Geochemistry
An Introduction

Introducing the essentials of modern geochemistry for students across the Earth and environmental sciences, this new edition emphasizes the general principles of this central discipline. Focusing on inorganic chemistry, Francis Albarède's refreshing approach is brought to topics that range from measuring geological time to the understanding of climate change. The author leads the reader through the quantitative aspects of the subject in a manner that is easy to understand.

The early chapters cover the principles and methods of physics and chemistry that underlie geochemistry, to build the students' understanding of concepts such as isotopes, fractionation, and mixing. These are then applied across many of the environments on Earth, including the solid Earth, rivers, and climate, and then extended to processes on other planets.

Three new chapters have been added – on stable isotopes, biogeochemistry, and environmental geochemistry. Student exercises are now included at the end of each chapter, with solutions available online.

Francis Albarède is Professor of Geochemistry at the Ecole Normale Supérieure de Lyon and a member of the Institut Universitaire de France. He had held visiting professorships at universities in the USA, Australia, and Japan. He has been President of the European Association of Geochemistry, Chief Editor of *Earth and Planetary Science Letters*, and the *Journal of Geophysical Research*. He has received numerous awards, including the Norman Bowen Award of the American Geophysical Union, the Arthur Holmes Medal of the European Union of Geosciences, and the Goldschmidt Award of the Geochemical Society. He is also author of *Introduction to Geochemical Modeling* (Cambridge University Press, 1995).

Geochemistry

An Introduction

Second Edition

Francis Albarède

Ecole Normale Supérieure de Lyon
Institut Universitaire de France

CAMBRIDGE UNIVERSITY PRESS
Cambridge, New York, Melbourne, Madrid, Cape Town, Singapore, São Paulo, Delhi

Cambridge University Press
The Edinburgh Building, Cambridge CB2 8RU, UK

Published in the United States of America by Cambridge University Press, New York

www.cambridge.org
Information on this title: www.cambridge.org/albarede

First published 2003
Second Edition 2009

Printed in the United Kingdom at the University Press, Cambridge

A catalog record for this publication is available from the British Library

ISBN 978-0-521-88079-4 hardback
ISBN 978-0-521-70693-3 paperback

Contents

Preface to the second edition

The material of several chapters has been deeply revised and rewritten for better intelligibility and to account for some major recent scientific developments. Many figures have been redrawn. I stuck to the black-and-white option to make it easier for teachers to distribute photocopied material for educational purpose: with due credit, I will gladly provide the figures as eps files that can be dressed with colors for classes. New chapters have been added, one on stable isotope fractionation, one on biogeochemistry, and one on paleo-environments; and existing chapters have been complemented with new material. Overall, the new edition is 50 percent longer than the first one.

Although I used graphic analogies whenever I thought it could spare the reader a difficult mathematical derivation without losing the substance of basic concepts, some of this new material will certainly be felt as a turn-off and I apologize for that. Boxes have been added for very specific material, such as the derivation of some equations, common misconceptions, or more anecdotal material, all of which can be left out without interrupting the main flow of the presentation. Quoting too much old work is pedantic but we can't really ignore the papers that created the basic concepts we use every day. I have marked as "must read" with a ♠ sign some references which laid the groundwork of entire fields.

Finally, at the end of most chapters I have incorporated an additional section of exercises, which have been found previously only on my website. In addition, the References and Further reading sections have been comprehensively updated.

Francis Albarède
Ecole Normale Supérieure, Lyon

Modern geochemistry is a discipline that pervades nearly all of Earth science, from measuring geological time through tracing the origin of magmas to unravelling the composition and evolution of continents, oceans and the mantle, all the way to the understanding of environmental changes. It is a comparatively young discipline that was initiated largely by Goldschmidt in the 1930s, but its modern development and phenomenal growth started only in the 1950s. Although there are many journals dedicated to geochemical research, there have been remarkably few general geochemical textbooks that cover more than a limited segment of the full scope of modern geochemistry. This is one reason why Francis Albarède's new book is most welcome. It is written by the author of the authoritative and widely acclaimed *Introduction to Geochemical Modeling* (Cambridge University Press, 1995), and it is intended as an undergraduate introductory course in geochemistry. Its scope is large, though not all-inclusive, concentrating on the inorganic chemistry of the condensed part of our planet. Although it started out as a translation from the original French book, the new English-language edition is much more than just a translation. The entire text has been substantially revised and in some parts expanded, and it is really a new book. Yet it retains a distinctly French flavor, particularly in the way many subjects are addressed via mathematical description. This approach is entirely normal for a student of the French Ecole Normale or a French university, but it will surprise many American teachers and students of geology alike. So if you are teaching or taking a course in "Rocks for Jocks," this book is not for you. But if you are interested in an introduction to modern geochemistry as a quantitative science, this book is definitely for you. Francis Albarède often uses a light touch, not taking the subject (or himself) excessively seriously, he uses refreshingly surprising analogs to approach important principles or processes, and his style is often informal. Look, for example, at the "Further reading" list. The books are classified into three categories, A, armchair reading, B, for students, and C, serious stuff, and each book is given a one- or two-line thumbnail characterization. Very nice. And by the way, the book itself, while clearly aimed at B, does contain material in all three categories!

What I particularly like about this book is its scope and choice of subjects, combined with sometimes bold brevity, which I hope will leave the student with an appetite for more. The emphasis is always on general principles rather than specific geochemical results or observations, and this should give the book a long residence time and keep it from becoming outdated. We are led from an introduction to the atomic and nuclear properties of the chemical elements to the principles of chemical and isotopic fractionation and mixing, geochronology and the use of radiogenic tracers to characterize source reservoirs, geochemical transport by advection and diffusion, the concept of closure temperature, chromatography, reaction rates to the treatment of large-scale systems, such as the oceans, the crust, and the mantle. The approach is initially mostly theoretical, focusing on the mathematical description of the behavior and interaction of single and multiple reservoirs. This is followed by a wide-ranging chapter on "Waters present and past," which covers topics from solution chemistry, water–rock interactions, erosion, rivers and oceans, to climate development during the Pleistocene. From there, we move to the "solid Earth," which deals with the evolution of mantle and crust, but also with the geochemistry of magmas. Finally, or almost finally, we are taken to phenomena of even much larger scale, the formation of the chemical elements in stars, the formation of the Solar System, the age and composition of Earth, Moon, and Mars.

The message the student should take from this is that geochemistry is a quantitative science that has made decisive contributions to the understanding of all these subjects. It has thus become one of the central disciplines of Earth science, a fact that is not always reflected in undergraduate curricula. This book should help to correct this common deficiency in the training of Earth scientists. It is an inspired book; I hope you will enjoy reading it as much as I did.

Albrecht W. Hofmann
Max Planck Institute for Chemistry, Mainz

Foreword to the French edition

I am specially happy to preface this book. First, because it is always a pleasure to be able to speak well of a friend's work; and Francis Albarède is a friend of long standing! We both embarked on our academic careers at about the same time. After some solid grounding in geology at the University of Montpellier, we were fortunate enough to begin our doctoral research in geochemistry in the 1970s in Professor Claude Allègre's laboratory at the Paris Institut de Physique du Globe, at a time when the discipline was really taking off in France. We also helped set up degree courses in geochemistry at the recently founded University of Paris 7, where we were appointed Assistant Lecturers. Our work together resulted in the publication of a short book in 1976, primarily for students, which quickly sold out and curiously enough was never reprinted! Few universities in those days offered specialist courses in geochemistry.

Times have clearly changed since then! Geochemistry is now taught in most universities and it is needless to recall here the fundamental contribution that this discipline has made to all areas of Earth sciences and cosmochemistry. It is always helpful, though, for students and for non-specialist faculty to have a textbook that provides a review of the basic concepts and the most recent contributions to the discipline. And this is the second reason why I am happy to present this book; because Francis Albarède's work fulfills both these requirements. The basic principles of the use of the chemical elements and their isotopes are set out clearly, together with their major applications in such varied domains as cosmochemistry (the formation of the chemical elements, of the Solar System, and of the planets), the internal dynamics of the Earth (with its various reservoirs and interaction among them, convection within the mantle, etc.), and its surface processes (hydrosphere, atmosphere, and climate change). Francis Albarède is particularly well qualified to deal with the diversity of geochemical applications because his own research has covered most of these major fields. A number of aspects that are sometimes overlooked in geochemistry books feature here, such as the processes of transport of elements (Chapter 4) or the concepts of residence time (in Chapter 5 on geochemical systems), and there is an overview of analytical techniques, which have proved so fundamental to the development of geochemistry. Moreover, the presentation is often novel (e.g. the presentation of geochronology, which is not just a catalog of the different methods), and the text is copiously illustrated with instructive diagrams and graphs.

One last reason why I particularly like this book is the method that Francis Albarède has chosen for setting out the principles underlying the main geochemical models: most of the relations describing these principles are demonstrated here, and the argument is invariably accompanied and supported by mathematical equations that unquestionably help the reader follow the reasoning. Advancing from one equation to the next is not always effortless, but the (slight) exertion required is well worth the trouble. Students need to discover or redis-cover the satisfaction to be gained from working out the equations describing a particular process from what are often intuitive relationships and, above all, from understanding that such equations are a short-hand representation of an underlying physico-chemical model that it is often easy to symbolize through a simple diagram: in short, they need to call on their faculties of understanding and their aptitude for model-making rather than their mem-ory. Reading this geochemistry textbook should encourage them to do just that. Readers may rest assured though, there is no need to know any advanced mathematics to understand this book; it is within the reach of any good college student. Nor is it devoid of humor: the allegory of dogs, and black and white cats to explain chemical fractionation and the absence of isotopic fractionation is a prime example! Through its resolutely model-based approach to processes and its concise and up-to-date explanations of the main contributions of the discipline, this book should become a standard text for college courses and a very valuable source of information for non-specialists eager to learn more about geochemistry. It comes out at a particularly fortunate time, just as new analytical instruments are about to widen the scope of geochemical tools substantially and give geochemistry a renewed impetus. I wish Francis Albarède's book swift and sustained success.

Professor Michel Condomines
Université de Montpellier II

Acknowledgments

I would like to thank those who made the first edition a success. Advice by Philippe Bonté, Dominique Boust, Hervé Cardon, Bill McDonough, Mireille Polvé, Yannick Ricard, Simon Sheppard, and Pierre Thomas is gratefully acknowledged. Careful reading of the original French manuscript by Janne Blichert-Toft, Fréderic Chambat, Michel Condomines, Don Francis, John Ludden, and Philippe Vidal, and by graduate students at ENS Lyon, has weeded out many errors of form and substance. Dave Manthey allowed me to reproduce graphics from his great Orbital Viewer. Agnès Ganivet kindly and effectively tidied up a first manuscript littered with syntactic errors. I would like to thank Nick Arndt, Edouard Bard, Janne Blichert-Toft, Marc Chaussidon, Al Hofmann, Dan Mckenzie, Bruce Nelson, Simon Sheppard, and Jacques Treiner, for reviewing the English manuscript. Chris Sutcliffe did an immense and wonderful job with translation and editing. Lesley Thomas was a clear-headed and efficient copy-editor.

In addition, the second edition owes also a great deal to new friends and colleagues. Detailed reviews by Janne Blichert-Toft and Simon Sheppard led to significant corrections and improvements. Scientific and editorial comments on particular chapters by Vincent Balter, Gilles Dromart, Toshi Fujii, Stephane Labrosse, Bruno Reynard, and Doug Rumble were very much appreciated. Particular thanks for the second edition are due to Tsuyoshi Iizuka, Bruno Reynard, Doug Rumble, and Ivan Vlastelic, for making unpublished material available, and to Peter Kolesar, Ran Qin, and John Rudge for pointing out errors in the first edition. Careful copy-editing by Zoë Lewin identified additional issues. Special thanks are due to the Educational Technology Clearinghouse (University of South Florida, http://etc.usf.edu/clipart/) for permission to use their clipart.

Working as a faculty at the Ecole Normale Supérieure in Lyon is good luck, both for the prevailing intellectual standard among scientists and students, and for the time saved by the liberal enforcement of academic chores by the Directeurs, Bernard Bigot, Philippe Gillet, and now Jacques Samarut. This book, like the previous ones, owes its existence to this luck. I would also like to express my gratitude to my wife, Janne Blichert-Toft, for letting such an inconsiderate intruder devour so much of our private life without complaining and for her steady encouragements. In writing this book, and in particular the first edition, I have also sought to express my gratitude to the Institut Universitaire de France: since my appointment has allowed me to devote more time to research, I felt it only right that this should be requited by some concrete contribution to the teaching of geochemistry.

Finally, special thanks are due to the Fondation des Treilles: nowhere on Earth can a book be prepared in such a magnificent environment worthy of Plato and Virgil surrounded by caring and friendly people. The last stretch was completed at Rice University in Houston, where Cin-Ty Lee and Alan Levander arranged a very fruitful and friendly time away from the daily troubles.

Introduction

Geochemistry utilizes the principles of chemistry to explain the mechanisms regulating the workings – past and present – of the major geological systems such as the Earth's mantle, its crust, its oceans, and its atmosphere. Geochemistry only really came of age as a science in the 1950s, when it was able to provide geologists with the means to analyze chemical elements or to determine the abundances of isotopes, and more significantly still when geologists, chemists, and physicists managed to bridge the chasms of mutual ignorance that had separated their various fields of inquiry. Geochemistry has been at the forefront of advances in a number of widely differing domains. It has made important contributions to our understanding of many terrestrial and planetary processes, such as mantle convection, the formation of planets, the origin of granite and basalt, sedimentation, changes in the Earth's oceans and climates, and the origin of mineral deposits, to mention only a few important issues. And the way geochemists are perceived has also changed substantially over recent decades, from laboratory workers in their white coats providing age measurements for geologists or assays for mining engineers to today's perception of them as scientists in their own right developing their own areas of investigation, testing their own models, and making daily use of the most demanding concepts of chemistry and physics. Moreover, because geochemists generate much of their raw data in the form of chemical or isotopic analyses of rocks and fluids, the development of analytical techniques has become particularly significant within this discipline.

To give the reader some idea of the complexity of the geochemist's work and also of the methods employed, we shall begin by following three common chemical elements – sodium (Na), magnesium (Mg), and iron (Fe) – on their journey around system Earth. These three elements were created long before our Solar System formed some 4.5 billion years ago, in the cores of now extinct stars. There, the heat generated by the gravitational collapse of enormous masses of elementary particles overcame the repulsive forces between protons and triggered thermonuclear fusion. These reactions allowed particles to combine forming ever larger atomic nuclei of helium, carbon, oxygen, sodium, magnesium, and iron. This activity is still going on before our very eyes as the Sun heats and lights us with energy released by hydrogen fusion. When, after several billion years, the thermonuclear fuel runs out, the smaller stars simply cool: the larger ones, though, collapse under their own weight and explode in one of the stellar firework displays that nature occasionally stages, as with the appearance of a supernova in the Crab nebula in AD 1054. The matter scattered by such explosions drifts for a while in interstellar space as dust clouds similar to the one that can be observed in the nebula of Orion. Turbulence in the cloud and collisions between the particles make the system unstable and the particles coalesce to form small rocky bodies known as planetesimals. Chondritic meteorites give us a pretty good idea of what these are like. Gravitational chaos amplifies very rapidly and the planetesimals collide to form larger and larger bodies surrounding a new star: a Solar System is born.

As the planet forms and evolves, our three chemical elements meet different fates. Sodium is a volatile element with a relatively low boiling point (881 °C) and large amounts of it are therefore driven off into space by the heat generated as the planet condenses. Iron, which is initially scattered within the rock mass, melts and collects at the heart of the planet to form the core, which, on Earth, generates the magnetic field. Magnesium has a boiling point of 1105 °C and behaves in an un-extraordinary way assembling with the mass of silicate material to form the mantle, the main bulk of the terrestrial planets, and residing in minerals such as olivine (Mg_2SiO_4), the pyroxenes ($Mg_2Si_2O_6$ and $CaMgSi_2O_6$), and, if sufficient pressure builds up, garnet ($Mg_3Al_2Si_3O_{12}$).

The chemical histories of the planets are greatly influenced by their magmatic activity, a term that refers to all rock melting processes. On our satellite, the Moon, the formation of abundant magmas in the first tens of millions of years of its history has left its mark in the surface relief. The crust, composed of the mineral plagioclase ($CaAl_2Si_2O_8$), is so light that it floats on the magma from which it crystallized, forming the lunar highlands that rise above the Moon's marias. The sodium that did not vaporize migrated with the magma toward the lunar surface and combined readily in the plagioclase alongside calcium. On Earth, water is abundant because temperatures are moderate and because the planet is massive enough for gravity to have retained it. The action of water and dissolved carbonates is another significant factor in the redistribution of chemical elements. Water causes erosion and so is instrumental in the formation of soils and sedimentary rocks. The presence of chemically bound water in the continental crust promotes metamorphic change and, by melting of metamorphosed sedimentary rocks, the formation of granite that is so characteristic of the Earth's continental crust. A substantial fraction of the sodium that did manage to enter into the formation of the early crust was soon dissolved and transported to the sea, where it has resided for hundreds of millions of years. Some marine sodium entered sediment and then, in the course of magmatic processes, entered granite and therefore the continental crust.

Magnesium, by contrast, tends to remain in the dense refractory minerals. It lingers in solid residues left after melting or precipitated during crystallization of basalt at mid-oceanic ridges or at ocean island volcanoes. Where it does enter fluids, it subsequently combines with olivine and pyroxene, which precipitate out as magma cools. Magnesium is resistant to melting and is predominant in the mantle, which has ten to thirty times the magnesium content of the crust.

The Earth is a complex body whose dynamics are controlled by mechanisms that commonly work in opposing directions: differentiation mechanisms, on the one hand, maintained by fractionation of elements and isotopes between the phases arising during changes of state (melting, crystallization, evaporation, and condensation), and, on the other hand, mixing mechanisms in hybrid environments such as the ocean and detrital rocks that tend to homogenize components derived from the various geological units (rain water, granite, basalt, limestone, soil, etc.). By fractionation we mean that two elements (or two isotopes) are distributed in unequal proportions among the minerals and other chemical phases present in the same environment.

It can be seen then that the elements must be studied in terms of their properties in the context of the mineral phases and fluids accommodating them and in the context of the processes that govern changes in these phases (magmatism, erosion, and sedimentation). By mineral phase is meant all the crystals from a small neighborhood (mm to cm) that belong to the same mineral species. We will start by discussing some useful rules of organic and nuclear chemistry that illuminate the geochemical properties of elements. An understanding of transport mechanisms is very important in geochemistry. The term "cycle" is sometimes employed but we will see later that its meaning needs to be clarified. Conversely, the terminology used above, which emphasizes that sodium resides for a long time in the ocean or that magnesium lingers in solid residues of melting, illustrates that transfers among the different parts of the globe, such as the core, mantle, crust, and oceans, are to be considered kinetically, or, more loosely, dynamically, in terms of flows or transport rates. We will next turn our attention to the mysterious field of stable isotopes, with the idea of clarifying the principles of their fractionation in nature. Radioactive decay, which alters the atomic nucleus and therefore the nature of certain elements at rates that are unaffected by the physical, crystallographic, or chemical environment in which they occur, will allow us to include the incessant ticking of these "clocks" in our study of radioactive processes.

The book then moves on to the essential study of the dynamics and evolution of the mantle and continental crust, and the study of marine geochemistry and its implications for paleoclimatology and paleoceanography. Two new chapters on biogeochemistry and environmental geochemistry have been added to the second edition. A last chapter deals with the geochemical properties of a number of elements. Most students deplore the lack of such a systematic approach in the literature; what is supposed to be common knowledge is never taught, simply because the odds of being inaccurate, unbalanced, and superficial are too great. An appendix provides an overview of a number of methods for analyzing chemical elements and isotopes.

I have been asked so many times by genuinely motivated colleagues from other disciplines where a compact description of geochemistry could be found. I hope this short textbook will meet this demand. It will probably be found that this book relies more heavily on equations than most other geochemistry textbooks. I maintain that a proper scientific approach to our planet must use all the available tools, especially those of physics and chemistry, to supplement purely descriptive and analytical approaches. I therefore ask readers to persevere despite the superficial difficulty of some of the equations, which are, after all, nothing more than a means of encoding concepts that ordinary language is powerless to convey with adequate precision. Oscar Wilde said that "... nothing that is worth knowing can be taught." A collection of observations is no more science than a dictionary is literature. Above all, I have made every attempt to avoid turning the reader of this book into a stamp collector.

But the devil is in the detail and I occasionally had to cut corners: some concepts, definitions, and proofs could have benefited from more rigor and from more detailed supporting arguments. I very much wanted to keep this book short, so I hope that my specialist colleagues will forgive the short-cuts used to this effect.

Readers may e-mail any queries, criticisms and – who knows – words of encouragement to the author (albarede@ens-lyon.fr).

1   The properties of elements

The 92 naturally occurring chemical elements (90, in fact, because promethium and technetium are no longer found in their natural state on Earth) are composed of a nucleus of subatomic nucleons orbited by negatively charged electrons. Nucleons are positively charged protons and neutral neutrons. As an atom contains equal numbers of protons and electrons with equal but opposite charges, it carries no net electrical charge. The mass of a proton is 1836 times that of an electron. The chemical properties of elements are largely, although not entirely, determined by the way their outermost shells of electrons interact with other elements. Ions are formed when atoms capture surplus electrons to give negatively charged anions or when they shed electrons to give positively charged cations. An atom may form several types of ions. Iron, for example, forms both ferric (Fe^{3+}) ions and ferrous (Fe^{2+}) ions, while it also occurs in the Fe^0 elemental form.

A nuclide is an atomic nucleus characterized by the number Z of its protons and the number N of its neutrons regardless of its cloud of electrons. The mass number A is the sum of the nucleons $N + Z$. Different interactions act in the nucleus and explain its binding: the short-range (nuclear) strong force, the long-range electromagnetic force, and the mysterious intermediate weak force. Two nuclides with the same number Z of protons but different numbers N of neutrons will be accompanied by the same suite of electrons and so have very similar chemical properties; they will be isotopes of the same element. The "chart of the nuclides" (Fig. 1.1) shows that in order to be stable, nuclides must contain a specific proportion of neutrons and protons. The semi-empirical formula for the energy of a nucleus is:

$$E = aA - bA^{2/3} + c\frac{(N - Z)^2}{A} - d\frac{Z^2}{A^{1/3}} \qquad (1.1)$$

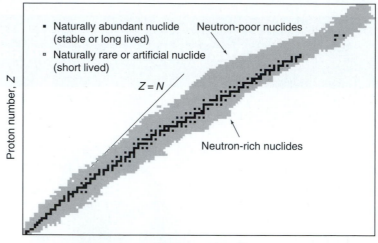

Neutron number, N

Figure 1.1   Chart of the nuclides (overview). The light stable elements have approximately the same number Z of protons and number N of neutrons but the heavier stable elements deviate towards the neutron-rich side according to (1.2): this relationship defines the valley of stability. Elements that depart significantly from this rule are unstable (radioactive).

and this describes the so-called "liquid drop" model of the nucleus. The constants a, b, c, and d can be adjusted to fit laboratory data. As a first approximation, the volume of the nucleus is proportional to A, its radius to $A^{1/3}$, and its surface area to $A^{2/3}$. The first term on the right-hand side expresses the volume energy, which is proportional to the number of nucleons; the second term is the surface energy, which subtracts the uncompensated attraction of the nucleons located near the surface of the nucleus; the third term expresses that, for a given A, the nuclear attraction between proton and neutron is slightly stronger than proton–proton and neutron–neutron attraction; the fourth term accounts for electrostatic energy which is inversely proportional to the distance between the neighboring charges of the protons. The locus of minimum energy, in the N, Z plot of Fig. 1.1, which is known as the "valley of stability," is obtained by minimizing (1.1) with respect to Z, and is conveniently represented by the equation:

$$Z = \frac{2A}{4 + 0.031A^{2/3}} \tag{1.2}$$

For light elements ($Z < 40$), the term in A at the denominator is very small, so $Z \approx A/2$ and therefore $N \approx Z$. At higher masses, electrostatic repulsion between protons gets stronger and $N > Z$. One easily finds that for ^{238}U, $Z = 92$ which is correct.

Nuclei with N and Z too far from this valley of stability are unstable and are said to be radioactive. An isotope is radioactive if its nucleus undergoes spontaneous change such as occurs, for instance, when alpha particles (two protons and two neutrons) or electrons are emitted. It changes into a different isotope, referred to as radiogenic, by giving out energy, usually in the form of gamma radiation, some of which is harmful for humans. Several internet sites provide tables of all stable and radioactive nuclides. The vast majority of natural isotopes of naturally occurring elements are stable, i.e. the number of their protons and neutrons remains unchanged, simply because most radioactive isotopes have vanished over the course of geological time. They are therefore not a danger to people.

1.1 The periodic table

The atomic number of an element is equal to the number of its protons. We have seen before that the atom's mass number is equal to the number of particles making up its nucleus. The Avogadro number $\mathcal{N}$ is the number of atoms contained in 12 g of the carbon-12 isotope. The atomic mass of an isotope is the weight of a number $\mathcal{N}$ of atoms of that isotope. Dimitri Mendeleev's great discovery in 1871 was to demonstrate the periodic character of the properties of elements when ordered by ascending atomic number (Fig. 1.2). Melting point, energy of formation, atomic radius, and first ionization energy all vary periodically as we work through Mendeleev's table. The geochemical properties of elements are reflected by their position in this table. The alkali metals (Li, Na, K, Rb, Cs), alkaline-earth metals (Be, Mg, Ca, Sr, Ba), titanium group elements (Ti, Zr, Hf), but also the halogens (F, Cl, Br, I), inert gases (He, Ne, Ar, Kr, Xe), rare-earths (lanthanides), or actinides (uranium family) all form groups sharing similar chemical properties; these properties are indeed sometimes so similar that it was long a challenge to isolate chemically pure forms of some elements such as hafnium (Hf), which was only separated from zirconium (Zr) and identified in 1922.

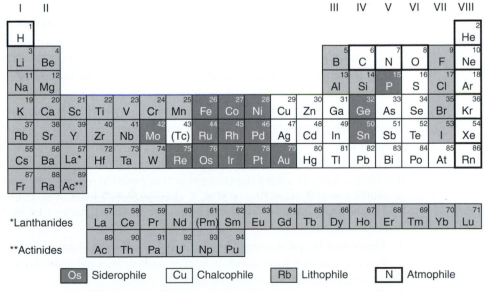

Figure 1.2   Mendeleev's periodic table of the elements and their geochemical classification after Goldschmidt. The elements in parentheses do not occur naturally on Earth. The atomic number of each element is given. Roman numerals over columns indicate groups.

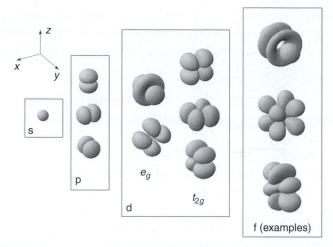

Examples of orbital geometry. Shown here are the surfaces of maximum probability of electron localization around the nucleus corresponding to various orbitals. s, p, d, f are the quantum numbers. Note the two types of d orbitals. Drawn using Orbital Viewer (Dave Manthey).

It is therefore very important to understand how elements are ordered in the periodic table. Put simply, an atom can be represented as a point-like nucleus containing the mass and charge of the nuclear particles and by mass-less electrons orbiting this point. The heavy nucleus is normally assumed to be immune to fluctuations in the more volatile electron clouds and is treated as stationary (Born-Oppenheimer approximation). Quantum mechanics requires the different forms of electron energy to be distributed discretely, i.e. at separate energy levels. It also requires that the different forms of momentum be quantized, not only the linear form of gas molecules bouncing around in a box, but also the angular momentum of the electrons on their atomic orbitals and around their spin axis. The angular momentum L plays the same role with respect to angular velocity ω as linear momentum $p = mv$ plays with linear velocity v: the familiar expression defining the linear kinetic energy as $p^2/2m$ becomes $L^2/2I$ for rotational kinetic energy, with the moment of inertia I playing for rotational energy the role of mass m for linear translation.

The Heisenberg principle states that the uncertainty of the position and the velocity of a particle vary inversely to one another and therefore prevents the exact calculation of electron orbits around the nucleus. An orbital is a complex function used by quantum mechanics to describe the probability of the presence of an electron around the nucleus but it is often reduced to a three-dimensional surface meant to represent the locus of maximum probability. It is denoted (Fig. 1.3) by a set of integers known as quantum numbers. The four levels of quantization are as follows:

1. The first (principal) quantum number n characterizes the total energy level of the electron and can take positive values 1, 2, 3, ... It defines the main electron shells, which are sometimes represented by the letters K, L, M, ...
2. The second (orbital) quantum number l characterizes the total orbital angular momentum L of the electron; it ranges from 0 to $n - 1$ and defines the number of lobes of the orbitals of each shell, which are usually designated by the letters s, p, d, f.

Table 1.1 Electronic configuration of the light elements

Element	Group	n	l	m	s	Configuration
H	I	1	0	0	$+1/2$	$1s^1$
He	VIII	1	0	0	$-1/2$	$1s^2 = [He]$
Li	I	2	0	0	$+1/2$	$[He]\,2s^1$
Be	II	2	0	0	$-1/2$	$[He]\,2s^2$
B	III	2	1	-1	$+1/2$	$[He]\,2s^2\,2p^1$
C	IV	2	1	-1	$-1/2$	$[He]\,2s^2\,2p^2$
N	V	2	1	0	$+1/2$	$[He]\,2s^2\,2p^3$
O	VI	2	1	0	$-1/2$	$[He]\,2s^2\,2p^4$
F	VII	2	1	$+1$	$+1/2$	$[He]\,2s^2\,2p^5$
Ne	VIII	2	1	$+1$	$-1/2$	$[He]\,2s^2\,2p^6 = [Ne]$
Na	I	3	0	0	$+1/2$	$[Ne]\,3s^1$

3. The third (magnetic) quantum number m $(0, \pm 1, \ldots, \pm l)$ gives the part L_z of the angular moment which points along the rotation axis; it defines the shape of the orbital.
4. The fourth quantum number s describes the momentum associated with the spin of the electron and gives the direction of spin of the electron around its own axis relative to its orbital movement.

The Pauli exclusion principle states that no two electrons can have the same quantum numbers.

The periodic table can be constructed by assigning a unique set of quantum numbers to each element (Table 1.1) and the filling of the successive orbitals can now proceed from lower to higher energy levels until the number of electrons matches the number of protons in the nucleus. This holds for n, by definition, but also for l because of the electrostatic screening by electrons on lower orbitals (see below): for example, orbital 2p is filled after orbital 2s. The filling order is shown in Fig. 1.4 and can be exactly matched with the periodic table.

A number of Internet sites provide detailed periodic classifications, of which I can recommend http://www.webelements.com/, while Dave Manthey's excellent site http://www.orbitals.com/orb/ov.htm provides software to create very professionally drawn orbital pictures (Fig. 1.4).

In the periodic table, groups I (alkali metals) and II (alkaline-earth metals) correspond to the filling of s orbitals, and groups III to VIII to that of the p orbitals. The intermediate groups (transition elements such as iron and platinum) differ in the occupation of their d orbitals. When occupied, these d orbitals are normally closer to the nucleus than the s orbitals of the next shell out. Occupation of the orbitals is noted nx^i, where x represents the type of orbital (s, p, d, f), n its principal quantum number and i the number of electrons it contains. Most elements of the first series (e.g. V, Cr, Mn, Fe, Co, Ni, Cu, Zn) have an electron formula of the type $[Ne]3s^2\,3p^6\,3d^i\,4s^2$, where [Ne] represents the fully occupied

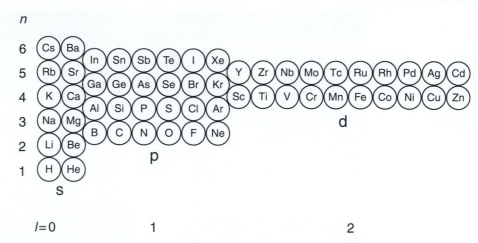

Figure 1.4 The filling order of the lowest orbitals of the elements in the periodic table. The vertical scale shows the energy levels.

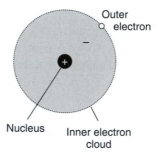

Figure 1.5 Shielding of the nuclear charge by the cloud of electrons orbiting between the outer electrons and the nucleus.

orbitals of a neon atom; and their divalent ions, such as Fe^{2+} and Cu^{2+}, have a configuration $[Ne]3s^2\,3p^6\,3d^i$. These transition elements differ only by the number i of electrons in orbital 3d but have an identical outer electron shell 4s, which explains why their chemical properties are so similar. This phenomenon is further amplified in the rare-earths (or lanthanides), such as La and Ce (shell 4f), and the actinides (5f), such as U and Th, where the s and p orbitals of the external shells are identical.

Simple rules hold for the prediction of atomic radii. First, the potential energy and atomic radius increase with n and therefore down each column of the periodic table. Second, the atomic radius decreases across each row. This is due to the reduction of electrostatic attraction of the outer electrons by the cloud of the inner electrons (Fig. 1.5), a phenomenon known as shielding. For the lanthanides and the actinides, the f electrons on their multi-lobate orbitals leave some parts of the nucleus exposed (Fig. 1.3) and therefore do not screen the increasing charge of the nucleus as efficiently as the more smoothly shaped lower-order orbitals. As a result, their atomic radii decrease smoothly with their increasing atomic number, a phenomenon known as lanthanide (and actinide) contraction.

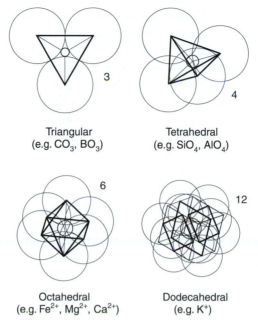

Triangular
(e.g. CO_3, BO_3)

Tetrahedral
(e.g. SiO_4, AlO_4)

Octahedral
(e.g. Fe^{2+}, Mg^{2+}, Ca^{2+})

Dodecahedral
(e.g. K^+)

Figure 1.6   The main ion coordination systems in naturally occurring minerals: triangular (three closest neighbors), tetrahedral (4), octahedral (6), and dodecahedral (12) coordination.

The energy stored in silicates as chemical bonds depends on the nature of the cations and the crystal sites accommodating them. Depending on their ionic radius, cations may occupy sites of varying size and the number of oxygen neighbors with which they bond (coordination number) increases with the size of the site. Carbon and boron atoms combine with oxygen in a triangular arrangement (i.e. threefold coordination, Fig. 1.6), while silicon and aluminum atoms combine with oxygen to form a tetrahedron (fourfold coordination). However, medium-sized cations, such as Fe^{2+}, Mg^{2+}, or Ca^{2+}, will take up vacant octahedral sites (sixfold coordination) between SiO_4 tetrahedra while the biggest ions, such as K^+ or OH^- hydroxyl anions, require the most spacious sites, normally of twelvefold coordination.

1.2 Chemical bonding

Atoms and ions combine to form matter in its solid, liquid, or gaseous states. The importance of occupation of the outer electron shell can be illustrated by comparing the interaction between two atoms of helium, where two electrons occupy orbital 1s, with the interaction between two hydrogen atoms, each with a single electron only. When the two helium atoms come close together and their electron clouds interpenetrate, one atom's electrons cannot be accommodated by the orbital of the other as this would infringe the Pauli exclusion principle. They must therefore jump to the 2s orbital, at a cost in energy that penalizes the formation of such bonds. Two hydrogen atoms, however, can lend one another a 1s electron. The resulting electron configuration is more advantageous than that of isolated hydrogen atoms and thus chemical bonding is favored. This is the essence of Pauling's valence bond theory, which, however, fell into disfavor for its inability to explain the spectroscopic properties of substances. In contrast, the crystal field theory sees fully ionized cations, such as Na^+ or Ca^{2+} being hosted in sites defined by negatively charged "ligands" such as the silicate network: the bond is assumed to be fully electrostatic. Although the valence theory accounts for many properties of the transition elements, it is now also considered as largely obsolete.

The dual character of most chemical bonds is accounted for by the concept of molecular orbitals: when atoms and ions come close to each other, their individual electronic orbitals merge into collective orbitals, which, however, in general remain difficult to calculate. The type of chemical bonding is determined by the probability of the presence of the electron next to one of the bound nuclei, which is exactly what the molecular orbitals are meant to describe. If the electron is transferred permanently, ionic bonding has occurred: a sodium atom in the presence of a chlorine atom will give up its isolated 3s electron because their outer shells will then be completely occupied, with sodium configured like neon and chlorine like argon. The ions formed in this way, noted as Na^+ and Cl^-, are particularly stable; their outer electron shell is largely spherical and these ions act like electrically charged spheres mutually attracted by their electrostatic fields to form ionic compounds such as table salt, NaCl. Conversely, when the number of electrons that can be exchanged fails to fill the outer shells of the two partners, a covalent bond is formed. Two hydrogen atoms lend one another their missing electron but must share the two 1s electrons over time by forming hybrid orbitals of complex geometry thereby allowing them to fill the outer shell of both atoms.

The transition elements (V, Fe, Cu, Zn, etc.) are particularly sensitive to their crystalline environment. They differ from each other by a different filling of their d orbital (Fig. 1.3). Two of these orbitals, called e_g, have their lobes lying along the axes of rectangular coordinates, while three orbitals, called t_{2g}, lie in each plane defined by two axes but with their lobes sticking out in between the axes. In the most common octahedral sites occupied by transition elements in silicates, with an oxygen atom at each apex along the axes, an electron occupying a t_{2g} orbital feels the repulsion by the electrons from the oxygen ions, the so-called crystal field, much less than an electron occupying an e_g orbital (Fig. 1.7). Calling Δ the crystal field stabilization energy (CSFE), i.e. the difference in bonding energy between t_{2g} and e_g, the energy shift of the three t_{2g} orbitals is $-2\Delta/5$, while the shift for the two e_g orbitals is $+3\Delta/5$, thereby ensuring that the mean energy shift with respect to a spherical environment is zero (Fig. 1.8).

Let us now give examples of how these concepts are relevant to the energetics of element partitioning in crystallographic sites:

1. Trivalent chromium ion Cr^{3+} has electronic formula $[Ar]3d^3\ 4s^0$ ([Ar] stands for the orbital filling of argon) in an octahedral site. One electron on each t_{2g} orbital gives this ion a bonding energy of $3 \times (-2\Delta/5) = -6\Delta/5$. In spite of a charge at odds with the major cations, such a large value makes Cr^{3+} an element abundant in some Fe–Mg minerals, most conspicuously pyroxene.

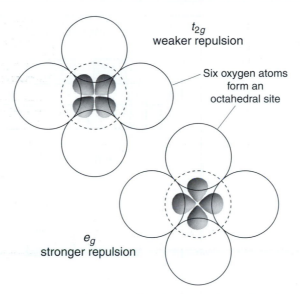

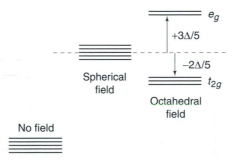

Figure 1.7 Crystal field effect on a transition element, such as Fe, Mn, Cr, in octahedral coordination. Six oxygen atoms form the apexes of the octahedron. The t_{2g} orbitals of the transition element lie between the oxygen atoms, while the e_g orbitals point toward them. The filling of an e_g orbital by electrons therefore has to overcome excess repulsion energy compared with a t_{2g} orbital.

Figure 1.8 Splitting of the d energy levels in the octahedral site of (Fig. 1.7). If the ion is inserted into a spherical site, the repulsion to overcome is symmetric for the electrons on all the d orbitals. In an octahedral site, which is particularly common in silicate rocks, the two e_g orbitals are subjected to stronger repulsion than the three t_{2g} orbitals. For the same overall interaction energy, and a difference Δ in bonding energy between t_{2g} and e_g (the crystal field stabilization energy), the energy shift of the t_{2g} orbitals is $-2\Delta/5$, while the shift for the e_g orbitals is $+3\Delta/5$. In a tetrahedral site, the situation would be reversed.

2. Ferrous iron Fe^{2+} has electronic formula $[Ar]3d^6 4s^0$. Once the lowest three t_{2g} orbitals are occupied by one electron each, there are two options: (i) if Δ is less than the repulsive energy of electron-electron pairing, the next electron will fit an upper e_g orbital or, (ii) if Δ is large, it will pair with an electron on the t_{2g} orbital. Two Fe^{2+} configurations are therefore possible in each case (Fig. 1.9) which receive their denomination from the way electronic spins add up. The most abundant iron in the mantle corresponds to the small Δ case and is referred to as "high-spin" Fe. The energy gain due to crystal

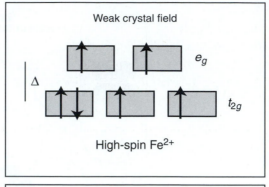

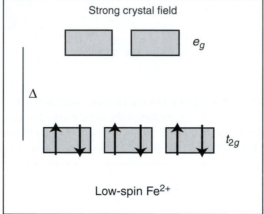

Figure 1.9 Two electronic configurations for the same ion Fe^{2+} with 6d electrons. When the crystal field is strong, the energy gap Δ between the t_{2g} and e_g orbitals increases, and the six electrons fill the three t_{2g} orbitals. This is the low-spin configuration. When the crystal field is weak, the energy cost of pairing electrons dominates and two electrons move to the upper e_g orbitals. This is the low-spin configuration. Only some ions of transition elements show such a dual configuration.

field is $4 \times (-2\Delta/5) + 2 \times (3\Delta/5) = -2\Delta/5$. Conversely, the large Δ case (full t_{2g}) is referred to as "low-spin" Fe and has a crystal field effect of $-12\Delta/5$.

3. For divalent nickel Ni^{2+} ($[Ar]3d^8\ 4s^0$), the configuration is unique: the three t_{2g} orbitals are fully occupied, while each of the e_g orbitals hosts one electron. The energy gain is $6 \times (-2\Delta/5) + 2 \times (3\Delta/5) = -6\Delta/5$; Ni^{2+} therefore snuggles in octahedral sites and is notably enriched in Fe–Mg silicates such as olivine and pyroxenes.

4. The common high-spin Mn^{2+} ($[Ar]3d^5 4s^0$) as well as Zn^{2+} ($[Ar]3d^{10}4s^0$) have symmetrical orbital configurations and therefore no energy gain: in general, they show little preference between silicate minerals or between silicates and melts.

The effect of the crystal field is essentially symmetrical when these ions occur in tetrahedral coordination.

In solutions, transition-element compounds form by interaction between a ligand and a cation, which is the basis of Lewis acid–base theory. Typical of such interaction is hydration, e.g. for Zn

$$Zn^{2+} + H_2O \Leftrightarrow Zn^{2+}(OH)^- + H^+ \qquad (1.3)$$

which shows the strong attraction of the Zn^{2+} cation for the lone electron pair carried by the oxygen. Water acts as the donor and Zn^{2+} is the acceptor. The reaction results in the liberation of a proton, which justifies the fact that Zn^{2+} is referred to as a Lewis acid and water as a Lewis base. In general, a species donating a pair of electrons is a Lewis base, whereas a species accepting this pair is a Lewis acid. The variable "hardness" of Lewis acids, not to be confused with their strength that measures the energy of the proton bond, helps us to understand the direction of inorganic and organic reactions. Ions are separated into hard "Class a" (groups I and II of the periodic table plus the lightest transition elements), soft "Class b" (the upper right of the periodic table), and borderline (Fe^{2+}, Cu^{2+}, Zn^{2+}, Pb^{2+}) elements as a function of their affinity for lone electron pairs. Why this additional criterion is important can be illustrated with the following reaction:

$$CoCl_4^{2-} + 6H_2O \Leftrightarrow Co\,(H_2O)_6^{2+} + 4Cl^- \qquad (1.4)$$

Adding the Lewis acid Ca^{2+}, which is a harder acceptor than Co^{2+} and therefore strongly binds to water molecules, displaces the reaction to the left and gives the strong blue color of the chloride to the solution. In contrast, adding Zn^{2+}, which is a softer Lewis acid than Co^{2+}, pushes the reaction to the right and shows the pink color of the hydroxide. Such reactions are essential to account for the role of metals with respect to organic ligands.

Pressure is especially important in the environment of ions. Oxygen is more compressible than smaller ions and, with its cation coordination number increasing, it plays an essential role. Thus, at depths of more than 660 km, pressure levels in the lower mantle are such that silicon coordination shifts from fourfold to sixfold thereby promoting the compact stacking of oxygen atoms. Likewise, under conditions of the lowermost mantle, site distortion becomes important and some minerals such as oxides show a transition from high-spin to low-spin iron. It remains, however, a real challenge to predict with full accuracy the chemical properties of elements at very great depths.

1.3 States of matter and the atomic environment of elements

Bonds formed by condensed materials are generally more complex than those formed by gases. In silicates, which are so important to our understanding of geological phenomena, a small silicon (or possibly aluminum) atom (Fig. 1.10) lies at the center of a tetrahedron of four oxygen atoms. As in carbon chemistry, SiO_4 tetrahedra may polymerize to varying degrees by sharing one or more oxygens at their apexes. The Si–O bond is a quite strongly covalent one. Other elements, such as Mg, Fe, or Na, may be accommodated within the silicate framework in their ionic forms Mg^{2+}, Fe^{2+}, Na^+.

The many crystallized silicate and alumino-silicate structures are classified according to the pattern formed by their tetrahedra. The most important ones in geology are:

1. Isolated-tetrahedra silicates: the most common minerals in this family are the various sorts of olivine, such as forsterite Mg_2SiO_4, and of garnet, such as pyrope $Mg_3Al_2(SiO_4)_3$.

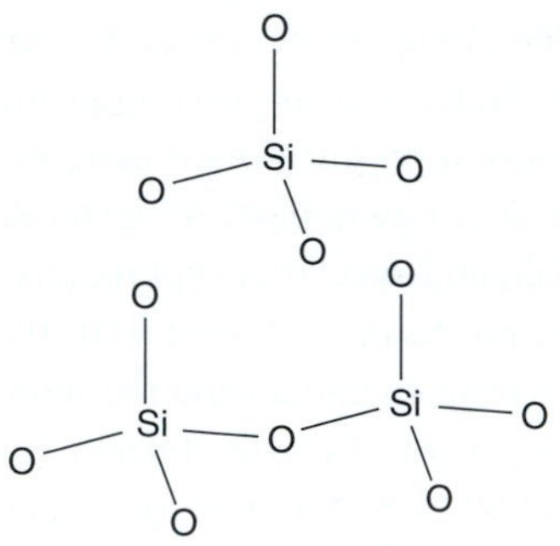

Figure 1.10 SiO_4 tetrahedra, which are the building blocks of silicate structures and their polymerization. Aluminum similarly forms AlO_4 tetrahedra.

2. Single-chain silicates: these are the pyroxenes, which fall into two groups with two different crystallographic systems; orthopyroxenes, such as enstatite $Mg_2Si_2O_6$, and clinopyroxenes, such as diopside $CaMgSi_2O_6$.
3. Double-chain silicates: amphiboles, such as tremolite $Ca_2Mg_5Si_8O_{22}(OH)_2$ or hornblende $Ca_2Mg_4Al_2Si_7O_{22}(OH)_2$. The formation of these hydroxylated minerals requires some degree of water pressure.
4. Sheet silicates: micas and clay minerals usually containing aluminum, potassium, and smaller ions such as Fe^{2+} and Mg^{2+}. A distinction is drawn between di-octahedral micas like muscovite (common white mica) $K_2Al_6Si_6O_{20}(OH)_4$ and tri-octahedral micas like biotite (ordinary black mica) $K_2Mg_6Al_2Si_6O_{20}(OH)_4$, the difference being the proportions of 2+ and 3+ cations and therefore site occupancy. This family is extremely diverse.
5. Framework silicates: these silicates are interconnected at each of their apexes. This family includes quartz SiO_2 and the feldspars, the most important of which are albite $NaAlSi_3O_8$, anorthite $CaAl_2Si_2O_8$, and the various potassium feldspars whose formula is $KAlSi_3O_8$.

Other significant minerals include iron oxides and titanium oxides, which are commonly cubic ionic solids such as magnetite Fe_3O_4 and ilmenite $FeTiO_3$. Corundum is the oxide of aluminum Al_2O_3. Calcium carbonate (calcite, aragonite) and magnesium carbonate (magnesite), formed by stacking of rhombohedral Ca^{2+} or Mg^{2+} ions, contain small carbonate groupings CO_3^{2-} able to rotate around the axis of symmetry.

The minerals cited above are only examples, albeit important ones, but alone they fail to provide a sufficiently precise representation of the chemical diversity of rocks. Elements are located within minerals at sites characterized by the number of oxygen atoms that they have as their immediate neighbors. We have already come across the tetravalent silicon ion Si^{4+} (i.e. carrying four positive charges) at a tetrahedral site and said that it could be replaced by the trivalent aluminum ion Al^{3+}. Electrical neutrality is maintained through paired substitutions such as $Al^{3+}Al^{3+}$ substituting for $Si^{4+}Mg^{2+}$. Such substitution is possible and even commonplace in pyroxenes, amphiboles, micas, and feldspars as the two ions have similar ionic radii (0.39 and 0.26 Å, respectively) and similar electrical charges.

Likewise, Mg^{2+} ions (0.72 Å) located at octahedral sites of naturally occurring olivine, i.e. those surrounded by six oxygen atoms, are commonly replaced by Fe^{2+} ions (0.61 Å) or Ni^{2+} ions (0.69 Å). Ions of the rare-earth element family, such as the Yb^{3+} ytterbium ion (0.99 Å), may substitute for the Ca^{2+} ion (1.00 Å) in clinopyroxenes or amphiboles. As these continuous substitution phenomena are analogous to those whereby various ions co-exist in aqueous solutions, the term solid solution is used.

Not all solid solutions are possible. The larger ions, such as the alkali metals (K^+, Rb^+) or the alkaline-earth metals (Sr^{2+}, Ba^{2+}) of the higher periods, have ionic radii that are too big for them to fit readily into the common silicate minerals. Ions with different charges manage to substitute for major ions if the electrical imbalance can be offset locally: a $(Ca^{2+})_2$ pair may thus be replaced by a $Yb^{3+}Na^+$ pair in a clinopyroxene, whereas its replacement by a Th^{4+} thorium ion requires the formation of a defect with a high energy cost. Ions that carry too high a charge or whose ionic radius is too small or too large are rejected by the lattice of essential minerals and concentrate either in accessory minerals, such as the phosphates or titanates, or in poorly characterized phases in grain fractures and interstices. These outcasts are termed incompatible elements. However, this is a relative concept: potassium and barium are incompatible in the feldspar-free mantle; yet, in the continental crust, where feldspar is abundant, K and Ba are compatible. The volatile elements and compounds, such as the rare gases, water, and carbon dioxide, call for some attention. As long as no gas phase is present, i.e. as long as the concentration of one of the most abundant volatiles in the solid or liquid in question does not exceed saturation level, they behave like any other element with varying levels of compatibility. In the absence of vapor, for example, the inert gases such as helium or argon need not be classified separately from the other trace elements.

1.4 Geochemical classifications

There are many ways to arrange elements by their geochemical properties. Such practice, reducing as it does the wide diversity of behavior of elements to limited ranges of behavior, might be thought pointless, but it does provide an overview of their chemical properties. The most widespread classification is probably that of Victor Goldschmidt (Fig. 1.2). Goldschmidt's scheme rests on the observation by Berzelius, a Swedish eighteenth-century chemist, that some elements tend to form oxides or carbonates whereas others form sulfides. The lithophile elements (Na, K, Si, Al, Ti, Mg, Ca) generally concentrate in the rock-forming minerals of the crust and mantle; the siderophile elements (Fe, Co, Ni, Pt, Re, Os) have an affinity for iron and therefore concentrate in the Earth's core; the chalcophile elements (Cu, Ag, Zn, Pb, S) readily form sulfides; the atmophile elements (O, N, H, and the inert gases) concentrate in the atmosphere. Certain elements in each group tend to be volatile; K is a more volatile lithophile than either Mg or Ti (see Chapter 12). Refractory elements such as Mg or Cr tend to concentrate in solid residues.

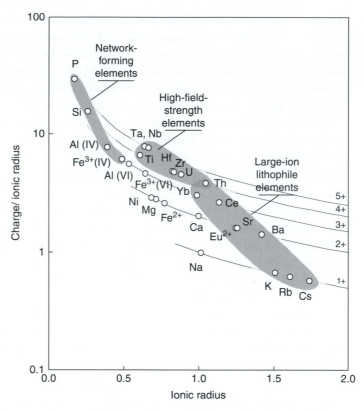

Figure 1.11 Plot of the electrostatic potential (charge/ionic radius) vs. ionic radius for different metallic ions. Ionic radii are given in Å (10^{-10}m). The curves of constant charge (1+ to 5+) are also shown. The roman numerals refer to coordination numbers. This plot is useful for the identification of groups of elements with coherent geochemical behavior: network-forming elements, high-field-strength elements (HFSE), and large-ion lithophile elements (LILE).

The ratio of the charge to the radius of an ion defines its electrostatic potential, i.e. its ability to modify the electrostatic field in which adjacent ions are bathing. This potential defines a large number of the properties associated with each element, such as its tendency to attract bond electrons (electronegativity) or to form compounds such as oxo-anions and hydrates. Plotting this potential vs. the ionic radius itself, which is a measure of the ion's ability to fit specific sites in minerals, is particularly informative of the geochemical properties of a particular ion (Fig. 1.11). Large ions carrying weak charges (K, Rb, Cs, Ba) are accommodated with difficulty by the main minerals of the mantle, except for K-feldspars, and so are concentrated in the continental crust. They are known as large-ion lithophile elements (LILE). Small ions carrying strong charges (Zr, Nb, Th, U) develop intense electrostatic fields (high-field-strength elements, HFSE), hence they do not readily substitute for the major elements in ordinary minerals. Relative positions in the periodic table remain, however, essential, and elements in the same column share common geochemical properties, as is the case of the alkali metals (Na, K, Cs), alkaline-earth metals (Mg, Ca, Sr), halogens (F, Cl, Br), or transition elements of the same row.

1.5 The different reservoirs and their compositions

In solids, elements are sequestered in minerals, minerals assemble as rocks, and rocks constitute the main geodynamic units of the mantle and the crust. Some elements occur in particularly large abundances in seawater or in the atmosphere (H, O, N, Ar). Terrestrial rocks fall into three main categories: igneous rocks, such as basalts and granites, produced by magmatic processes, i.e. from the melting of rocks; sedimentary rocks, formed by the accumulation of clastic and biological particles or by chemical precipitation on the floors of the oceans and other bodies of water; and metamorphic rocks, produced by "firing" existing rocks at high temperature, under pressure, and, for most of them, in the presence of aqueous or carbonic fluids. Extra-terrestrial rocks cannot be classified in quite the same way. Igneous rocks occur on the Moon, most probably on Mars, and as meteorites (achondrites). Another type of meteorite, known as a chondrite, has no terrestrial equivalent. Chondrites were formed by condensation of gases from the solar nebula and by droplets of silicate liquids called chondrules, from which their name is derived. Metamorphic transformations may affect planetary rocks and meteorites as well.

A reservoir is a loosely defined term referring to a very large body of rock (mantle, crust), water (ocean), or gas (atmosphere) whose mean composition stands in sharp contrast to the composition of other reservoirs. It may contain a variety of components, but its composition is normally very different from the composition of components present in other reservoirs: for instance, the Si-rich rocks making up the continental-crust reservoir are easily differentiated from the Mg-rich rocks making up the mantle. A reservoir may be spatially continuous (e.g. the ocean) or scattered over large distances in the Earth (e.g. recycled oceanic crust in the deep mantle). The most abundant chemical elements in each of the main terrestrial reservoirs are listed in Appendix A. However, a clearer understanding of these chemical distributions requires some idea of which minerals contain these elements (see next chapter). Oxygen is found just about everywhere, whereas silicon is confined to the silicates, which are by far the most abundant minerals. The silicon content of minerals is a particularly significant parameter because the SiO_2 (silica) concentration is a measure of the "acidity" of rocks: this obsolete term, which dates from the days when silicates were viewed as silicic acid salts, still pervades the literature. The silica concentration of minerals increases from olivine, pyroxene, amphibole, and mica through to feldspar and quartz. Magnesium and iron are particularly abundant in olivine, in pyroxene of igneous rocks, in amphibole, and in the sheet minerals (biotite, chlorite, and serpentine) of metamorphic rocks. A felsic (acidic) rock, such as a granite or a rhyolite, is rich in silicon and poor in magnesium and iron; a mafic rock, the archetype of which is basalt, has a high Mg and Fe content. Calcium is found, above all, in igneous pyroxene and calcitic feldspar (plagioclase), in sedimentary carbonate, and in metamorphic amphibole. Aluminum has many carriers: in igneous rocks it is located, by increasing order of pressure, in plagioclase, spinel (oxides), and garnet; it concentrates in the clay minerals of sedimentary rocks and in mica of metamorphic rocks (biotite and muscovite). Potassium and sodium are scarcely found outside mica and feldspar.

When rock melts, some elements (Na, K, Al, Ca, Si) are fusible, whereas others (Mg and to a lesser extent Fe) are more refractory; magmatic melting therefore contributes to geochemical fractionation among reservoirs. In the same way, some elements (Na, K, Ca, Mg) are more soluble in water than others, inducing further geochemical fractionation during erosion and sedimentation.

When the composition of the mantle is compared with that of the Earth as a whole it can be seen to have a high refractory-element content, especially of Mg and Cr, and a lower content of fusible elements, especially Na, K, Al, Ca, and Si, demonstrating its residual character with regard to melting. As might be expected, olivine and pyroxene are predominant in the mineralogy of the upper mantle: peridotite is the ubiquitous rock forming the upper mantle. The continental crust, on the other hand, is enriched in fusible elements (accommodated mainly in feldspar, quartz, and clay minerals) and exhibits a melt "liquid" character in contrast to the residual mantle. The oceans are obviously enriched in soluble Na, K, and Ca cations and anions (Cl^-, SO_4^{2-}), while elements that are both insoluble and fusible (Si, Fe, and Al) accumulate in clastic sedimentary rocks (clays).

The composition of the Sun, the Earth's crust and mantle, etc., is given in Appendix A. It is not always easy to determine these compositions; while observation of the solar spectrum and analysis of meteorites, of seawater, and of river water yield data that can be tabulated fairly directly, determining the composition of the Earth's crust calls for discussion of the nature of the lower crust, the lower mantle, the core, and of the Earth as a whole. The mechanisms responsible for forming these major, but not directly observable, geological reservoirs must therefore be reasonably well understood before their compositions can be estimated. This will be covered briefly in Chapters 11 and 12.

1.6 The nucleus and radioactivity

Although the forces holding the nucleus together are extremely powerful, observation shows that some nuclei are unstable, i.e. radioactive. Radioactivity is a property of the nucleus and involves gigantic energies of the order of a few MeV per nucleon (1 eV/molecule $= 1.6 \times 10^{-19} \times 6.022 \times 10^{23} = 96.5$ kJ mol^{-1}) (Table 1.2). The temperatures, which are a measurement of atomic and molecular excitation, required to break a nuclear bond are orders of magnitude higher than those required to exchange electrons (chemical reactions) or even to remove an electron from an atomic orbital (ionization energy) and only occur in stellar interiors like that of the Sun. These enormous energies explain the extreme efficiency of nuclear energy with respect to any other alternative energy form. Nuclear bonds are not dependent on the atom's electron suite and therefore not on any chemical reaction or mineralogical phase change which take place at much lower energies: removing the first electron from an atom typically requires only a few eV per atom (ionization potential). Mineralogical, temperature, and pressure processes involve even smaller energies. Radioactivity is therefore independent of the chemical and mineralogical environment of the element, of temperature and of pressure, which affects only the electron shells. For example, the rubidium-87 or uranium-238 decay probability per unit of time is

Table 1.2 The molar energy scale of different physical and chemical processes

Bond	Strength (eV/molecule)	Strength (kJ mol^{-1})	$T(K)^1$
Nuclear	6×10^6	5.8×10^8	7.0×10^{10}
Electronic	6	580	69 800
Chemical^2	4.1	393.5	47 400
Kinetic^3	4.2×10^{-5}	0.040	0.5

[1] Equivalent temperature (energy per mol/gas constant).
[2] Combustion of one mole (12 grams) of carbon in oxygen.
[3] One mole of air (22.4 liters at ambient pressure) at 60 km h^{-1}.

the same in the ocean or in a lake, in granite or limestone, in the Earth's crust or lower mantle, on the Moon or on Mars, etc. This probability has not varied measurably over the course of geological time. The remarkable consistency of radioactive isotopic chronometers of diverse geologic and planetary objects obtained by using different radioactive isotopes dispels any doubt about the validity of nuclear clocks. The extreme temperature conditions found in massive stars provide a few exceptions to this rule, but these are of no practical relevance to our Solar System.

There are a number of decay processes:

1. The α (alpha) process, which is the emission of a helium nucleus (two protons and two neutrons), is common at high mass. The nucleons (positively charged protons and neutral neutrons) are held together by the short-range attractive strong force. This force is essentially restricted to adjacent nucleons and therefore varies linearly with their number. In contrast, the electromagnetic force, which tends to pry the positively charged protons apart, is weaker but acts over a broader range, so that it involves the entire nucleus. It varies with the total number of proton pairs and therefore with the squared number of protons. Overall, for the heavier nuclei, the repulsive force therefore tends to compensate the attractive force and the mean energy holding the particles together decreases. A heavy nucleus become knobby and wobbly and some parts tend to detach. The α (alpha) process occurs when the sum of the masses of the daughter nuclide and the α particle is less than the mass of the parent nuclide. It is a consequence of a quantum property of nuclear particles known as the tunnel effect: although the potential barrier opposing the ejection of an α particle is much greater than the energy gain resulting from separation, there is a non-zero probability that an ejection will occur. For example $^{147}\text{Sm} \rightarrow {}^{143}\text{Nd} + \alpha$.

2. The β^- (beta minus) process involves the emission of an electron by the parent nucleus. In a vacuum, a free neutron does not survive more than 15 minutes before it turns into a proton and an electron. A neutron bound in a nucleus is definitely more stable but, as indicated above, the lower energy of the proton–neutron interaction with respect to that of similar nucleons favors a nucleus with an equal number of protons and neutrons. When $N > Z$, this condition is violated, and excess energy is released by converting a

neutron into a proton and an electron. This process calls for a special type of force, the weak interaction, which is analogous to electromagnetic forces but, as reflected by its name, much weaker. This interaction is of little relevance to our familiar environment and is difficult to explain simply. The continuous distribution of electron energy emitted during this process cast doubt on the quantification of nuclear energy until it was shown that there was a particle of zero mass, a neutrino, that practically does not interact with the matter through which it passes, but which can carry considerable energy. Collapsing stars lose energy through neutrino emission. For example, $^{87}Rb \rightarrow {}^{87}Sr + \beta^- + \bar{\nu}$, where β^- is the electron and $\bar{\nu}$ an antineutrino. Beta-minus radioactivity is a common process when the nucleus has a high neutron/proton ratio. A symmetrical process of β^+ (beta plus) positron emission (particle with the same mass as an electron but of opposite charge) takes place when the nucleus has a high proton/neutron ratio ($N < Z$), a situation rarely found for natural nuclides, which, as we will learn later in Chapter 12, form in neutron-rich environments.

3. Capture of an electron of the K shell is a less frequent process, and was identified by von Weizsäcker during his investigation of excess argon-40 in the Earth's atmosphere. For example, $^{40}K + e^- \rightarrow {}^{40}Ar$. Electron capture affects the nuclide in much the same way as positron emission does.

4. Spontaneous fission of some heavy atoms like uranium-238 or plutonium-244 is a rather rare and very slow process; it forms the basis of the fission-track dating method.

A nuclide may be unstable with respect to two decay processes simultaneously and the probabilities of decomposition by each process are additive. For example, potassium-40 (^{40}K) decays dually into ^{40}Ca by β^- emission and into ^{40}Ar by electron capture; this is known as a branched mode of radioactive decay.

Exercises

1. Give the first three quantum numbers of an electron in a 3p orbital.

2. Find the electronic formula of the elements K, Hf, and F. Compare the orbital electronic configuration for K and Rb (s block), Hf and Zr (d block), F and Cl (p block). For each element, which ionic configuration should be the most stable?

3. Plot the ionic radius and the first ionization energy as a function of the number Z of protons.

4. Which ion of Na^+, K^+, Rb^+, Mg^{2+}, Ca^{2+} has the smallest radius? Why?

5. Why is the ionic radius of Yb (0.99 Å) smaller than that of La (1.16 Å)?

6. Calculate the crystal field stabilization energy (CFSE) in units of Δ for the common ions Sc^{3+}, Ti^{4+}, V^{5+}, Cr^{3+}, Mn^{2+}, Fe^{2+}, Co^{2+}, Ni^{2+}, Cu^{2+}, and Zn^{2+} in octahedral vs. tetrahedral environments in high-spin configuration.

7. The PetDB database (http://www.petdb.org/) collects geochemical data on mid-ocean ridge basalts (MORB). Table 1.3 gives the concentrations of the transition elements of the first row in some glassy samples from the East Pacific Rise (EPR) and also the

Table 1.3 Concentrations of transition elements of the first row in some glassy samples from the East Pacific Rise (EPR) and also the mean value of these concentrations calculated in MORB from a much larger set of samples. Data in ppm except for TiO_2, MnO, and FeO, which are given in weight%

Sample	Sc	TiO_2	V	Cr	MnO	FeO	Co	Ni	Cu	Zn
1	26.6	2.07	176	5.99	0.22	11.8	26.8	10.6	45.9	131
2	41.0	2.27	385	23.4	0.22	12.7	46.3	33.5	74.9	114
3	44.1	3.21	361	81.7	0.25	15.3	45.8	72.7	74.3	115
4	43.6	1.77	308	161	0.21	10.8	45.8	64.4	86.8	92.4
5	38.1	2.08	277	204	0.17	10.2	45.4	104	72.7	88.2
6	32.4	0.96	180	383	0.13	8.5	45.3	145	93.4	50.5
7	33.7	1.34	207	665	0.17	9.2	52.7	632	81.2	68.4
Mean EPR	38.7	1.90	270	199	0.20	10.8	43.7	84.4	74.4	94.1

mean value of these concentrations calculated in MORB from a much larger set of samples. Normalize the data for each sample to the mean value, plot the normalized values against the atomic number, and use the results from the previous exercise to infer whether fractionation is due to the precipitation of minerals in which these elements are in octahedral sites (e.g. olivine, pyroxene) or to minerals in which they could also reside in tetrahedral sites (oxides).

8. The platinum-group elements (PGE) consist of the following transition elements: Ru (44), Rh (45), Pd (46), Os (76), Ir (77), and Pt (78) with their proton numbers given in parentheses. Discuss how to apply the crystal field theory to these elements and decide what you need to know to understand their crystal chemistry.

9. Let the length of a polyhedron edge be a and the radius of the circumscribed sphere be R. For a tetrahedron it can be shown that $R = (1/4)a\sqrt{6}$, while for an octahedron $R = (1/2)a\sqrt{2}$. The ionic radius of O^{2-} is 1.4 Å. Calculate the radius of the cations that closely fit in a tetrahedral or octahedral site. Compare with a table of ionic radii (e.g. http://www.webelements.com).

10. Using the data listed in Table 1.4, draw a semi-logarithmic plot of the composition of:
    (i) ordinary chondrites normalized to CI carbonaceous chondrites for decreasing 50 percent condensation temperature
    (ii) continental crust normalized to the BSE or primitive mantle by decreasing values of the normalized value
    (iii) the primitive mantle normalized first to CI carbonaceous chondrites, then to ordinary chondrites using the same order
    (iv) a Hawaiian basalt and a mid-ocean ridge basalt normalized to primitive mantle using the same order.

11. Using the first plot of the previous exercise, try to evaluate the condensation temperature of the elements for which this parameter is missing.

12. Using all the plots and a periodic table, find which group of elements trivalent yttrium (Y) is homologous to. Then do the same for thorium (Th) and uranium (U).

Table 1.4 Concentration of various trace elements (in ppm) in different materials. CI = Orgueil-type chondrites, BSE = Bulk Silicate Earth. T_{cond} is the 50 percent condensation temperature

	T_{cond}	Carbonaceous chondrites (CI)	Ordinary chondrites	BSE	Continental crust	MORB basalt	Hawaiian basalt
K	1000	558	798	240	15772	848	12200
Ti	1549	436	617	1205	4197	8513	1.7
Rb	1080	2.3	3.0	0.600	58	1.26	2.1
Sr	–	7.8	10.7	19.87	325	113.2	221
Y	1592	1.56	2.10	4.30	20	35.82	19.5
Zr	1780	3.94	6.03	10.47	123	104.24	108
Nb	1550	0.246	0.38	0.658	12	3.51	8.3
Ba	–	2.34	4.23	6.600	390	13.87	60.2
La	1520	0.2347	0.31	0.648	18	3.90	6.808
Ce	1500	0.6032	0.88	1.675	42	12.00	17.56
Pr	1532	0.0891	0.126	0.254	5	2.074	2.81
Nd	1510	0.4524	0.66	1.250	20	11.18	13.74
Sm	1515	0.1471	0.193	0.406	3.9	3.75	3.665
Eu	1450	0.056	0.076	0.154	1.2	1.335	1.246
Gd	1545	0.1966	0.304	0.544	3.6	5.077	3.93
Dy	1571	0.2427	0.353	0.674	3.5	6.304	3.61
Er	1590	0.1589	0.236	0.438	2.2	4.143	1.738
Yb	1455	0.1625	0.215	0.441	2.0	3.90	1.426
Lu	1597	0.0243	0.032	0.0675	0.33	0.589	0.203
Hf	1652	0.104	0.167	0.283	3.7	2.97	2.548
Ta	1550	0.0142	0.023	0.0372	1.1	0.192	0.679
Pb	–	2.47	0.305	0.150	12.6	0.489	0.633
Th	1545	0.0294	0.043	0.0795	5.6	0.187	0.527
U	1420	0.0081	0.013	0.0203	1.42	0.071	0.181

13. Equilibrium of silicate melts can be described by the reaction

$$2O^- \Leftrightarrow O^{2-} + O^0 \tag{1.5}$$

in which O^0, O^-, and O^{2-} stand for bridging (Si-O-Si), singly bonded (Si-O) and free oxygen, respectively. Assuming that one mole of melt is produced from X_{SiO_2} moles of silica and $(1 - X_{SiO_2})$ mole of metal oxide MO, calculate the mole fractions of O^0, O^-, and O^{2-} with the assumption that the reaction above can be described by the equilibrium constant.

2 Mass conservation and elemental fractionation

Before discussing the composition of the different systems of geological interest and the exchanges of matter that make those systems evolve relative to each other, it is worth recalling the principles governing the geochemical differentiation of our planet. These seemingly simple principles conceal what are often daunting complexities. They are: the principle of conservation of mass, elementary and isotopic fractionation induced by phase changes, kinetic fractionation, and radioactivity.

In the Introduction, we alluded to the contrast between mixing processes and differentiation processes. We will now look at a number of examples.

1. Partial melting of the mantle beneath mid-ocean ridges produces basaltic liquids whose chemical composition is different from the ultramafic chemical composition of the source peridotite. This chemical fractionation of elements between the molten fluid and its parent medium can be described by thermodynamic rules. The former makes up the oceanic crust while the latter forms the refractory base of the lithosphere (located in the oceanic plates beneath the crust). It must be kept in mind that there is no chemical or isotopic fractionation in the system unless at least two phases co-exist (solid/liquid, vapor/liquid, mineral A/mineral B, . . .) each to host a different share of the initial inventory. Conversely, when the oceanic crust and oceanic lithospheric mantle plunge at subduction zones, they begin a long journey within the mantle, where convection folds and stretches them, in much the same way as a baker kneads dough, progressively eliminating the differences that initially existed between the two constituent parts. Convective mixing, or better, stirring, therefore undoes the effect of magmatic differentiation.

2. The erosion of a granite yields clay minerals and quartz grains, and adds to the dissolved load of run-off water; the clay accumulates on the ocean floor, the quartz is deposited as sandstone on the continental shelf or slope, and the river water mixes with

the ocean waters. Through the process of erosion, a granite body of initially homogeneous composition is separated into three different products that meet very different fates. Over the course of time, the sediments laid down are scraped off along subduction zones, where they are buried under considerable thicknesses of rock. There tectonic deformation and metamorphic mineralogical transformations at high temperature and high pressure, combined with fluid action, provide conditions amenable to the partial melting of crustal rocks (anatexis), and the siliceous fluids formed escape toward the surface producing granite intrusions. The work of differentiation performed by erosion is therefore destroyed by metamorphic and anatectic recombination.

3. In the ocean, chemical differentiation occurs at the surface, as biological activity draws on nutrient elements as well as calcium carbonate and silica. The redissolution of dead organisms and waste materials, together with bottom-water circulation, remixes the components that were separated out by biological activity. Once again the opposing processes of differentiation and mixing are at work.

We turn now to the methods of mass balance analysis, which are the essence of these differentiation and mixing processes. By such methods we can determine the relative abundances of the constituent parts mixed together in a composite product and evaluate the proportions of phases that have separated out in a differentiation product. First some essential definitions. Any system, whether natural or artificial, is made up of components and species. Components are chemical entities (atoms, ions, or, for a rock, metal oxides, such as Na, Na^+, Na_2O) from which the chemical composition of the rock can be described fully and uniquely. Rock compositions are commonly given in weight percent of its constitutive oxides SiO_2, Al_2O_3, Na_2O, etc., while the composition of solutions is normally given as moles of ionic component, Na^+, Ca^{2+}, Cl^-, etc., per kilogram of water. One unit system is as good as another, provided consistency is maintained. The set of components itself may not be unique: a rock can be described as proportions of atoms or as proportions of oxides. Components are not normally found as such in the system; a familiar but often misunderstood example being that of oxygen whose pressure (fugacity) can always be defined even when no gas phase is present. Components are preserved during reactions and phase changes, and their abundances are independent of the physical conditions in which the system finds itself. Species are all the forms of chemical associations of components, whether present or not (expressed or virtual), in the system. A species may be a $KAlSi_3O_8$ feldspar in a rock or an HCO_3^- ion in a solution. Species abundances are altered by temperature, reactions, and phase changes.

2.1 Conservation of mass

Let us begin with a few simple questions of geochemistry.

1. A river receives the input of a tributary. Given the flow rate and chemical composition of the two streams ahead of their confluence, what is the composition of the river water downstream of the confluence?

2. What is the chemical composition of a sediment for which the composition and abundance of the clay and quartz fragments are known?
3. A certain percentage of olivine of known composition precipitates from a basalt lava. What is the composition of the residual melt?

All three questions draw on the same principle: the whole is the sum of its parts. The elements of the river downstream are a combination of all of the upstream elements; the rock combines all of the elements present in the minerals; the parent basalt can be reconstructed by reincorporating the olivine in the residual lava. This principle is perfectly applicable, provided simply that all the components forming the aggregate system have been identified for certain. It is applicable whether the initial components have lost their physical identity during mixing (case 1) or whether their identity has been preserved, as in sand (case 2).

Let us try to determine the silicon content of the sediment in Question 2. We have a mass M_{sed} of sediment composed exclusively of M_{clay} kilograms of clay and M_{qz} kilograms of quartz. The concentration of silicon in the clay fraction is C_{clay}^{Si}, that of the quartz fraction C_{qz}^{Si}, and that of the total sediment C_{sed}^{Si}. The applicable conservation equations are first that of the total mass of material:

$$M_{sed} = M_{clay} + M_{qz} \tag{2.1}$$

and that of silicon:

$$M_{sed} C_{sed}^{Si} = M_{clay} C_{clay}^{Si} + M_{qz} C_{qz}^{Si} \tag{2.2}$$

Dividing (2.2) by (2.1) yields:

$$C_{sed}^{Si} = f_{clay} C_{clay}^{Si} + \left(1 - f_{clay}\right) C_{qz}^{Si} \tag{2.3}$$

where the weight fraction, $f_{clay} = M_{clay}/M_{sed}$, of the clay component is shown. Notice that the conservation of mass has been written in the trivial form $f_{clay} + f_{qz} = 1$. A similar expression may be obtained for each element, e.g. for aluminum:

$$C_{sed}^{Al} = f_{clay} C_{clay}^{Al} + \left(1 - f_{clay}\right) C_{qz}^{Al} \tag{2.4}$$

Subtracting the concentration of each corresponding element in the quartz from both sides of (2.3) and (2.4) and dividing one by the other gives a single equation no longer dependent on f_{clay}. This indicates that in a $\left(C_{sed}^{Al}, C_{sed}^{Si}\right)$ diagram (Fig. 2.1), the silicon and aluminum concentrations of sediments obtained by varying the proportion of quartz and clay form a straight line (mixing line) through the points representing the compositions of the two components. A mixture like this with linear properties is said to be conservative.

More generally, we are interested in a component j of mass M_j and an element i whose concentration in the component j is C_j^i. Here, i might refer to silicon and j to, say, the clay content of the sediment. The properties of the entire system, here the sediment, can be denoted by index 0, and the conservation equations can be written in the form:

$$\sum_j f_j = 1 \tag{2.5}$$

$$C_0^i = \sum_j f_j C_j^i \tag{2.6}$$

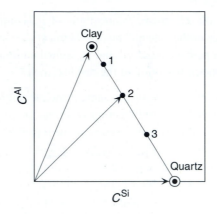

Figure 2.1 Graphic rules for conservation of mass and for mixing (arbitrary units). A mineral or rock composition can be represented by a vector in the chemical concentration space (here Al and Si). Points 1, 2, and 3 represent three compositions of mixtures of clay and quartz. Notice the linear character of the mixing relationships.

Equation 2.5 is known as the closure condition and indicates that the list of constituents is exhaustive. Equation 2.6 indicates that the concentration of the system as a whole is the weighted mean concentration of the composition of its components. The fraction by mass of each component in the mixture is the "weight" by which the concentration of each component must be multiplied to represent its contribution to the total inventory of the element in the entire system. An example of this principle, taken from everyday life, is the simple concept of rock. A rock is a sample, typically a piece of boulder or cliff, which has been just knocked off by the geologist's hammer. There is no such ideal thing as a rock in nature: it is a man-made sample, an artefact, whose composition results from a chance combination of all crystals collected by a particular hammer blow.

Vectors provide a useful means of describing the relationships of conservation. A mineral, a solution, or a rock in which the concentrations of n elements ($i = 1, \ldots, n$) have been measured are represented by a point (or a vector) in an n-dimensional space. In the example above, a vector equation could be written formally in the two-dimensional space of Si and Al concentrations:

$$\begin{pmatrix} C_{\text{sed}}^{\text{Si}} \\ C_{\text{sed}}^{\text{Al}} \end{pmatrix} = f_{\text{clay}} \begin{pmatrix} C_{\text{clay}}^{\text{Si}} \\ C_{\text{clay}}^{\text{Al}} \end{pmatrix} + f_{\text{qz}} \begin{pmatrix} C_{\text{qz}}^{\text{Si}} \\ C_{\text{qz}}^{\text{Al}} \end{pmatrix} \qquad (2.7)$$

supplemented, of course, by the closure condition $f_{\text{clay}} + f_{\text{qz}} = 1$. Equation 2.7 shows the sediment vector as a linear combination of the vectors representing the components. If the sediment is composed of three components (let us add, for the sake of illustration, a carbonate component), then a vector formed by the concentration of the three elements in the rock will lie within the triangle formed by the equivalent vectors representing the clay, quartz, and carbonate components. Generally, when trying to interpret the chemical composition of a mixture in terms of constituents that are sometimes too numerous to be shown graphically, statistical methods are used, the simplest and most effective of which is principal component analysis (PCA).

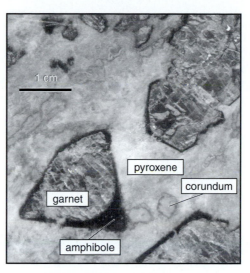

Figure 2.2 In this ancient metamorphosed basalt (eclogite), the fall in pressure when the rock rose to the surface destabilized the mineral assemblage. Pyroxene and garnet reacted with fluids no longer visible to give amphibole and corundum (sample courtesy of P. Thomas).

Representing compositions in vector form also sheds light on the concept of reaction assemblage, or mineral reaction, that is so useful in metamorphic petrology. Figure 2.2 shows a natural assemblage of minerals where pyroxene and garnet have reacted to produce amphibole and corundum.

Let us illustrate these principles through a simpler mineral assemblage, i.e. the quartz (SiO_2), forsterite (Mg_2SiO_4), and enstatite ($Mg_2Si_2O_6$) assemblage. A rock composed of two components or, as is customary in petrology, two oxides, cannot have more than two minerals co-existing at equilibrium. Three minerals cannot be stable simultaneously because, in (SiO_2, MgO) space, any "rock" vector could be represented by an infinite number of combinations of "mineral" vectors whose equivalence is represented by the reaction forsterite + quartz ⇔ enstatite (Fig. 2.3). Thermodynamics tells us that this reaction evolves spontaneously from left to right, causing complete consumption of either quartz or forsterite. The stable mineralogical assemblage therefore includes only two minerals: the enstatite and, depending on the composition of the rock, either the quartz or the forsterite.

Geochemists are fond of ratios of elements or isotopes. There are two reasons for this: ratios are generally measured more precisely than concentrations and they are less sensitive to dilution phenomena. For example, the abundance of olivine in a basalt does not alter the ratio of those elements that are segregated in the melt (incompatible elements). The evaporation of seawater concentrates Cl^-, Na^+, ions, etc., but does not alter their concentration ratios. It is a serious mistake to try to apply the rules for absolute concentrations to ratios. The results stated below are explained in Appendix B. For the ratio A/B (e.g. the Si/Al ratio), the conservation equations become:

$$\sum_{\text{compos. } j} \varphi_j^{B} = 1 \tag{2.8}$$

$$\left(\frac{C^A}{C^B}\right)_0 = \sum_j \varphi_j^{B}\left(\frac{C^A}{C^B}\right)_j \tag{2.9}$$

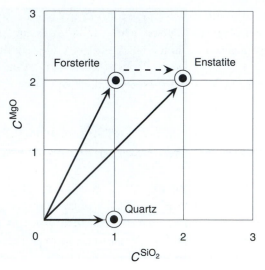

Figure 2.3 There cannot be three independent vectors in a two-dimensional space. A third mineral, therefore, requires there to be a reaction. Here forsterite + quartz ⇔ enstatite. This eliminates the deficient mineral, either quartz or forsterite. Units are numbers of oxide molecules per gram-formula weight.

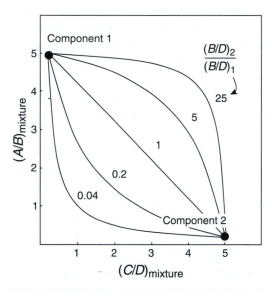

Figure 2.4 When concentration ratios or isotope ratios are plotted, the mixture or fractionation is not generally given by straight lines but by hyperbolas whose concavity depends on the abundance ratios of the elements in the denominator in the components of the mixture.

where the weight $\varphi_j^B = f_j C_j^B / C_0^B$ represents the fraction of element B residing in phase j. Equation 2.8 is a closure condition equivalent to (2.5), but this time restricted to element B. Let us now compare the combination of ratios in a mixing process using (2.9). It can be seen that, unlike the weights f_j assigned to the concentration balance (2.6), weights φ_j^B assigned to the ratio balance are dependent on the element in the denominator of the ratio. Therefore, in ratio/ratio diagrams, mixtures do not generally form straight lines but hyperbolas (Fig. 2.4). A mixture relation like this is linear only if the denominator element of each ratio is the same (or in a constant ratio). This is why the conservation relations in ternary petrological diagrams, whose coordinates are normalized to the sum of the three variables depicted (as in the famous AFM plots), are linear. By contrast, plots involving fractional parameters, such as the magnesium number mg# = Mg/(Fe + Mg), which is widely used in petrology, must be handled with care.

2.2 Elemental fractionation

It is very useful to introduce partition coefficients when studying the substitution of minor elements or trace elements in the lattice of minerals in equilibrium with magmatic fluids or natural solutions from which they precipitate. When an element i is in solution in two co-existing phases j and J (e.g. j stands for seawater and J for a carbonate that precipitates out), the Nernst law can be written:

$$\frac{x_J^i}{x_j^i} = K_{J/j}^i (T, P, x) = k_0 \exp \left(-\frac{\Delta G_0}{RT} \right) \qquad (2.10)$$

where x_j^i is the molar proportion of element i in phase j, R is the gas-law constant, and ΔG_0 a measure of the energy of exchange of this element between the two phases j and J. The partition (or distribution) coefficient $K_{J/j}^i$ depends on the temperature T, pressure P, and composition of the phases. The pre-exponential factor k_0 is a measure of the non-ideality of the solutions. After a simple adjustment that takes molecular weights into account, the concentration ratios between phases may also be described by partition coefficients. The ratios of trace element concentrations (typically less concentrated than 1000 parts per million) in two phases under comparable conditions of temperature, pressure, and aggregate composition of the system (acid, basic, aqueous) are usually constant and referred to as partition coefficients. It is a common usage in geochemistry to restrict the term partition coefficients to mineral/liquid coefficients, but this choice is arbitrary and occasionally misleading. A liquid/mineral partition coefficient is a perfectly valid concept. Figure 2.5 gives a few examples of mineral/liquid partition coefficients in magmas.

The term "reasonably" justifies the activity of many experimental and analytical laboratories. To a fairly good degree of approximation, the dependence of K on temperature and pressure can be described by:

$$d \ln K = \frac{\Delta H}{RT^2} dT + \frac{\Delta V}{RT} dP \qquad (2.11)$$

(Appendix C) where ΔH and ΔV measure differences in the enthalpy and molar volume of element i between the two phases. The dependence of K on composition is more complex: a satisfactory model for solids is Onuma's elastic model, in which energies are elastic energies required by the substituting element to replace that element which normally makes up the mineral. In order to insert an atom of ionic radius r_0 into a crystallographic site, which we consider as a spherical cavity of radius r, work must be done against the electrostatic forces F. As a first approximation, these forces can be taken as proportional to the contraction or expansion of the ionic radius from r to r_0, i.e. $F \approx k (r - r_0)$ (Hooke's law).

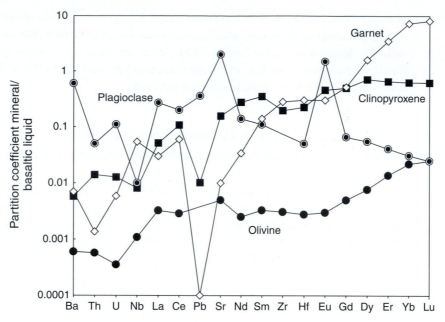

Figure 2.5 Typical partition coefficients for some important trace elements between the main minerals and the liquid for a basalt composition.

In this equation, k is a constant related to certain elastic properties of the medium known as Young's modulus and the Poisson ratio. A measure of the change in elastic energy U upon compression or expansion is:

$$dU = -PdV = -\frac{F}{4\pi r^2}4\pi r^2 dr = -F dr \qquad (2.12)$$

which is integrated between r_0 and r as:

$$\Delta U = -\frac{k}{2}(r - r_0)^2 \qquad (2.13)$$

Most ions of identical charge fit in with such a parabolic relationship between the binding energy, and therefore $\ln K$, and their squared radius (Fig. 2.6). Measurement of the different ΔH values and evaluation of elastic energies are daunting tasks. Although fractionation theory applies remarkably well to both low- and high-temperature systems, it is essential to remember that it requires approximations, however good they may be, and that it cannot be compared with a zero-order principle such as the conservation of mass.

If one phase plays a particular role, such as seawater or a magmatic liquid from which minerals precipitate, the fractionation of elements can be described by taking this phase as a reference and introducing an aggregate (bulk) solid/liquid partition coefficient $D^i_{s/l}$ such as:

$$\sum_{\text{solids } j} f_j = 1 \qquad (2.14)$$

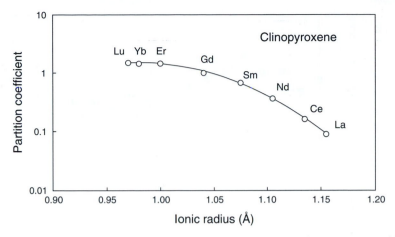

Figure 2.6 Variation of the partition coefficients of rare-earth elements between clinopyroxene and basaltic liquid by ion radius (Blundy *et al.*, 1998). The parabolic shape of the curve reflects a physical control of the partition coefficient by the elastic properties of the crystalline lattice.

$$D^i_{s/l} = \sum_{\text{solids } j} f_j K^i_{j/\text{liq}} \tag{2.15}$$

described as the weighted mean of individual partition coefficients $K^i_{j/\text{liq}}$ of element i between mineral j and the liquid, the weights being the abundance of each mineral in the precipitate. Although (2.15) is exact, this equation does not suppose that $D^i_{s/l}$ remains constant when the mineral fractions f_j change during the process. Later we will come across geological settings where these conditions are fulfilled. The coefficient $D^i_{s/l}$ provides a measure of the element compatibility (or of its refractory character) in a given environment, and it is standard practice in igneous geochemistry to represent the concentrations of elements of different magmatic rocks in order of increasing compatibility as the mantle melts, the mean order being established by experiment. These "spidergrams" entail normalization, i.e. division by concentrations of a reference rock or reservoir (chondrites, Bulk Silicate Earth, mid-ocean ridge basalts = MORB) and have become a standard graphic tool in geochemistry (Fig. 2.7).

Let us apply the simple theory of chemical fractionation to the change in the concentration of an element i during partial melting of a source rock made up of several minerals. This theory leads to the so-called equilibrium- or batch-melting equations. If the proportion of liquid in the molten rock is F (melt fraction), the mass balance condition is written:

$$C^i_{\text{source}} = F C^i_{\text{liq}} + (1 - F) \sum_{\text{solids } j} f_j C^i_j = \left[F + (1 - F) D^i_{s/l} \right] C^i_{\text{liq}} \tag{2.16}$$

or

$$C^i_{\text{liq}} = \frac{C^i_{\text{source}}}{F + (1 - F) D^i_{s/l}} \tag{2.17}$$

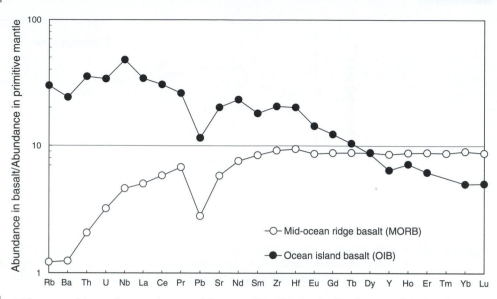

Figure 2.7 Spidergram of trace elements in essential types of basalts standardized to a model of concentration of the primitive mantle. The incompatibility of elements, i.e. the tendency of crystals to reject them to the melt, diminishes toward the right. MORBs are "depleted" in incompatible elements, while ocean island basalts (OIB) are "enriched" relative to the "primitive mantle."

This equation separates two regimes of melting that are best understood by plotting the enrichment factor $C^i_{\text{liq}}/C^i_{\text{source}}$ during melting as a function of the bulk solid/liquid partition coefficient (Fig. 2.8). For $F < D^i_{\text{s/l}}$, the enrichment factor is nearly constant and equal to $1/D^i_{\text{s/l}}$, while for $F > D^i_{\text{s/l}}$, the enrichment factor varies approximately as $1/F$. In the mantle or crust, the melt fraction F is generally low, of the order of 0.1 to 15%. For incompatible elements such as Th, Nb, La, or Ba, $D^i_{\text{s/l}}$ is virtually negligible and the elements are very rapidly extracted from the residue: their concentration in the liquid is inversely proportional to the melt fraction and so varies very rapidly at low values of F. For a highly compatible element like nickel, $D^i_{\text{s/l}}$ is very high, while F remains low compared with unity. For the compatible elements, concentration C^i_{liq} varies very little with F. Concentrations of primary basalt liquids in nickel, but also in magnesium, chromium, and osmium, are therefore buffered by the melt residue. Buffering of elements with large $D^i_{\text{s/l}}$ by the solid residue is clearly visible in Fig. 2.9, which shows that the variability of basalt compositions falls very markedly as the level of compatibility of the element increases.

However complex the distributions observed in rocks, and especially in magmatic rocks, it is useful to understand that certain elements are the unmistakable fingerprint of certain minerals. For example, olivine concentrates Ni, garnet concentrates heavy rare-earth elements, the pyroxenes concentrate Sc and Cr, and the feldspars concentrate Eu and Sr. The trained observer can quickly spot from an analysis of trace elements the influence of certain minerals during the formation of a rock.

Separate treatment is reserved for elements that reach the limit of saturation leading to phase changes. This is what happens with zirconium in felsic lavas and granites when zircon ($ZrSiO_4$) precipitates, with chromium in basalts whenever chromite forms, and more

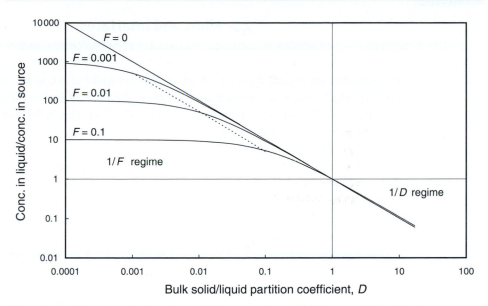

Figure 2.8 Enrichment/depletion factor during melting as a function of the bulk solid/liquid partition
coefficient D for different degrees F of melting. Note the two regimes separated by the dashed
line $F = D$: for $F < D$, the enrichment factor remains constant and equal to $1/D$, while for $F > D$,
it varies as $1/F$. Efficient fractionation between very incompatible elements requires small melt
fractions.

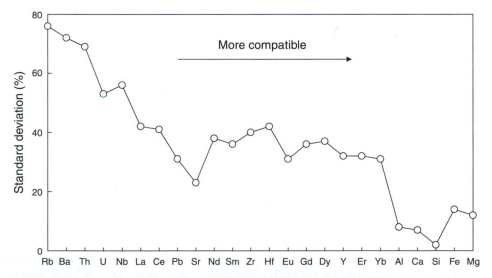

Figure 2.9 The variability of concentrations of elements in basalts (measured by the relative standard
deviation) decreases with their compatibility (after Hofmann, 1988). This reflects the buffering of
the more compatible element concentrations in melts by the residual solid.

importantly still with water or CO_2, which separate out in a vapor phase when magmas
ascend. At this point, the concentration of the element forming the major phase (e.g. Zr
or CO_2) remains fixed, we say buffered, by either the zircon or the vapor. The balance
equations have to be re-written to cover this new situation.

2.3 Films and interfaces

Although they are essentially made of minerals, most rocks also contain some interstitial material, which may evade precise analysis when the scale of the constituting phases is too small, typically below 10 nanometers (1 nm = 10^{-9} m). This is the case for cement in sedimentary rocks, but also for common interstitial phases of igneous rocks and metamorphic rocks, such as apatite, oxides, carbonates, clay minerals, which have been deposited by the percolation of aqueous or carbonated fluids. Peridotite xenoliths in basaltic rocks often look shiny because of the presence of an interstitial film of magmatic glass. It is also now well understood that a cryptic mechanism of diffusive segregation often concentrates elements incompatible with the structure of olivine from mantle rocks, Ti, Ca, Al, as ill-defined mineral phases no larger than a few nm.

Two major questions arise with interfaces and films. First, the energy of small minerals should be understood through the sum of two terms: volume energy, which is the only form we have been considering until now, and surface energy, which is the extra energy needed to produce the dangling bonds. It is therefore inconvenient to treat phases that take up hardly more than a few lattice units as we would treat large crystals. Second, we may wonder whether minute phases and films account for a substantial proportion of the elementary budget of a rock sample. Let us take an ideal rock formed of spherical minerals from a single species, e.g. olivine, and with homogeneous radii a. We assume that these minerals are homogeneously covered with a film of thickness δa. Let us consider an element with a whole-rock concentration C_0, a concentration C_R in the mineral, and a partition coefficient D_f between the interstitial film and the mineral. In order to calculate the contribution of the film to the total inventory of the rock for this element, we first note that the mass fraction F_f of the film is

$$F_f = \frac{4\pi a^2 \delta a}{\frac{4}{3}\pi a^3 + 4\pi a^2 \delta a} = \frac{3\delta a/a}{1 + 3\delta a/a} \tag{2.18}$$

and then write the mass balance as:

$$C_0 = (1 - F_f)C_R + F_f D_f C_R \tag{2.19}$$

Since $F_f \ll 1$, it is clear that the impact of the film on the rock budget becomes noticeable only when $F_f D_f \approx 1$. For a film of 100 nm wrapped around olivine crystals with a radius of 1 mm, the effect is only visible for elements with $D_f > 10^4$. This is potentially harmful for large-ion lithophile elements (LILE) such as U, Th, and Ba: the external part of minerals must be carefully taken away before analysis. It is standard practice for this type of analysis to remove mineral grains with exposed interface under the optical microscope and then to submit the residue to leaching (= partial dissolution in strong acids) before complete dissolution. The ground truth, however, is that situations in which most of the element inventory resides in the interstitial phase ($D_f F_f \gg 1$), with sample chemistry being driven by the interstitial phase, are exceedingly rare.

2.4 Distillation processes

The distillation of alcohol, whereby we seek to enrich the ethyl-alcohol content of the condensate resulting from the boiling of wine or any other low-grade fermented beverage, is a familiar enough example. When a finite quantity of a substance changes state and the product of the transformation is isolated, the progressive chemical or isotopic fractionation associated with the progress of this transformation defines the phenomenon of distillation. This process occurs naturally in various geological settings:

- fractional crystallization of magmas during which solids, termed cumulates, are successively isolated from the residual magma;
- fractional melting of the mantle, producing liquids extracted immediately from the molten source;
- progressive condensation of atmospheric water vapor, in the course of which precipitation loses contact with the high atmosphere H_2O;
- boiling of hydrothermal solutions.

It can be shown (see box) that the change in concentration C_{res}^i of an element i in the residual phase during formation of a new phase obeys Rayleigh's law:

$$\mathrm{d} \ln C_{res}^i = \left(D^i - 1 \right) \mathrm{d} \ln f \tag{2.20}$$

where D^i is the bulk partition coefficient between the new phase and the parent residual phase (the order is essential) and f the fraction by mass of the residual phase relative to the stock of original material. The most common form of this equation is:

$$C_{res}^i = C_0^i f^{D^i - 1} \tag{2.21}$$

where C_0^i is the concentration in element i of the parent phase of the original material, i.e. for $f = 1$.

A first application of this theory is the fractional crystallization of magmas, for which we obtain:

$$C_{liq}^i = C_0^i f^{D^i - 1} \tag{2.22}$$

Here, the parameter f represents the residual liquid fraction and C_0^i the composition of the parent magma. The concentration of the solid at equilibrium with the liquid is obtained by multiplying (2.22) by the partition coefficient of the element. For incompatible elements, like Th, Ba, and La, the partition coefficients D^i, which measure solid/liquid fractionation, are low: their concentrations in the residual magma and, consequently, in the extracted mineral phases are therefore inversely proportional to f. Over most of the range of magma differentiation, liquid is the dominant phase. Parameter f is therefore close to unity and the concentrations of incompatible elements vary little during crystallization. Moreover, the ratios of incompatible elements (e.g. Th/La) in differentiated lavas are virtually unaffected by fractional crystallization. By contrast, for compatible elements, such as Ni and Cr, high D^i values cause very high variations in concentration in the residual

The Rayleigh distillation equation

Let us consider a liquid system at high temperature and sufficiently well stirred for its composition to remain homogeneous. As the liquid cools, a solid phase appears and we are going to look at how the concentration of an element i changes as all of the liquid in the reservoir progressively crystallizes. We designate the properties of the newly crystallized solid by the subscript sol, those of the residual liquid by the subscript liq, and those of the liquid at the beginning of crystallization by the subscript 0. We write the mass of each phase (solid or liquid) as M and the mass of element i accommodated by a phase as m^i. We apply two conditions: that of the closed system (transformation of liquid to solid, without external inputs or outputs) and that of equilibrium fractionation of element i between the solid and liquid. The closed-system condition is written in its incremental form:

$$dM_{liq} + dM_{sol} = 0 \tag{2.23}$$

$$dm^i_{liq} + dm^i_{sol} = 0 \tag{2.24}$$

The equilibrium condition between the last solid formed and the residual liquid is written using the partition coefficient $D^i_{s/l}$ in the form:

$$\frac{dm^i_{sol}}{dM_{sol}} = D^i_{s/l} \frac{m^i_{liq}}{M_{liq}} \tag{2.25}$$

Let us replace the properties of the liquid by those of the solid from (2.23) and (2.24):

$$\frac{dm^i_{liq}}{dM_{liq}} = D^i_{s/l} \frac{m^i_{liq}}{M_{liq}} \tag{2.26}$$

or

$$\frac{dm^i_{liq}}{m^i_{liq}} = D^i_{s/l} \frac{dM_{liq}}{M_{liq}} \tag{2.27}$$

Taking the logarithm of concentration $c^i_{liq} = m^i_{liq}/M_{liq}$ of element i in the liquid and differentiating gives:

$$\frac{dc^i_{liq}}{c^i_{liq}} = \frac{dm^i_{liq}}{m^i_{liq}} - \frac{dM_{liq}}{M_{liq}} \tag{2.28}$$

By combining (2.27) and (2.28), and introducing the fraction of residual liquid $F = M_{liq}/M_0$, we obtain:

$$\frac{dc^i_{liq}}{c^i_{liq}} = \left(D^i_{s/l} - 1 \right) \frac{dF}{F} \tag{2.29}$$

which is the form sought, applied to progressive crystallization of a liquid, and which leads us to (2.20).

	Parent	Evolving	Partition	
Process[a]	phase	phase	coefficient	Environment
Crystallization	Liquid	Solid	Solid/liquid	Magmatic
	Magma	Cumulate	Magma/cumulate	Magmatic
	Brine	Salt	Salt/brine	Evaporitic
Melting	Source rock	Magma	Liquid/solid	Magmatic
Evaporation	Solution	Vapor	Vapor/liquid	Hydrothermal
Condensation	Vapor	Liquid water	Liquid/vapor	Atmospheric

Table 2.1 Distillation processes

[a] Add "fractional" to the term describing the process.

magma for low variations of f. Thus the precipitation of 10–20% olivine removes most of the nickel initially present in the magma (Fig. 2.10).

Conversely, the coefficients of interest for the fractional melting in which the liquid is progressively drawn off will be the liquid/solid fractionation coefficients. This gives the equation:

$$C_{liq}^i = \frac{C_0^i}{D^i} (1 - F)^{\frac{1}{D^i} - 1} \qquad (2.30)$$

where $F = 1 - f$ is the liquid fraction extracted, but C_0^i is now the concentration of the source (mantle, crust, etc.). In a system with a low degree of melting, which is the usual case in nature, the value of F is close to 0. By symmetry with the foregoing reasoning, it can be seen that the concentrations of incompatible elements, which have very high liquid/solid partition coefficients (note the reverse order of phases), will vary greatly with the degree of melting. As in the case of partial fusion at equilibrium, concentrations of compatible elements remain buffered, however, by the residual solid. A simple, but crucial, principle is inferred: melting should be preferably investigated with incompatible elements, crystallization with compatible elements (Fig. 2.10).

Dividing (2.20) written for A by the ratio of the same equation written for B, the variable f can be eliminated:

$$\frac{d \ln C_{res}^A}{d \ln C_{res}^B} = \frac{D^A - 1}{D^B - 1} \qquad (2.31)$$

In a logarithmic plot, the so-called log–log plot, $(\ln C^A, \ln C^B)$, concentrations of the different phases present therefore form two-by-two alignments. Examples of this type of relation abound in the geochemical literature (Table 2.1). An important application is the demonstration that mid-ocean ridge basalt chemistry can be accounted for by fractional melting of the mantle, in which case the differential form of (2.30) applies:

$$\frac{d \ln C_{MORB}^A}{d \ln C_{MORB}^B} = \frac{d \ln C_{res}^A}{d \ln C_{res}^B} = \frac{\left(1/D^A\right) - 1}{\left(1/D^B\right) - 1} \qquad (2.32)$$

where res stands for residue and D for the bulk-solid partition coefficients during MORB melting.

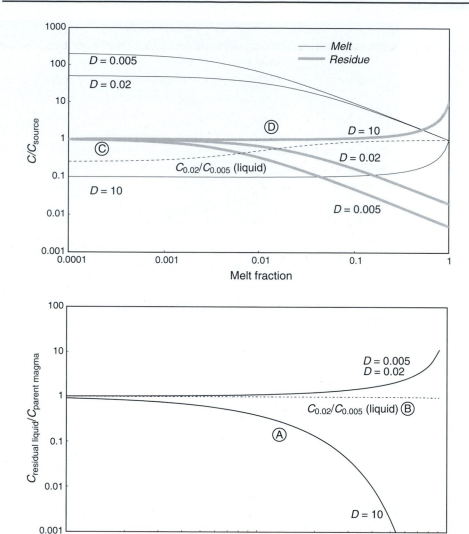

Figure 2.10 Comparison of the effect of fractional crystallization (bottom) and partial melting (top) on the concentration in magmatic melts of elements with different compatibilities. The diagrams show the case of bulk partition coefficients for elements with $D = 0.005$, 0.02, and 10. Curve A: Even the removal of a few percent of a mineral hosting a very compatible element, e.g. Ni in olivine ($D \approx 10$), changes concentration in the residual melt drastically, whereas this change is hardly noticeable with highly incompatible elements such as Th, Ba ($D \approx 0.005$), or La ($D \approx 0.02$). Curve B: Even extreme mineral fractionation does not change the ratio of two incompatible elements such as Th/La in the residual melt. Curve C: For small to moderate fractions of melt, compatible elements remain buffered by the solid. Curve D: In contrast, the ratio of two incompatible elements is fractionated until the fraction of melt exceeds the largest of the two partition coefficients. At this stage, the bulk of the incompatible elements has been transferred to the melt and their relative concentrations are proportional to their abundances in the source. The dashed lines labelled $C_{0.02}/C_{0.005}$ represent the ratio of two elements with partition coefficients of 0.02 and 0.005, respectively.

Exercises

1. Table 2.2 shows the major element compositions of the glass (= gl) and mineral phases (olivine = ol, clinopyroxene = cpx, and plagioclase = plag) in a basaltic lava from the Réunion Island hot spot (Indian Ocean). Calculate the composition of a bulk lava made of this glass with 15 wt% olivine and 10 wt% clinopyroxene phenocrysts.
2. Using the same data, calculate the major element composition of a wehrlite cumulate (80 wt% olivine and 20 wt% clinopyroxene), of a gabbroic cumulate (10 wt% olivine, 40 wt% clinopyroxene, 50 wt% plagioclase).
3. Using the same data, we consider how much cumulus olivine may change whole-rock compositions. Keeping the same interstitial glass and assuming variable proportions of olivine phenocryst from 0 to 40 percent, calculate the major element contents of the whole rock and turn the MgO and FeO weight percents into moles per 100 grams of rock using the following atomic weights: 55.847 for Fe, 24.305 for Mg, and 16.00 for O. It is known that, at equilibrium, $(FeO/MgO)_{ol}/(FeO/MgO)_{gl} = 0.3$ and this condition can be drawn as a line in the same diagram. Plot the molar FeO/MgO ratio of the olivine vs. this ratio in the whole rock for olivine with different forsterite contents (e.g. different FeO/MgO ratios). Can you suggest a method to estimate whether olivine megacrysts present in a lava are phenocrysts in equilibrium with their host glass or xenocrysts? This method should not involve the analysis of the glass itself.
4. Table 2.3 shows the partial analysis (wt% of oxide) of a deep-sea sediment composed of quartz, clay, and carbonates. Calculate the abundance of each mineral phase in this sample.
5. Let us evaluate the concentration of the siderophile elements Fe and Ni and the proportion of other elements in the Earth's core. We assume that the Earth is made of a silicate mantle (we neglect the crustal contribution), which is often referred to as the Bulk Silicate Earth, and a metallic core. The core makes one third of the total mass of the Earth. We further assume that the Earth is made of ordinary chondrites and we will

Table 2.2 Concentration of major elements (in weight percent of oxide) in the different phases of a basaltic lava from the Réunion Island (Indian Ocean)

	Glass	Olivine	Clinopyroxene	Plagioclase
SiO_2	48.81	38.79	46.63	52.49
TiO_2	2.68	0.047	3.8	0.153
Al_2O_3	14.40	0.026	6.27	29.05
FeO	11.07	20.02	8.25	0.893
MnO	0.17	0.269	1.03	0.027
MgO	6.65	40.56	13.5	0.095
CaO	11.65	0.297	19.64	12.27
Na_2O	2.74	0.04	0.59	4.33
K_2O	0.79	0.024	0.05	0.372

Table 2.3 Composition of a deep-sea sediment and its mineral phases

Oxide	Sediment	Quartz	Clay	Carbonate
SiO_2	45.7	100	51.4	0
Al_2O_3	13.2	0	26.4	0
CaO	16.5	0	0	55

Table 2.4 Concentration of Fe and Ni (wt%) in H and LL ordinary chondrites and for the Bulk Silicate Earth (BSE).

Element	H	LL	BSE
Fe	27.5	18.5	6.3
Ni	1.6	1.0	0.020

Table 2.5 Concentrations of various cations in water samples (1–3) taken from an estuary (μmol l^{-1})

Element	River	1	2	3	4	Seawater
Cl	2.20×10^2	1.07×10^5	2.13×10^5	3.20×10^5	4.26×10^5	5.33×10^5
Mg	1.69×10^2	1.05×10^4	2.08×10^4	3.11×10^4	4.15×10^4	5.18×10^4
Al	1.85	4.93×10^{-1}	1.34×10^{-1}	4.0×10^{-2}	1.5×10^{-2}	8.3×10^{-3}
Si	2.32×10^2	1.49×10^2	1.18×10^2	1.07×10^2	1.02×10^2	1.01×10^2
Ca	3.75×10^2	2.36×10^3	4.35×10^3	6.34×10^3	8.33×10^3	1.03×10^4
Fe	7.14×10^{-1}	9.75×10^{-2}	1.41×10^{-2}	2.77×10^{-3}	1.24×10^{-3}	1.03×10^{-3}

test two extreme compositions, those of the H chondrites and LL chondrites. Use data in Table 2.4. For each case, find the abundances of Fe, Ni, and the unaccounted-for elements.

6. Using the data of Appendix A and F and a spreadsheet, calculate the abundances of various elements in the mean mantle by removing the amounts held in the continental crust from the BSE inventory.

7. Table 2.5 gives the composition of water samples taken along an estuary. We assume that, because of its high solubility, Cl is a good mixing indicator between freshwater and seawater. Plot the concentration of each element vs. [Cl] down the estuary. Elements Al, Si, and Fe are considered to be non-conservative. Explain why.

8. Explain why moderate contamination of a basalt by granitic rocks does not greatly change the FeO/MgO ratio of the hybrid magma with respect to the ratio of the original basalt.

9. Black smokers from mid-ocean ridges spout waters resulting from the mixing of hydrothermal solutions with seawater. Which of the following plots do you expect to produce binary mixing arrays that are straight lines:

Table 2.6 Concentration C_0^i of Ni, La, and Yb in the peridotitic source and mineral/liquid partition coefficients $K_{min/liq}^i$

Element i	Ni	La	Yb
C_0^i	2500	0.5	0.4
$K_{ol/liq}^i$	10	0	0
$K_{opx/liq}^i$	1	0	0.05
$K_{cpx/liq}^i$	1	0.01	0.3

    a. $^{87}Sr/^{86}Sr$ vs. [Mg]
    b. $^{87}Sr/^{86}Sr$ vs. [Sr]
    c. $^{87}Sr/^{86}Sr$ vs. [Sr]/[Mg]
    d. $^{87}Sr/^{86}Sr$ vs. [Mg]/[Sr]
    e. $^{87}Sr/^{86}Sr$ vs. 1/[Sr]
    f. $^{87}Sr/^{86}Sr$ vs. [Mg]/([Mg]+[Sr])

10. Plot the Sr concentrations and $^{87}Sr/^{86}Sr$ ratios vs. the Mg concentrations in mixtures of hydrothermal solutions and seawater (0–100 wt% mixing ratios) using the following data: $^{87}Sr/^{86}Sr = 0.709$, [Sr] = 8 ppm, [Mg] = 1260 ppm in seawater; $^{87}Sr/^{86}Sr = 0.703$, [Sr] = 16 ppm, [Mg] = 0 ppm in the hydrothermal end-member. The atomic weight of Sr is 87.62 and the atomic abundance of isotope 86 is 9.86%.

11. Partial melting of a peridotite produces basaltic liquids. We assume that the mineral composition of the residue is invariable and made of 70 wt% olivine, 20 wt% orthopyroxene and 10 wt% clinopyroxene. Calculate the concentrations of the following elements: nickel (Ni), lanthanum (La), ytterbium (Yb), and the La/Yb ratio in the melt for melt fractions F equal to 0.002, 0.01, 0.02, and 0.1. Concentrations in the peridotitic source and mineral/melt partition coefficients are given in Table 2.6.

12. The melt calculated in the previous exercise for $F = 0.1$ rises and fractionates minerals in conduits before the residual liquids are erupted as differentiated lavas. The cumulates contain 70 wt% olivine and 30 wt% clinopyroxene. Calculate the Ni, La, and Yb concentrations and the La/Yb ratios for a fraction crystallized X equal to 0.01, 0.1, and 0.2. Compare the relative evolution of these parameters resulting from partial melting and fractional crystallization.

13. Very low partition coefficients are particularly difficult to assess experimentally because their mineral concentrations are very low and prone to contamination by co-existing phases. This is the case for rare-earth elements between olivine and melt. Let us call r the ionic radius of the rare-earth elements in octahedral coordination in this mineral. Use the partition coefficients of Sm ($r = 0.96$ Å, $K_{ol/liq}^{Sm} = 0.000424$), Gd ($r = 0.94$ Å, $K_{ol/liq}^{Gd} = 0.000973$), and Yb ($r = 0.87$ Å, $K_{ol/liq}^{Yb} = 0.0196$) to evaluate the olivine/liquid partition coefficients of La ($r = 1.03$ Å), Ce ($r = 1.01$ Å), and Nd ($r = 0.98$ Å). (Hint: calculate the three coefficients of the parabola relating ln $K_{ol/liq}^i$ with r.)

References

Blundy, J. D., Robinson, J. A. C., and Wood, B. J. (1998) Heavy REE are compatible in clinopyroxene on the spinel therzolite solidus. *Earth Planet. Sci. Lett.*, **160**, 493–504.

Hofmann, A. W. (1988) Chemical differentiation of the Earth: the relationship between mantle, continental crust, and oceanic crust. *Earth Planet. Sci. Lett.*, **90**, 297–314.

3 Fractionation of stable isotopes

Tracing natural processes using isotopic abundances is probably the most successful aspect of modern geochemistry. The methods relying on isotopic data are analytically intensive since they depend on complex separation procedures and expensive equipment, but overall they are conceptually simple and robust. The natural variability of the abundances of isotopes in nature results from different processes:

1. Under a variety of thermodynamic and kinetic conditions, isotopes distribute themselves unevenly among co-existing phases (minerals, liquids, gases). These effects are in general very subtle as are the differences between the isotopes that create them, which explains why they escaped detection until the 1950s.
2. Radioactivity removes the parent isotope from an element and adds the decay product to another: ^{87}Rb becomes ^{87}Sr. We will see in the next chapter that the rate of removal of the parent isotope is identical anywhere and at any time in the universe, so this process only affects the relative abundances of the radiogenic isotopes.
3. Cosmic rays are particles produced outside the Solar System and their origin is still not quite understood. The energy of some of these particles, mostly protons and α particles, occasionally exceeds the nuclear binding energy. In the upper atmosphere, some nuclei, such as nitrogen and oxygen, get chipped by the collision, a process known as "spallation." Most particles reaching the ground are actually secondary and can be accounted for by spallation.
4. For all we should be concerned about at this point, the effect of nucleosynthetic processes stopped with the collapse of the solar nebula. We will see that some rare objects found in meteorites and known as refractory inclusions still keep the memory of a pre- or early-solar isotopic variability. Turbulence in the solar nebula and mixing upon planetary accretion and differentiation have for all practical purposes erased – with the remarkable exception of oxygen – most traces of these heterogeneities in planetary bodies.

Thermodynamic modes of isotopic fractionation are collectively referred to as mass-dependent processes and are the topic of this chapter. Kinetic fractionation will be returned to in Section 5.4. The logic underlying thermodynamic fractionation can be applied to isotopes with the added advantage that the behavior of two isotopes of the same element will be more similar than that of two distinct elements, and, indeed, the chemistry of two isotopes of a heavy element, e.g. lead, is virtually unchanged. The applications of isotopic fractionation to the mapping of low- and medium-temperature phenomena are fundamental. The principles are rather similar to those set out in the previous chapter for chemical fractionation.

3.1 Principles of stable isotope fractionation

Let us imagine how isotopes ^{16}O ($\approx 99.8\%$ of natural oxygen) and ^{18}O are distributed between liquid water and vapor. The reaction can be written:

$$H_2^{18}O_{liq} + H_2^{16}O_{vap} \Leftrightarrow H_2^{16}O_{liq} + H_2^{18}O_{vap} \tag{3.1}$$

The four compounds appearing in this equation are chemically identical two by two, but differ by their isotopic abundances: they are known as isotopomers of H_2O. The mass action law governing this equilibrium is written:

$$\left(\frac{H_2^{18}O}{H_2^{16}O}\right)_{vap} \bigg/ \left(\frac{H_2^{18}O}{H_2^{16}O}\right)_{liq} = \left(\frac{^{18}O}{^{16}O}\right)_{vap} \bigg/ \left(\frac{^{18}O}{^{16}O}\right)_{liq} = \alpha^O_{vap/liq}(T, P) \tag{3.2}$$

The notation $\alpha^O_{vap/liq}(T, P)$ for the fractionation coefficient now replaces that of partition coefficients K to signal that the exchange concerns just a single nucleus. Fractionation coefficients are defined likewise for many stable isotope pairs, such as deuterium/hydrogen $^2H/^1H$ (D/H), $^{13}C/^{12}C$, $^{15}N/^{14}N$, $^{34}S/^{32}S$, etc., between such varied phases as gases, natural solutions, magmatic liquids, and minerals. The isotopes exchanged have the same outer electron configuration. Their chemical properties are therefore very similar and their molar volumes almost identical, making α very close to unity and virtually independent of pressure.

The Second Principle of Thermodynamics states that, because the stability of a system is only achieved when its energy is at its minimum, somewhere behind any fractionation factor looms a term that accounts for the exchange of energy between the different participants of a reaction. Before we set out to discuss the physics of stable isotope fractionation, let us therefore first summarize a few simple ideas about the different forms of internal energy in a system. Energy can be stored in different ways:

1. Translational energy. Even when a gas as a whole is at rest, its atoms and molecules move freely and therefore have a mean kinetic energy $\frac{1}{2}m\overline{v^2}$, where the bar indicates that the squared-velocity is averaged and m is the mass of the atom. Translational energy is the unique form of energy of monatomic species such as rare gases (He, Ne, etc.).
2. Rotational energy. Molecules have a momentum of inertia I which measures the mean squared-deviation of the mass from the main rotation axis and, for a given angular velocity ω, the rotational energy is $\frac{1}{2}I\omega^2$. Rotating charges, such as electrons, can be seen as little current loops, which act as little magnets, and the energy of the magnetic field is proportional to the magnetic moment.

3. Vibrational energy. Because they are subjected to combined attractive and repulsive forces, the individual atoms of a diatomic molecule, such as H and Cl in HCl, vibrate around their position at rest. The energy of this movement is the most important cause of isotopic fractionation. Acoustic waves in solids provide a collective form of vibrational energy. Vibrating charge pairs can be seen as little antennas. The energy that these antenna can emit or receive is proportional to the electric dipole of the pairs.

4. Electronic energy. Electrons in atoms and molecules can rise to energy levels above their ground state, normally from low-energy to high-energy orbitals by absorption of light. This phenomenon is, however, of very short duration and involves energies too large to be of concern in this chapter.

The total energy of a system is the sum of these four components. Vibrational and rotational energy are gained or lost by absorption or emission of photons, i.e. by interaction of matter with electromagnetic radiation, such as infrared, visible, or ultraviolet light. In much the same way as "kicks" (shocks) redistribute translational energy among gas molecules, the electromagnetic field carried by photons kicks the electronic dipole and magnetic moments. We are going to see that all these kicks must carry discrete amounts of energy.

Quantum mechanics states that energy is measured by quanta, that we can see as "grains," which means that it is distributed over discrete energy levels. These quanta characterize the spacing between successive energy levels. For the conditions of temperature of interest to us, the translation quanta are much smaller than the rotational quanta, which themselves are much smaller than vibrational quanta (Fig. 3.1). The "classical" thermal energy RT (or kT for an isolated molecule with k being the Boltzmann constant) represents the equivalent of an enormous number of translational and rotational quanta, which indicates that the corresponding levels are well populated and can be treated as a continuum via the methods of classical physics. For reasons that will appear later, they do not produce temperature-dependent effects. At temperatures of geological interest, both translational and rotational energies are therefore unimportant for isotope fractionation in solids and liquids, with rare exceptions such as the CO_3 spinner of carbonate minerals. In contrast, the more energetic vibrational levels are more scantily populated, which is the reason for the temperature dependence of isotopic effects, and they therefore will need special attention.

When two atoms come close to one another, their orbitals can interact with one another to form a single molecular orbital. The energy of a chemical bond, e.g. O–H for the hydroxyl radical, varies with the distance between the bonding atoms (Fig. 3.2). Although this plot does not describe the situation as accurately as true molecular orbitals, one can see the optimum position as resulting from a trade-off between the repulsion of the positively charged nuclei and the attraction between each nucleus and the orbiting electrons. Energetically, the most favorable position of the electrons is between the nuclei, which leads to a mutual attraction between the O and H atoms. These competing effects tend to confine the two atoms to an optimum distance at the lowest point of a potential "well." Any movement changing the length of the bond will be opposed by a counteracting force, which leaves the system in a perpetual state of vibration. We can represent such a system with two balls

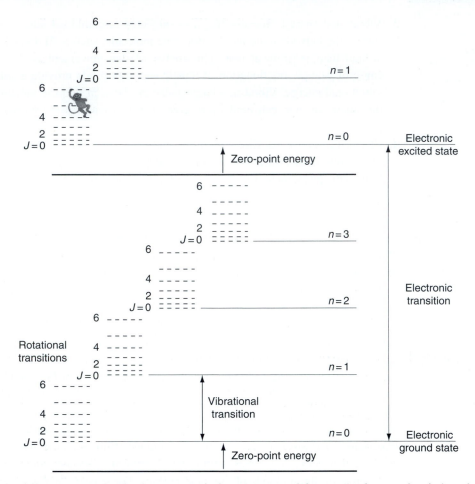

Figure 3.1   The different energy levels of a compound. The separation of the rotational energy levels (J is the number of rotational quanta) is very small with respect to the separation of the vibrational levels and that of electronic levels between different orbitals. Note the zero-point energy, a consequence of the uncertainty principle, which only appears for the vibrational energy levels.

attached by a spring: this is the model of the harmonic oscillator. College physics tells us that if such a system does not lose energy by friction or radiation, the sum of the potential and kinetic energies remains constant. The potential energy $V(r)$ of a diatomic molecule is a function of the separation distance r between the nuclei and goes through a minimum for $r = r_0$. The potential energy $V(r)$ can be expanded about the value at the minimum as

$$V(r) = V(r_0) + \left(\frac{dV}{dr}\right)_{r_0} (r - r_0) + \frac{1}{2}\left(\frac{d^2V}{dr^2}\right)_{r_0} (r - r_0)^2 + \cdots \qquad (3.3)$$

which assumes that terms with degrees > 2, known as anharmonic, are small enough to be neglected: this is the parabolic or harmonic approximation of the potential well. The first term on the right-hand side is the energy at the minimum and the second is zero because the derivative is evaluated at the minimum. Let us call k the second-order derivative calculated

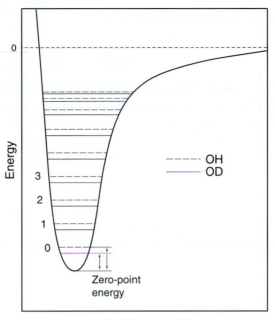

Figure 3.2 Energy of the oxygen—hydrogen bond. The potential of the O—H bond as a function of the distance between the two nuclei is shown as a continuous curve. The zero represents the energy of the system when O and H are fully dissociated. Because the heavy nucleus hardly budges when the light electrons wiggle, this curve is independent of the mass of the nuclei (Born–Oppenheimer approximation) and is therefore the same for all isotopes. The quantized energy levels ($n = 0$, 1, 2, ...) are shown for the O—H (oxygen—hydrogen) and O—D (oxygen–deuterium) bonds. The first energy levels are nearly equally spaced. The isotopically heavier molecules are in a lower state of energy than the lighter molecules. Zero-point energy is the elevation of the ground state ($n = 0$) above the minimum of the potential well. Differences in zero-point energies between molecules with different isotopes (e.g. O—H and O—D) account for all stable isotope fractionation effects.

at r_0 and show that it measures the force of the spring. Stretching the bond by applying a force F increases its potential energy and this takes the form:

$$F = -\frac{dV}{dr} \quad (3.4)$$

with the minus sign ensuring that energy increases when the force stretches the bond. Comparing the two equations leads to Hooke's law, which we already met in the previous chapter:

$$F = -k(r - r_0) \quad (3.5)$$

The constant k is therefore a measure of the "hardness" of the spring, i.e. of how much energy is needed to achieve a particular change in the bond length. Now, let us summon our college physics and remember one of Newton's great discoveries: $F = M d^2r/dt^2$, or force and acceleration are proportional, and mass is the ratio of the former by the latter.

Bringing everything together, we get:

$$\frac{d^2 (r - r_0)}{dt^2} + \frac{k}{M} (r - r_0) = 0 \tag{3.6}$$

in which we use the property that the derivative of r_0 is zero. Periodic functions of the form $\sin \sqrt{k/M}\, t$ clearly satisfy this equation. The most important parameter of this solution is the frequency of the oscillation $\nu = (1/2\pi) \sqrt{k/\mu}$. The harmonic mean mass (known as the reduced mass) μ such as $1/\mu = (1/M_O) + (1/M_H)$ arises because each atom actually carries out only part of the work it would have to do if it was attached to a steady support. As expected, heavy atoms do not jump around very quickly!

Upon formulation of black-body radiation theory, Planck laid the groundwork of quantum mechanics by hypothesizing that a system undergoing a periodic movement with frequency ν can only occupy the discrete energy states $E_n = nh\nu$, in which h is the Planck constant equal to 6.63×10^{-34} J s and $n = 0, 1, 2, \ldots$ From observations on the photoelectric effect, Einstein postulated that the energy of an electromagnetic wave is actually bundled with the properties of a particle, the photon. Photons have no mass and no electric charge. Upon emission or absorption of photons, material shifts from one energy level to another, but the energy carried by each photon is always $h\nu$.

For the harmonic oscillator, the situation is slightly more complicated by the existence of the zero-point energy, without which, however, no fractionation of stable isotopes would exist. The relationship between the energy of our ball-and-spring system and frequency is:

$$E_n = \left(n + \frac{1}{2} \right) h\nu \tag{3.7}$$

with $n = 0, 1, 2, \ldots$ Again, moving up and down the energy levels involves the absorption or the emission of photons with energy $h\nu$. This does not happen, however, for homonuclear, symmetrical diatomic molecules, such as H_2 and O_2, which cannot change their vibrational and rotational energy level and explains why these gases are transparent (see Sections 9.1 and 12.9). The existence of zero-point energy E_0 for $n = 0$ is a result of the Heisenberg uncertainty principle: if the energy of the system was allowed to reach, as it is in classical physics, its minimum value, both the position ($r = r_0$) and the velocity ($\sqrt{2 (E_0 - V_0)/m}$) of each atom would be precisely known, which is not permitted by the uncertainty principle. Such a limitation does not arise for rotational energy because the position of the atom on its orbit remains undetermined at all times. Vibrational zero-point energy is important for natural systems because at nearly up to ambient temperature, most oscillators are precisely at that energy level.

Not all the natural systems are simple diatomic harmonic oscillators. More complex molecules have more complex patterns of vibration. Fortunately, group theory allows us to make the situation more tractable and demonstrates that any arbitrary vibration of a molecule can be broken down into the superposition of simple "normal modes." Most molecules made of N atoms have $3N - 6$ normal modes of vibration, whereas this number reduces to $3N - 5$ for linear molecules. Figure 3.3a shows some of the normal modes for some of the common gaseous species, OH, H_2O, CO_2, CH_4. The frequencies shown are well known to spectroscopists (infrared, Raman, ultraviolet) and are used to identify

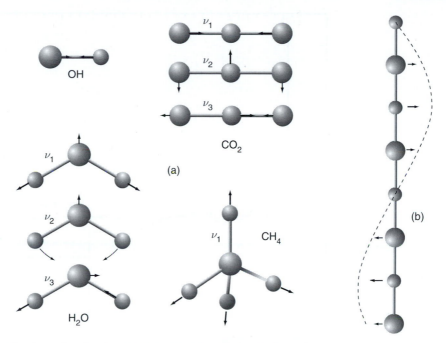

Figure 3.3 (a) Vibrations of molecules, OH, H_2O, CO_2, CH_4. All the vibrations, however complex, can be described as the superposition of a small number of normal modes (shown here as arrows) corresponding to different frequencies v_1, v_2, ... measurable by infrared or Raman spectroscopy. (b) For solids, the vibrations take the form of traveling waves.

the presence of these gases in the molecular clouds of distant galaxies as well as in the terrestrial atmosphere.

Crystalline solids behave in yet another way (Fig 3.3b): because ions are regularly distributed and interact with their neighbors in a continuous way, vibrations take the form of traveling waves. The pseudo-particles associated with these waves are known as phonons. Models related to the statistical theory of the heat capacity C_V of solids – the increment of energy associated with a small temperature step – assume either a single vibration frequency (Einstein solid) or a frequency continuum increasing up to a cut-off value imposed by the lattice (Debye elastic solid). The Einstein and Debye models are reasonably successful for simple crystals, but the wealth of spectroscopic data and the advent of large computers have now made it possible to evaluate exact isotopic properties using so-called "ab initio" models. Models of fractionation involving liquid phases require particularly intensive numerical computing.

Where are the isotope effects in all this? Heavy atoms react more slowly than light ones and therefore tend to occupy lower energy levels. Bond energy varies with v and therefore with $1/\sqrt{\mu}$, where μ is the harmonic mean mass of the atoms that form the molecule. A consequence of this rule is that the quantized energy levels and the zero-point frequencies are lower for a bond involving the heavier isotopes: the molecule OD is more stable than OH (Fig. 3.1). This can be seen for the substitution of H by D in magnesium hydroxide $Mg(OH)_2$ (brucite), which drastically reduces the frequency of the OH vibration (Fig. 3.4).

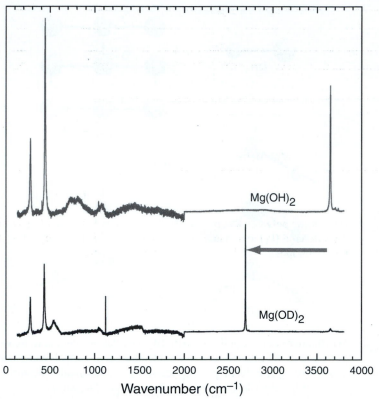

Figure 3.4 Energy shift in brucite ($Mg(OH)_2$) induced by the substitution of D for H in the OH radical (courtesy of Bruno Reynard). Spectroscopists do not normally work with frequency but with wavenumbers, in cm^{-1}, which are frequencies divided by the velocity of light.

Let us go back to the exchange of ^{16}O and ^{18}O between water vapor and liquid water (3.1) and consider a vibrational mode with different hardness parameters $k_{vap} \ll k_{liq}$ indicative of loose bonds in the gaseous state (Fig. 3.5). The variation ΔE of energy during this exchange is

$$\Delta E = \left(E_{H_2^{18}O_{vap}} + E_{H_2^{16}O_{liq}} \right) - \left(E_{H_2^{18}O_{liq}} + E_{H_2^{16}O_{vap}} \right) \tag{3.8}$$

or

$$\Delta E = \left(E_{H_2^{18}O_{vap}} - E_{H_2^{16}O_{vap}} \right) - \left(E_{H_2^{18}O_{liq}} - E_{H_2^{16}O_{liq}} \right) \tag{3.9}$$

For the sake of illustration, let us assume that most of the atoms will occupy their ground, zero-point energy level ($n = 0$) so that:

$$\Delta E = \frac{1}{2}h \left[\left(v_{H_2^{18}O_{vap}} - v_{H_2^{16}O_{vap}} \right) - \left(v_{H_2^{18}O_{liq}} - v_{H_2^{16}O_{liq}} \right) \right] \tag{3.10}$$

$$= \frac{h}{4\pi} \left[\left(\sqrt{k_{vap}} - \sqrt{k_{liq}} \right) \left(\frac{1}{M_{H_2^{18}O}} - \frac{1}{M_{H_2^{16}O}} \right) \right] \tag{3.11}$$

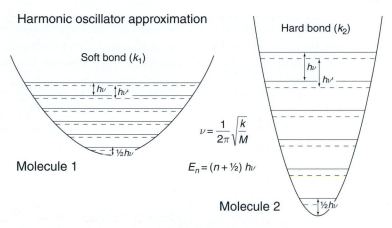

Harmonic oscillator approximation

$$\nu = \frac{1}{2\pi}\sqrt{\frac{k}{M}}$$

$$E_n = (n + \tfrac{1}{2})\, h\nu$$

Figure 3.5 Comparison between two bonds, one soft (left) and one hard (right) with the force constant of the soft bond (k_1) less than that of the hard bond (k_2). The total energy is at a minimum when heavy isotopes preferentially populate the hard bond.

This quantity is clearly > 0 which indicates that, in order to achieve energy reduction, reaction (3.1) must proceed from right to left (backward). The heavy ^{18}O therefore fractionates preferentially into the liquid phase. It is a general result that heavy isotopes fractionate in favor of the stiffest bonds. Since transforming solid or liquid into vapor requires addition of energy to break bonds (latent heat), solids and liquids, where most of the energy is stored in vibrations, lower the total energy of the system by concentrating heavier isotopes. In contrast, vapor offers sites that are higher on the energy scale than the corresponding liquid and therefore tend to concentrate the lighter isotopes relative to the liquid. In a liquid–vapor equilibrium, such as H_2O liquid and vapor below $220\,°C$, the liquid is enriched in the heavier isotope (e.g. ^{18}O, ^2H), while the vapor preferentially concentrates the lighter isotope (e.g. ^{16}O, ^1H).

The Boltzmann distribution

Let us consider N identical particles, atoms or molecules, which can occupy a large number of energy levels, n_1 being at the level E_1, n_2 at the level E_2, etc. We assume that each individual particle has an equal probability to land on any energy level. Without loss of generality, we will assume that N is the Avogadro number (1 mole). We are going to justify to some extent that the fractional proportions $f_i = n_{i/N}$ of these atoms or molecules that have the energy E_i at temperature T is:

$$f_i = \text{const} \times e^{-\frac{E_i}{RT}} \tag{3.12}$$

Our assumptions are that the occupancy of a particular energy level, say E_i, is independent of the occupancy of another level, say E_j. The function f therefore requires that $f_{i+j} = f_i \times f_j$. The only function with such a property is the exponential function, which suggests the form:

$$f_i = \text{const} \times e^{-\beta E_i} \tag{3.13}$$

The positive constant β is to be determined and the minus sign ensures that f_i remains finite. From the condition that the fraction proportions f_i sum to unity, we get the constant equal to Q^{-1}, where $Q = \sum_{i=0}^{\infty} e^{-\beta E_i}$ is known as the partition function. This can be recast as:

$$f_i = \frac{e^{-\beta E_i}}{Q} \tag{3.14}$$

In order to demonstrate that β is actually equal to $1/RT$, we need to pay a quick visit to the concept of entropy. Boltzmann formulated a statistical definition of the entropy S of a system as $R \ln \Omega$, where Ω is the number of indistinguishable microscopic configurations (how atoms distribute themselves among the possible states). The second principle then states that the entropy of an isolated system increases when it spontaneously evolves towards the state with the largest number of accessible configurations. The number Ω can be found using a simple argument: the total number of permutations of N atoms is $N!$, where the exclamation mark stands for the function factorial (e.g. $3! = 3 \times 2 \times 1$). Among the $N!$ permutations, $n_1!$ correspond to indistinguishable occupations of energy level E_1 with a similar outcome for other energy levels. The number of distinguishable configurations is therefore:

$$\Omega = \frac{N!}{n_1! n_2! \ldots} \tag{3.15}$$

This expression would be a nightmare if it was not for the handy Stirling approximation $\ln N! \approx N \ln N - N$, valid for large N, and which gives a new expression for the entropy (remember that the logarithm of a product of numbers is the sum of the logarithms of these numbers):

$$\frac{S}{R} = +N \ln N - n_1 \ln n_1 - n_2 \ln n_2 - \cdots - N + n_1 + n_2 + \cdots \tag{3.16}$$

We now recall that $N = \sum_i n_i$: the second part of the right-hand side therefore vanishes and we get:

$$S = -R \sum_i f_i \ln f_i \tag{3.17}$$

Taking the differential of the entropy S at constant T:

$$dS = -R \sum_i df_i \ln f_i - R \sum_i f_i \frac{df_i}{f_i} \tag{3.18}$$

or, using (3.14)

$$T dS = RT \beta \sum_i E_i df_i - RT \ln Q \sum_i df_i \tag{3.19}$$

At constant volume, $\sum_i E_i df_i = dU = T dS$, which requires that $RT\beta = 1$. This ends the demonstration of one of the most important equations in science.

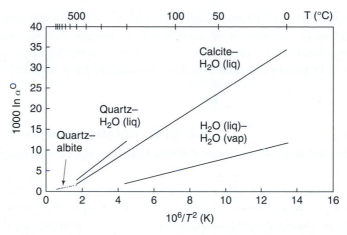

Figure 3.6 Fractionation of oxygen isotopes between a mineral phase and water, or between different mineral phases. Notice that preferential incorporation of ^{18}O is normally favored by the mineral structure and that isotopic fractionation decreases rapidly with increasing temperature.

Temperature may be understood through the Boltzmann distribution (see box), which determines the fraction of a population of atoms that populates a certain energy level. At moderately low temperatures, typically below ambient temperature, there is not enough energy to open up many vibrational levels and all the atoms are in their ground state. The exchange energy is constant and fractionation temperature dependent. In this case, which only finds useful applications when hydroxyl groups are involved, the law of mass action requires the temperature dependence of $\ln \alpha$ in $1/T$. Temperature in this case only attests to the total available vibrational energy and therefore indirectly constrains the relative occupation of the available levels. Higher temperatures have an additional effect: they open up new grounds for occupation. The more energy available, the more vibrational quanta can wander up the energy levels. In this case, which is a straight outcome of the quantum theory, we observe (Fig. 3.6) fractionation laws of the type:

$$\ln \alpha = \frac{A}{T^2} + C \tag{3.20}$$

where A and C are constants. These laws, to whom the names of Bigeleisen, Mayer, and Urey are attached for their 1947 papers, are widely used for thermometry of magmatic, metamorphic, or hydrothermal systems. The symmetry constant C is usually zero for isotopic exchange between anhydrous isotopomers so that fractionation at magmatic temperatures is normally small or negligible.

Since bond energy varies with the inverse square-root of the mean mass of the atoms that form the molecule, the thermodynamic fractionation factor between isotopes generally falls very quickly with increasing mean atomic mass as it varies with the difference between smaller and smaller quantities. At 500–800 °C, fractionations of 20–80 per mil (‰) are commonly observed for the D/H ratio, of 2–8‰ for oxygen isotopes, and fall to less than 1‰ for zinc and copper isotopes.

A last effect deserves attention when we deal with molecules with very different degrees of symmetry. For example, the molecule CO has only one single possible configuration, whereas the methane molecule can be reproduced identically to itself by rotating the CH_4

tetrahedron of $2\pi/3$ around any C–H bond: $4 \times 3 = 12$ indistinguishable configurations are therefore accessible to the same molecule. Methane thus has access to 12 times as many rotational states with respect to CO and, everything else being equal, its relative stability will be greatly enhanced with respect to that of carbon monoxide.

3.2 Delta notation and stuff

Using raw isotopic ratios to describe natural variability is very inconvenient because of their small amplitude. Also there is a common difficulty with interlaboratory biases. The same sample processed and analyzed by different groups on different instruments will come out differently and even sometime more differently than the actual natural variability. This is not a real issue because the knowledge of "true" isotopic ratios is practically pointless. Reproducibility is the magic word, not accuracy. The practice for all laboratories is to measure the same reference material so that each and every analyst can report his or her own data on unknown samples with respect to the same standard. Reference samples must be multiple, broadly available, homogenous, and they should be analyzed using the same analytical procedure as the unknown samples. Under these conditions, we expect that deviations of the isotopic composition from that of the reference, usually small numbers, should be known with great precision, typically 0.02 to 1‰ depending on the element and sample size.

The delta notation with respect to a particular reference material is generally used for isotopic ratios of stable nuclides; $\delta^{18}O$ and $\delta^{17}O$ represent the deviations of the isotopic $^{18}O/^{16}O$ and $^{17}O/^{16}O$ ratios in the sample in parts per thousand relative to the same ratio in the reference material. $\delta^{18}O$ is defined as:

$$\delta^{18}O = \left[\frac{\left(^{18}O/^{16}O\right)_{\text{sample}}}{\left(^{18}O/^{16}O\right)_{\text{ref}}} - 1 \right] \times 1000 \qquad (3.21)$$

with the reciprocal expression of the isotopic ratio of a sample as a function of its $\delta^{18}O$ value:

$$\left(\frac{^{18}O}{^{16}O} \right)_{\text{sample}} = \left(\frac{^{18}O}{^{16}O} \right)_{\text{ref}} \left(1 + \frac{\delta^{18}O}{1000} \right) \qquad (3.22)$$

It is common practice to use a light but abundant isotope in the denominator. Jargon refers to heavy oxygen for high $^{18}O/^{16}O$ ratios and to light oxygen otherwise.

A useful approximation of the fractionation properties is to compare the difference of $\delta^{18}O$ values between two co-existing phases 1 and 2 with the fractionation coefficients α of the same phases. The following notation is used:

$$\delta^{18}O_2 - \delta^{18}O_1 = 1000 \frac{\left(^{18}O/^{16}O\right)_2 - \left(^{18}O/^{16}O\right)_1}{\left(^{18}O/^{16}O\right)_{\text{ref}}} \qquad (3.23)$$

Since α is very close to one, we can use the approximation $\ln \alpha \approx \alpha - 1$ and insert it in the previous equation rewritten as:

$$\delta^{18}O_2 - \delta^{18}O_1 = 1000 \frac{(^{18}O/^{16}O)_1}{(^{18}O/^{16}O)_{ref}} (\alpha - 1) \approx 1000 \ln \alpha \tag{3.24}$$

in which we have taken into account that the ratio of two $^{18}O/^{16}O$ ratios is always very close to one. The difference in delta values between co-existing phases (minerals, liquids) therefore decreases as $1/T^2$, which implies that the sensitivity of stable-isotope thermometers decreases with increasing T, hence their important applications to low- and medium-temperature processes.

For elements with more than two isotopes, at constant temperature, the amplitude of fractionation α increases with the difference in mass of the isotopic ratios. For example, the difference in the $^{18}O/^{16}O$ ratio between two samples is twice the difference in the $^{17}O/^{16}O$ ratio. This is because bond energy varies with the mass of the bonding atoms. Therefore $\ln \alpha$ can be expanded to the first order relative to the difference in mass Δm in the isotope ratio, giving:

$$\ln \alpha (\Delta m) = \ln \alpha (\Delta m = 0) + f \Delta m + \mathcal{O} \left(\Delta^2 m \right) \tag{3.25}$$

where f, which is the derivative of $\ln \alpha$ relative to Δm for $\Delta m = 0$, is a coefficient that is independent of mass, termed mass discrimination. In this equation, $\mathcal{O}$ means "of the order of" and we will neglect terms of higher than first order. Moreover, when the mass difference is zero, there is no fractionation between a mass and itself ($\alpha = 1$), and the first term on the right-hand side is therefore zero, giving for fractionation of ^{18}O and ^{16}O between phases 1 and 2 ($\Delta m = 2$):

$$\alpha^O_{2/1} = \left(\frac{^{18}O}{^{16}O} \right)_2 \Big/ \left(\frac{^{18}O}{^{16}O} \right)_1 = e^{2f} \tag{3.26}$$

When fractionation is small, i.e. when f tends toward 0, we utilize the linear approximation obtained by developing the logarithm of the left-hand side of (3.25) to the first order:

$$\alpha^O_{2/1} \approx 1 + 2f \tag{3.27}$$

We will have, for example, the two linear equations:

$$\left(\frac{^{17}O}{^{16}O} \right)_2 = \left(\frac{^{17}O}{^{16}O} \right)_1 (1 + 1f) \tag{3.28}$$

$$\left(\frac{^{18}O}{^{16}O} \right)_2 = \left(\frac{^{18}O}{^{16}O} \right)_1 (1 + 2f) \tag{3.29}$$

Alternatively, (3.27) can be re-written as

$$\frac{1 + \delta^{18}O_2/1000}{1 + \delta^{18}O_1/1000} \approx 1 + 2f \tag{3.30}$$

We use the approximation $(1 + x)^{-1} \approx 1 - x$ valid for small x and neglect the products of deltas to obtain:

$$\delta^{18}O_2 - \delta^{18}O_1 \approx 2000 f \tag{3.31}$$

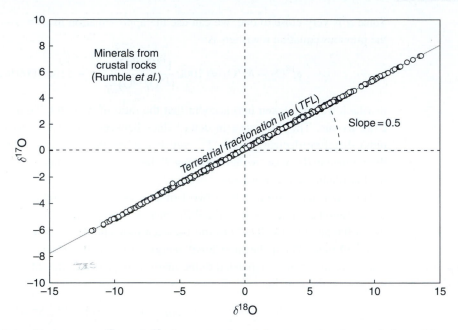

Figure 3.7 Relationship between $\delta^{17}O$ and $\delta^{18}O$ for >500 minerals from crustal rocks, mostly from China (Rumble *et al.*, 2007). The values define a nearly perfect straight line (errors are smaller than symbol size) going through (0,0) and with a slope of 0.5 demonstrating the perfect dependence of oxygen isotope fractionation on the mass difference between masses at the numerator and denominator. Laser ablation data by courtesy of Doug Rumble. Meteoric waters and sediments fall on the same line, called the terrestrial fractionation line.

Likewise for $\delta^{17}O$:

$$\delta^{17}O_2 - \delta^{17}O_1 \approx 1000f \tag{3.32}$$

In a plot with $x = \delta^{17}O$ and $y = \delta^{18}O$, isotope fractionation makes the isotope composition of sample 2 move on a straight line going through sample 1 with a slope of $\approx 1/2$ (see below for details on this value). This pattern is actually quite general and, for a planet starting with an isotopically homogeneous reservoir of oxygen, all the differentiation and mixing processes produce samples with $\delta^{17}O$ and $\delta^{18}O$ plotting on this very same straight line (Fig. 3.7). As we have not really assumed that fractionation takes place at equilibrium (we just used an expansion of $\ln \alpha$ with respect to Δm), kinetic and biological processes must also follow the same pattern. The $\delta^{17}O$ and $\delta^{18}O$ values of essentially all terrestrial samples plot on the terrestrial fractionation line (Fig. 3.7). It will be seen later that one very important application of this simple theory is in the discrimination of planetary material. The same principle holds for the isotopic variability of all the elements with more than two stable isotopes, such as Fe, Zn, and Mo: if Δm_x and Δm_y are the mass differences relative to the isotopic ratios used for x and y, respectively, the mass fractionation line has a slope of $\Delta m_y / \Delta m_x$. Some rare outliers exist, due to photodissociation reactions (different isotopes of the same element absorb radiations with slightly different wavelengths) or autocatalytic effects. Atmospheric sulfur with isotope ratios $^{34}S/^{32}S$ and $^{33}S/^{32}S$ and ozone are prime examples: these exotic processes are known by the name of mass-independent

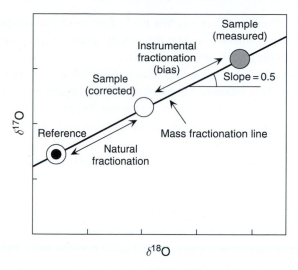

Figure 3.8 Fractionation depending on the mass of isotopes ^{16}O, ^{17}O, and ^{18}O is consistent whether its cause is natural or analytical. The $^{17}O/^{16}O$ and $^{18}O/^{16}O$ ratios are shown in delta notation. All samples from Earth lie on the same straight line of mass fractionation whose slope is equal to the ratio of the mass difference (17 − 16) by difference (18 − 16), i.e. 0.5. Natural fractionation is a deviation determined relative to an arbitrary reference sample, hence the δ notation. The isotopic bias is introduced by the mass spectrometric analysis and follows a similar mass-dependence law: it must be corrected for each sample by appropriate techniques.

fractionation. It is important to emphasize at this early stage that isotope measurements made in the laboratory report isotopic fractionation values that are the sum of fractionation related to the action of all natural processes *and* of instrumental fractionation, sometimes referred to as instrumental mass bias (Fig. 3.8). The true instrumental bias will never be known, except by artificially mixing pure isotopes into certified standard materials, but since every laboratory places – by definition – reference material at the same point in the diagram, this is something we can live with.

Let us finally note that, because we restricted ourselves to a first-order approximation, all the isotope fractionation laws seem to be giving a unique value for the slope of the fractionation line, e.g. 0.5 for oxygen. To higher orders, however, the data are good enough to tell one process from another, in particular whether isotope fractionation is controlled by quantum mechanics effects, with a dependence in $1/M$, or by kinetic effects, with a dependence in $\exp\left(-1/\sqrt{M}\right)$, and to recognize the effects of Rayleigh distillation during phase change. This complex topic is beyond the scope of this textbook.

3.3 Hydrogen

It is important to remember that hydrogen is largely held in the hydrosphere, ocean, ice caps, and groundwater. The mantle may contain a large fraction of the terrestrial hydrogen inventory but its local concentrations are always small (a few hundred parts per million or ppm). The consequence is that the hydrogen isotope compositions of aquifers are hardly modified by interaction with the ground, unless the proportion of groundwater held in the pores of the rock is extremely small.

On Earth, the average D/H ratio is about 1.4×10^{-4}, or equivalently, the isotopic abundance of D is 140 ppm. On Mars, the Viking landers measured a D abundance of 780 ppm. These numbers are much larger than the 20 ppm value of the solar nebula inferred from spectroscopic data on Jupiter and the reason is the faster escape rate of H from planetary atmospheres with respect to the twice-as-heavy D.

The reference material is the standard mean ocean water (SMOW), a man-made composite of various samples of desalinated seawater and δD usually refers to SMOW. Hydrogen isotopes fractionate very significantly between vapor and liquid and between vapor and ice under the temperature conditions prevailing at the surface of the Earth, which makes δD values universally employed tracers of the hydrological cycle. The factor 2 difference between the two isotopes is unique in the periodic table and accounts for the isotopic variability of δD that ranges up to several hundreds of per mil. Ice at the South Pole only contains 85 ppm D ($\delta D = -400$). At the temperatures of interest for environmental studies, vapor/liquid D/H fractionation varies from $-170‰$ at $-25\,°C$ to $-110‰$ at $0\,°C$, and $-75‰$ at $+25\,°C$. As will be discussed in the next section, the isotope variability is amplified by the progressive distillation of rain and snow from the vapor transported by the atmosphere.

Fractionation of D from H is also observed between water and hydrous minerals. Although the experimental data are few and particularly difficult to interpret and to reproduce, hydrous minerals are typically depleted in deuterium by 30–50 per mil with respect to co-existing water and δD values of -40 may be found in minerals in equilibrium with seawater and in magmatic rocks altered by seawater. As expected, the effect of temperature is to reduce isotopic fractionation. The δD values of metamorphic and igneous hydrous minerals (chlorite, biotite, muscovite, amphibole) are typically in the range of -40 to $-95‰$. Hydrogen isotope geochemistry therefore represents a powerful means of investigating the dehydration reactions associated with the burial of hydrous rocks, such as schists, serpentinites, and amphibolites, at subduction zones in particular, and the deep water cycle in general.

3.4 Oxygen

Oxygen has three stable isotopes at masses 16, 17, and 18, with average abundances of 99.76, 0.037, and 0.204 percent, respectively. For oxygen, as for hydrogen, the most broadly used reference material is the SMOW but low-temperature carbonate data are also commonly reported with respect to the carbonate of a belemnite from the Pee Dee formation (PDB), which is about 30 per mil heavier than SMOW. The $\delta^{18}O$ value of the Sun, as measured by the solar wind implanted in metal grains from the Moon regolith (Hashizume and Chaussidon, 2005) and by the cometary mission Stardust is about -40 to $-60‰$. The mean terrestrial value is recorded by mantle rocks, which are the largest terrestrial reservoir of oxygen, and is $\delta^{18}O \approx +5.5\%_0$. If atmospheric oxygen was in equilibrium with seawater, its $\delta^{18}O$ would be very similar to SMOW. Actually, its value is more like $\approx +23.5\%_0$. This is the Dole effect, which reflects the control of photosynthesis and respiration.

Fractionation of ^{16}O from ^{18}O between liquid water and vapor is about one order of magnitude smaller than for hydrogen isotopes, which is what is roughly expected from the comparison of $\sqrt{18/16} - 1$ and $\sqrt{2/1} - 1$. At the temperatures of interest for environmental studies, vapor/liquid $^{18}O/^{16}O$ fractionation is only mildly temperature dependent: $-15\%_0$ at $-25\,°C$, $-12\%_0$ at $0\,°C$, and $-9\%_0$ at $+25\,°C$. Because temperature is the only variable and is common to both systems, it is expected that δD and $\delta^{18}O$ should be strongly correlated in rain waters (and snow). The linear relation $\delta D = 8\delta^{18}O + 10$ discovered by Craig in 1961 and called the Global Meteoric Water Line (GMWL) can be explained by a process of fractional condensation of water vapor upon migration of moist equatorial air toward the poles. The SMOW does not plot on this line simply because water vapor is isotopically fractionated with respect to surface seawater. Let us write the Rayleigh equation for the number of moles n of each isotope left in the atmosphere with D^i being the partition coefficient of isotope i between water and vapor as:

$$d\ln n_{vap}^{^{18}O} = \left(D_{liq/vap}^{^{18}O} - 1\right) d\ln f \tag{3.33}$$

$$d\ln n_{vap}^{^{16}O} = \left(D_{liq/vap}^{^{16}O} - 1\right) d\ln f \tag{3.34}$$

in which f is the mass fraction of vapor left; and subtract the second equation from the first. We obtain:

$$d\ln \left(\frac{^{18}O}{^{18}O}\right)_{vap} = \left(D_{liq/vap}^{^{18}O} - D_{liq/vap}^{^{16}O}\right) d\ln f \tag{3.35}$$

$$\approx D_{liq/vap}^{^{16}O} \left(\alpha_{liq/vap}^{^{18}O/^{18}O} - 1\right) d\ln f \tag{3.36}$$

where we have simply introduced the original definition (3.2) of α. A similar expression holds for D/H. We now observe that, because ^{16}O and H are the dominant oxygen and hydrogen species in both the liquid and the vapor, $D_{liq/vap}^{^{16}O}$ and $D_{liq/vap}^{H}$ are both very close to unity and introduce the approximation $d\ln D/H = d\ln(1 + \delta D/1000) \approx d\delta D/1000$. After this replacement has been made, we divide the same expression for D/H by the last equation, and we get the slope of the GMWL as:

$$\left(\frac{d\delta D}{d\delta^{18}O}\right)_{vap} \approx \frac{\alpha_{liq/vap}^{D/H} - 1}{\alpha_{liq/vap}^{^{18}O/^{16}O} - 1} \tag{3.37}$$

From the equation describing the isotopic evolution of the vapor, we now proceed to derive the δD–$\delta^{18}O$ relationship in precipitations. Equation (3.24) shows that rain and snow samples are just shifted from the vapor line towards more positive values by the factors 100 $\ln\alpha_{liq/vap}^{D/H}$ for x and 1000 $\ln\alpha_{liq/vap}^{^{18}O/^{18}O}$ for y but the slope of the precipitations is identical to that of the vapor. This fully justifies the correlation between δD and $\delta^{18}O$ observed by Craig.

Among the processes that fractionate oxygen isotopes at low temperature, probably the most important is between water and the different forms of dissolved and crystallized carbonates. This was understood in 1947 by Urey and his students who first identified the potential of this technique for paleothermometry. Kim and O'Neil (1997) made a very careful investigation of $^{18}O/^{16}O$ fractionation between calcite and water and suggested the expression:

$$1000 \ln \alpha_{calcite/water}^{^{18}O/^{16}O} = \frac{18030}{T} - 32.42 \qquad (3.38)$$

Fractionation is prone to effects of salinity. The different carbonate ions H_2CO_3, HCO_3^-, and CO_3^{2-} have very different $\delta^{18}O$ values, decreasing in that order. Today, the oxygen isotope geochemistry of foraminifera has become the essential tool for establishing temperature and salinity in the ancient oceans. Seawater $^{18}O/^{16}O$ changes only by addition of melt water, which is meteoric water with very different isotopic properties, and we will see that this is actually used to trace the evolution of the ice caps. If this effect can be neglected, the $\delta^{18}O$ of a marine calcite fossil may be used to infer the temperature of the seawater in which the animal grew. Fractionation of oxygen isotopes in phosphates (e.g. especially tooth enamel) are used in the same way.

Because of prominent temperature effects, oxygen isotope geochemistry provides unique constraints on the origin of igneous and metamorphic rocks:

1. Magmatic differentiation of basaltic rocks takes place at temperatures of about $1100\,^{\circ}C$ and therefore does not entail any significant fractionation of oxygen isotopes (< 0.4 per mil) (Fig. 3.6). The $\delta^{18}O$ values of most fresh basaltic rocks derived from either glass or olivine fall in the range 5.2–5.8‰ which is within 0.3 per mil of mantle values.
2. Partial melting at temperatures in excess of $800\,^{\circ}C$ creates no or minimal isotope fractionation.
3. Large deviations of $\delta^{18}O$ values from the mantle values require that either some source rocks were once involved in low-temperature processes (they contain sediments or altered rocks) or that the samples was exposed to hydrothermal alteration or weathering.
4. The order of decreasing $\delta^{18}O$ at equilibrium should be quartz, feldspar, Fe–Mg silicates, magnetite. The commonest rocks in the continental crust derive largely from the metamorphic transformation of sediments (schist, gneiss) and their melting (granite).

Fig 3.6 also shows the $^{18}O/^{16}O$ fractionation between albite and water, which cuts across all the other curves. A first implication is that exchanged silicates at hydrothermal and surface temperatures have $\delta^{18}O$ values much higher than the co-existing water: clay minerals are enriched by about 12–25 per mil with respect to seawater in the temperature range of 150–25 °C. This shift is clearly visible in the high $\delta^{18}O$ values of altered oceanic basalts. It also explains the role that the sedimentary cycle plays with respect to the continental crust. The consistently high $\delta^{18}O$ values (typically +7 to +15‰) of gneiss and mica-schist, but also of the anatectic granites which form by melting of these metamorphic rocks reflects this so-called metasedimentary origin. Such high $\delta^{18}O$ values of silicic metamorphic and magmatic rocks with respect to the mantle value ($\approx +5.5‰$) require the

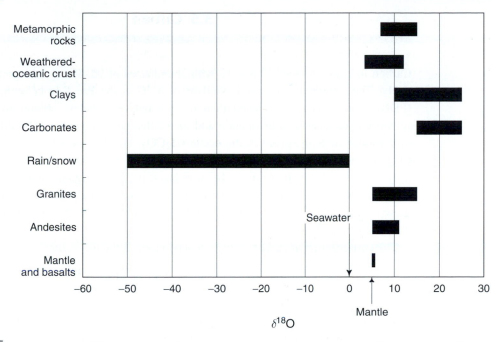

Figure 3.9 Distribution of $\delta^{18}O$ values in rocks and natural water. Notice seawater at 0‰ and the Earth's mantle at 5.2‰. It is the equilibration of sedimentary and metamorphic rocks with seawater at low temperature that raises the $\delta^{18}O$ of these rocks and lowers that of seawater. The broad isotopic variation of meteoric water results from atmospheric precipitation, which occurs at low temperature.

presence in their source material of rocks and minerals that have been subjected to low-temperature interaction with a hydrous fluid, such as seawater, groundwater, or meteoric water.

Conversely, upon interaction of rocks with seawater at temperatures of $\approx 300\,°C$, the reversal of albite–water $^{18}O/^{16}O$ fractionation drags these rocks towards lower $\delta^{18}O$ values. This is also clearly visible in the low $\delta^{18}O$ values of the gabbroic section of ophiolites, widely regarded as the deep section of the oceanic crust exposed to circulating seawater at such temperatures.

The distribution of oxygen isotopes in the natural environment is summarized in Fig. 3.9. Marine alteration, especially the interaction of seawater with mid-ocean ridges, is thought to be responsible for the current $^{18}O/^{16}O$ ratio of seawater: throughout the Earth's history, seawater has circulated in submarine hydrothermal systems and its oxygen has been brought into isotopic equilibrium at average temperatures of the order of 275 °C with the oxygen of basalts erupted by the volcanic systems of mid-ocean ridges. To these purely thermal effects must be added that of interaction with ^{18}O- and D-depleted meteoric water, which, as discussed below, is quite spectacular in modern and fossil peri-magmatic geothermal fields. Oxygen and hydrogen isotope evidence on shallow intrusions indicates that even the most fresh-looking plutonic rocks may have interacted with fluids contaminated by meteoric fluids.

3.5 Carbon

Carbon has two isotopes 12 and 13 with abundances of 98.89 and 1.11 percent, respectively. The universally accepted standard of $\delta^{13}C$ is the Pee Dee belemnite carbonate (PDB). Carbon is less abundant than oxygen but nevertheless ubiquitous. The major reservoirs are located in the mantle and in sedimentary carbonates. The different forms of carbonate ions dissolved in the ocean (H_2CO_3, HCO_3^-, and CO_3^{2-}) and atmospheric CO_2 are minor reservoirs with respect to carbonates and mantle carbon but are particularly important to us. Carbon can be in reduced (C, CH_4, organic material) or oxidized form (CO, CO_2). It is under different forms in the mantle and the relative abundances of these forms are not well constrained, but the "terrestrial" value of $\delta^{13}C$ should not be very different from $-7‰$.

The core reaction of carbon isotope geochemistry is the following:

$$CH_4 + 2O_2 \Leftrightarrow CO_2 + 2H_2O \qquad (3.39)$$

Contrary to all the substitutions we have seen so far, the two carbon molecules involved in this reaction have very different configurations and therefore very different normal modes of vibration (Fig. 3.3). In addition, the number of symmetrical configurations of the reactants and the products is very different. Isotope fractionation induced by such a reaction is therefore particularly strong, especially at low temperatures: the $\delta^{13}C$ of CO_2 is higher than that of CH_4 by 80 per mil at $0\,°C$, 33 per mil at $200\,°C$, and by 10 per mil at $700\,°C$. This is a strong indication that biological processes must have a profound impact on the carbon isotope compositions of carbon-bearing systems. From left to right, the previous chemical reaction produces oxidative energy and is a mockup of respiration, which liberates CO_2 with high $\delta^{13}C$. Solar energy is needed to activate the non-spontaneous reaction from right to left: this is the essence of photosynthesis, which liberates oxygen and stores low-$\delta^{13}C$ reduced carbon in plants. Why is it that the final products do not have the same isotope compositions as the initial product such as suggested by the reaction? It is simply that the reactions are much more complicated and involve repeated exchanges of carbon inside the cell during which intermediate products are lost. For respiration, the excreted CO_2 is isotopically buffered by the biological material.

In addition to fractionation between reduced carbon and CO_2, $^{12}C/^{13}C$ also fractionates at low temperature between atmospheric CO_2, the dissolved carbonate species, and $CaCO_3$ carbonates with $\delta^{13}C$ being typically 10 per mil higher at $15\,°C$ in calcite than in atmospheric carbon dioxide.

The fractionation just referred to concerns the distribution of isotopes between phases in equilibrium, but another type of fractionation (kinetic isotope effect or KIE) takes place when species with different oxidation states and coordination (here CO_2 and CH_4) fail to achieve equilibrium, in particular during biological reactions. For example, carbonic anhydrase enzyme found in mammal blood speeds up the conversion of CO_2 into an HCO_3^- ion by five orders of magnitude, thus preventing bubbles of excreted CO_2 from forming. Light isotopes having higher vibrational frequencies enter into reaction paths more often than heavy isotopes and are therefore exchanged more readily. They are also bound less

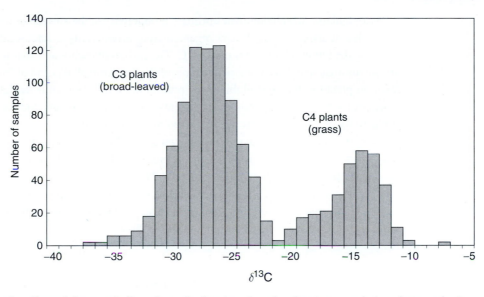

Figure 3.10 The effect of the metabolic cycle on the fractionation of carbon isotopes during photosynthesis. The C3 cycle (broad-leaved plants) and C4 cycle (grasses) correspond to very different photosynthetic processes within the cell. Past climatic conditions can therefore be inferred from the $\delta^{13}C$ value of organic remains (Deines, 1980).

strongly than heavy isotopes. In particular for fast reaction rates, equilibrium is therefore not necessarily achieved. Thus, by comparison with the environment within which they formed, organic products are significantly depleted in ^{13}C relative to ^{12}C. Such isotopic effects are extensively used in studying the genesis of fossil fuels and low-temperature mineralization in which biological processes are important, and in fingerprinting traces of early life in Archean sediments.

Since the isotopic standard of this element is the PDB marine carbonate, marine carbonates, as a whole, have $\delta^{13}C$ values close to 0‰. The carbon in atmospheric CO_2 has a $\delta^{13}C$ value close to -7‰. The continental biomass (dominated by plants), in contrast, has very negative $\delta^{13}C$ values. The distribution has two maxima, one at about -14‰, corresponding generally to grasses, and a larger one at about -25‰, corresponding to most broad-leaved plants (Fig. 3.10). This contrasted fractionation demonstrates that photosynthesis does not take place at equilibrium. It is brought about by two different CO_2 fixation mechanisms by very different plants, with what are known as the C4 pathway, predominating for grasses, and the C3 pathway, for broad-leaved plants and conifers. The very negative values of the C3 plants are essentially due to intracellular kinetic fractionation during C fixation mediated by the enzyme known as "rubisco." The variation in $\delta^{13}C$ of fossil plant matter collected from bores in peat bogs is thus utilized for tracing changes in vegetation between glacial and interglacial periods. Increased $\delta^{13}C$ levels of organic matter are interpreted as marking the advance of grasslands, while very negative $\delta^{13}C$ levels indicate a return of broad-leaved plants. The $\delta^{13}C$ of ≈ -25‰ of bitumen is inherited from their vegetal precursors. Carbon in oil tends to be even lighter.

Phytoplankton, which dominates primary productivity at the surface of the oceans, has a $\delta^{13}C$ value of about $-25\permil$. Where organic-matter productivity is intense, there is a corresponding depletion of surface water in ^{12}C. The carbon of carbonates that form the tests of calcareous algae (such as, in particular, coccoliths that form chalk) and of foraminifera that graze on plankton (such as globigerina) reflects the intensity of this depletion. The $\delta^{13}C$ values of foraminifera in carbonate sediments serve as a measuring rod for the biological productivity of ancient oceans where the sedimentation occurred.

The $\delta^{13}C$ values of most diamonds (known as peridotitic), carbonatites (magmas essentially made of carbonates such as those produced by the Oldoinyo Lengai volcano, Tanzania) and volcanic gases cluster around $-7\permil$. This isotope composition is considered to be representative of deep-seated carbon. It is coincidentally very similar to atmospheric carbon. Undegassed mid-ocean ridge basalts also have similar values, but CO_2 outgassing drives the $\delta^{13}C$ of the residual carbon towards $-20\permil$.

3.6 Sulfur

Sulfur has four isotopes of masses 32, 33, 34, and 36 with mean abundances of 95.0, 0.75, 4.21, and 0.02 percent, respectively. Because of their small abundances, isotope 33 is rarely measured while isotope 36 is left out, so that the bulk of the geochemical information comes from the $^{34}S/^{32}S$ ratio only. Most of the S element inventory is confined to the mantle and the core, but sulfate is one of the major components of seawater while calcium sulfate and iron sulfide make up an important component of sediments. For decades, the accepted standard was the Canyon Diablo Troilite (CDT). Troilite (FeS) is a sulfide found as large blebs in the Canyon Diablo iron meteorite, which excavated Meteor Crater in Arizona. After evidence was found that CDT was not truly homogeneous, it has been replaced by a synthetic silver sulfide (Ag_2S) produced by the International Atomic Energy Agency. Because the surface reservoirs are presumably small with respect to the deep ones, the mean terrestrial $\delta^{34}S$ is probably very close to zero.

As for carbon, the most interesting aspects of sulfur stable isotopes are related to competing dominant states of oxidation, H_2S, SO_2, and SO_3 in the gas phase, S^{2-} (sulfide), SO_3^{2-} (sulfite), and SO_4^{2-} (sulfate) in solutions and ionic compounds. As for carbon, isotopic equilibrium is rarely attained and kinetic isotope effects are important whenever redox and biological processes prevail. The trend is the same as with carbon: oxidized species are heavy (high $\delta^{34}S$ values), whereas the reduced species are light. It is easier to discuss fractionation in the gas phase; SO_2 is heavier than H_2S by 38 per mil at $0\,^\circ C$, 15 per mil at $200\,^\circ C$, and 4 per mil at $700\,^\circ C$. Under the same conditions, SO_3, which hydrates into sulfate, is heavier than H_2S, which hydrates into different sulfide ions, by 80, 33, and 4 per mil, respectively. In general sulfide and sulfate in equilibrium at ambient temperature precipitate from solutions with a difference of about $+20$ in favor of the sulfate. Today's marine sulfate has a $\delta^{34}S$ of about $+20\permil$, and so do modern marine evaporites: this value reflects the fact that the dominant mass of sulfur is in the mantle. We can also expect that diagenetic reduction of marine sulfate, whether inorganic or biologically mediated, results in sulfides (often present in organic-matter-bearing sediments such as pyrite FeS_2) with $\delta^{34}S$ values a few tens of per mil below the marine value largely due to kinetic isotope effects. The $\delta^{34}S$ value of marine sulfate has fluctuated by several per mil over the Phanerozoic (see Chapter 9), which reflects changes in the burial of sedimentary pyrite and atmospheric oxygen pressure.

Sulfur isotopes have developed into a tracer of sulfide ore genesis thanks to the complex speciation of this element in solution. First, the diprotic acid H_2S dissociates into HS^- and S^{2-} as a function of pH and the dissociation constants vary with, among other parameters, temperature. Second, fractionation of sulfur isotopes between sulfate and sulfide species, either at equilibrium or kinetically controlled, is assumed. Depending on the pH and redox potential of the solution, different proportions of these species co-exist and each of them has a different $\delta^{34}S$ value. Sulfides, such as pyrite, and sulfate, such as barite, can precipitate with a range of $\delta^{34}S$ that reflects these conditions. Using these principles, it can be demonstrated that the sulfur from black smokers is largely – although non exclusively – contributed by reduced basaltic sulfide and not by marine sulfate.

True igneous sulfides are rare because they are quickly oxidized by hydrothermal fluids. Nickel sulfide (pentlandite) is an all-too-rare truly magmatic sulfide in peridotites and basalts and its $\delta^{34}S$ is often close to zero.

Recently, the minor isotopes 33 and 36 of sulfur have found novel applications. First, photochemical reactions between solar ultraviolet radiations and SO_2 in the upper atmosphere create strong deviations from the mass-dependent fractionation relationships $\delta^{33}S = (1/2)\delta^{34}S$ and $\delta^{36}S = 2\delta^{34}S$. We will return to this point upon discussion of the rise of atmospheric oxygen. Second, because isotope fractionation between sulfides and sulfates is often large, the linear approximation of the Rayleigh law, which we met for O and H isotopes in meteoric waters, breaks down at the level of precision obtained by modern mass spectrometers. Small non-mass-dependent effects can thus be detected that seem to be powerful tracers of microbiological activity.

3.7 Nitrogen

Nitrogen has two isotopes 14 and 15, with mean abundances of 99.63 and 0.37 percent, respectively. The standard is, not surprisingly, atmospheric nitrogen (AIR). Although N_2 is the main component of the atmosphere, it has been argued that a substantial proportion is in the mantle. In addition, $^{15}N/^{14}N$ in extra-terrestrial material varies a lot as a result of loss from planetary atmospheres. We will therefore abstain from proposing a mean terrestrial value of $\delta^{15}N$.

As with carbon and sulfur, fractionation is dominated by nitrogen species of different oxidation states, N_2, NH_3, NO, NO_2 in the gaseous state and, for the last three, their dissolved equivalents, ammonium NH_4^+, nitrite NO_2^-, and nitrate NO_3^-. At equilibrium under surface temperature conditions, nitrites and nitrates are isotopically heavier, and NH_3 is lighter than N_2. Again, kinetic isotope effects are important. Biological processes play a key role in defining the nitrogen isotope compositions of surface material. Fixation converts atmospheric nitrogen into ammonium needed for amino acids with variable depletions in ^{15}N, i.e. $\delta^{15}N = 0$ to $-27‰$. Nitrification is an exothermic reaction that turns organic ammonium from mostly reduced amino (NH_2) groups into nitrite then nitrate. Nitrification reactions being nearly quantitative, the end products are variably depleted in ^{15}N. Reduction of marine nitrate into N_2, called denitrification, is a very active process taking place in interstitial waters below the water–sediment interface: isotopically light N_2 is lost to the atmosphere, while the heavier residual nitrates are flushed back into the ocean. This explains the $\delta^{15}N$ of $\approx +5‰$ of deep-sea nitrates.

Nitrogen is surprisingly abundant in diamonds in which its $\delta^{15}N$ is $\approx -7‰$. There is a clear difference in nitrogen isotope compositions between mid-ocean ridge (-4) and ocean island basalts ($+3$), yet with significant overlap. The interpretation of these variations is still unclear but they seem to be related to how much recycled lithosphere is present in the source of each type of basalt.

3.8 Other elements

Recent progress in analytical techniques has opened up new grounds for stable isotope geochemistry. Besides boron, which we will see later can be used as a pH-meter of ancient oceans, and silicon, which is currently revived as a paleoceanographic tracer, the use of the isotope compositions of some metals has recently attracted considerable attention. Lithium has an interesting cycle giving information on fluid–rock interactions and subduction-zone processes. Magnesium and calcium are relevant to weathering and erosion cycles, molybdenum to redox conditions in the ocean and in diagenesis, Fe, Cu, Zn to biological processes and ore-body genesis, and selenium to environmental studies in changing redox conditions.

Exercises

1. Consider oxygen gas and the reaction $^{16}O_2 + {}^{18}O_2 \Leftrightarrow 2{}^{16}O^{18}O$. What is the number of distinguishable configurations of each of the molecules involved in the reaction? If oxygen isotopes were distributing themselves at random in proportion to their abundances in the system and without any other effects, calculate the α value of this reaction.

2. The $\delta^{18}O$ value of modern seawater is $0‰$ while the average value of the polar ice cap is $-45‰$. The ice cap holds 2 wt% of the oceanic water. Calculate the $\delta^{18}O$ value of an ice-free ocean. Other water reservoirs can be neglected.

3. The $\delta^{18}O$ values of benthic carbonates decreased from $-1.2‰$ to $-2.5‰$ between the Early Eocene and the Oligocene. Assume an ice-free ocean (no salinity variation) and determine the cooling of the deep ocean over the same period using

Table 3.1 Temperature coefficients of 1000 ln $\alpha_{min}^O = A_{min}T^{-2} + C_{min}$ for different minerals. Third column: values of $\delta^{18}O$ in minerals from the same gneiss samples

Mineral	$10^{-6}A_{min}$	C_{min}	$\delta^{18}O$ ‰
Feldspar	3.38	−2.92	9.2
Plagioclase	2.76	−3.49	7.6
Magnetite	−1.47	−3.70	0.3
Muscovite	2.38	−3.89	6.6

Table 3.2 Coefficients of 1000 ln $\alpha_{min}^O = AT^{-2} + BT^{-1} + C$ for the liquid water–vapor oxygen and hydrogen isotope exchange reactions

	$10^{-6}A$	$10^{-3}B$	C
α^O	1.137	−0.4156	−2.0667
α^D	24.844	−76.248	52.612

the equation $\delta^{18}O_{calcite} - \delta^{18}O_{seawater} = 2.78 \times 10^6 T^{-2} - 2.91$, where T is the absolute temperature.

4. Explain why δD and $\delta^{18}O$ in ice cores can be used to trace local precipitation temperatures.

5. Oxygen isotope thermometry of a metamorphic rock. The oxygen isotope compositions of minerals from a gneiss sample have been measured and reported in Table 3.1. Let us call α_{min}^O the value of the $^{18}O/^{16}O$ fractionation coefficients between minerals and water. The temperature dependence of this coefficient can be written as $1000 \ln \alpha_{min}^O = A_{min}T^{-2} + C_{min}$ in which A_{min} and C_{min} are mineral-specific constants and T is the absolute temperature. The values of A and C are reported in Table 3.1 for the different minerals. What is the quartz–magnetite apparent equilibration temperature? Plot $\delta^{18}O - C$ for each mineral vs. A: what is the slope of this alignment? What is the $\delta^{18}O$ value of the water in equilibrium with this mineral assemblage?

6. Explain why even subtle variations of $\delta^{18}O$ values in fresh basaltic glasses are said to indicate a source component processed at low temperature.

7. Let us call α^O the value for $^{18}O/^{16}O$ vapor/liquid fractionation in water and α^D the ratio for D/H fractionation. Using the values listed in Table 3.2 calculate these coefficients at $5\,°C$ and $25\,°C$. Calculate the $\delta^{18}O$ and the δD values of water vapor in equilibrium with seawater at $25\,°C$. This water vapor now condenses as rain at $5\,°C$. Using the Rayleigh fractionation law, calculate the $\delta^{18}O$ and the δD values of rain water when 10, 20, 50, and 80 wt% of the water vapor have condensed. Plot the corresponding δD vs. $\delta^{18}O$ values and compare the slope of the alignment with the meteoritic water line.

References

♠ Bigeleisen, J. and Mayer, M. G. (1947) Calculation of equilibrium constants for isotopic exchange reactions. *J. Chem. Phys.*, **15**, 261–267.

♠ Craig, H. (1961) Isotopic variations in meteoric waters. *Science*, **133**, 1702–1703.

Deines, P. (1980). The isotope composition of reduced organic carbon. In *Handbook of Environmental Isotope Geochemistry*, ed. P. Fritz and J. C. Fontes, vol. 1, pp. 329–406. Amsterdam: Elsevier.

Hashizume, K. and Chaussidon, M. (2005) A non-terrestrial ^{16}O-rich isotopic composition for the protosolar nebula. *Nature*, **434**, 619–621.

Kim, S. T. and O'Neil, J. R. (1997) Equilibrium and nonequilibrium oxygen isotope effects in synthetic carbonates. *Geochim. Cosmochim. Acta*, **61**, 3461–3475.

Rumble D., Miller M. F., Franchi I. A., and Greenwood R. C. (2007) Oxygen three-isotope fractionation lines in terrestrial silicate minerals: An inter-laboratory comparison of hydrothermal quartz and eclogitic garnet. *Geochim. Cosmochim. Acta*, **71**, 3592–3600.

♠ Urey, H. C. (1947) The thermodynamic properties of isotopic substances. *J. Chem. Soc. (London)*, 562–581.

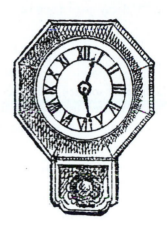

We have seen that radioactivity is not dependent on the chemical bonding of atoms, or on temperature, or on pressure. Radioactivity can be described as an event whose probability of occurrence is independent of time and depends only on the duration of measurement. The probability that a radioactive nuclide will decay per unit of time is denoted λ. This probability, better known as the decay constant, is specific to the radioactive nuclide under consideration. Radioactive decay, like incoming calls at a telephone exchange, is a prime example of a Poisson process, in which the number of events is proportional to the time over which the observation is made. In the absence of any other loss or gain, the proportion of parent atoms (or radioactive nuclides) disappearing per unit of time t is constant:

$$\frac{\mathrm{d}P}{P\,\mathrm{d}t} = -\lambda \tag{4.1}$$

For a number of parent atoms $P = P_0$ at time $t = 0$, this equation integrates as:

$$P = P(t) = P_0 \mathrm{e}^{-\lambda t} \tag{4.2}$$

In this form, Eq. 4.2 is not generally useful for measuring ages. The half-life $T_{1/2}$, which is the time it takes for half of the parent nuclides to decay, is $\ln 2/\lambda \approx 0.69/\lambda$. After five half-lives, 97% of the radioactive isotopes have decayed away, and 99.6% have decayed after eight half-lives. The product λP measures the number of decay events per unit time. It is commonly referred to as the activity of the radioactive nuclide P and denoted $[P]$. The becquerel (Bq) is a unit equal to one decay event (count) per second. A liter of seawater has an activity of 12 Bq, mostly because of the potassium-40 and uranium naturally dissolved in it. The human body contains enough potassium-40 and carbon-14 to register an activity of 5000–10,000 Bq as perfectly natural in origin. A granite cobblestone normally produces a radioactivity of several thousand Bq.

In order to determine the age of a system using (4.2) from the measurement of the number of parent atoms at the present time, we must also know P_0. If we do, then we have a chronometer based on the *decay of a radiogenic nuclide* and the age will be given by:

$$t = \frac{1}{\lambda} \ln \frac{P_0}{P}$$ (4.3)

Dating using the ^{14}C decay gives a good example:

$$t = \frac{1}{\lambda_{14}C} \ln \frac{^{14}C_0}{^{14}C}$$ (4.4)

If we don't, a different solution is in order. For each parent atom, a daughter atom (or radiogenic nuclide) is created, usually of a single element, whose number can be denoted D. In a closed system and for a stable daughter nuclide D, the number of parent and daughter atoms is constant. Therefore:

$$D = D_0 + P_0 - P = D_0 + P\left(e^{\lambda t} - 1\right)$$ (4.5)

The term $P\left(e^{\lambda t} - 1\right)$ is a measure of the accumulation of the radiogenic nuclide during time t. Even if D and P are measured, this equation is no more a timing device than the previous one unless we know the number of daughter atoms D_0 at time $t = 0$. A simple case where this condition applies is when the initial number of daughter nuclides is small enough so as to be negligible. We will refer to such chronometers as *systems with high parent/daughter ratios*. This approximation is fairly generally valid for the potassium–argon dating method and for uranium–lead dating of zircons described later, and the time t will be given by:

$$t = \frac{1}{\lambda} \ln \left(1 + \frac{D}{P}\right)$$ (4.6)

For example, for the ^{238}U–^{206}Pb chronometer:

$$t = \frac{1}{\lambda_{238}U} \ln \left(1 + \frac{^{206}Pb}{^{238}U}\right)$$ (4.7)

When the previous methods fail (typically D_0 cannot be neglected with respect to D), a different assumption can often be used, which is the basis of the isochron method. This method dates an event of isotopic homogenization, typically the precipitation of minerals from the same melt or solution, and removes the ambiguity arising from our ignorance of the initial state of the system. Isotopic homogenization results from the fact that the chemical properties of different isotopes from the same element are very similar, though as seen in Chapter 3 not identical. To help our understanding, this principle is illustrated in Fig. 4.1 by a playful comparison. A fenced yard with a tree in the center represents two crystalline sites with different energy levels. In the first case, we release a few dozen cats and dogs into the yard–tree system and we can well imagine that after some brisk movement among our elements, they will arrange themselves in appropriate sites, cats in

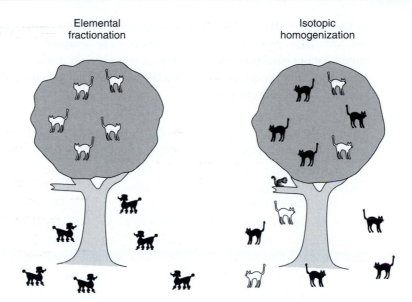

Elemental
fractionation

Isotopic
homogenization

Figure 4.1 Left: Cats and dogs interact vigorously affecting site occupation (tree or yard). Just like two elements with different chemical properties, they arrange themselves so as to achieve the most stable configuration. Right: White cats and black cats have very similar properties and like isotopes of the same element are arranged randomly among the available sites. The most likely arrangement is an identical proportion of isotopes in each site.

the tree and dogs on the ground in the yard, under the tree. Cats and dogs are two different elements in competition for the same sites: they arrange themselves spontaneously so as to move to the most stable configuration! Any alternative configuration is either intrinsically unstable (dogs in the tree, cats on the ground) or out of equilibrium (cats and dogs on the ground). Now, we clear these animals out and then release into the yard a few dozen cats that only differ by the color of their fur, either black or white. The probability that a cat will take to the tree or the ground is independent of the color of its coat, the energy of interaction is low, and the most likely arrangement is one of maximum entropy where the proportion of white and black cats is the same at each site. Our black and white cats are isotopes with very similar properties that share the sites evenly regardless of their energy level.

If elements and their isotopes are allowed to move easily between sites in crystals, liquids, and gases, either because the liquid states enable effective mixing or because thermal diffusion allows atoms to move rapidly, the elements with variable properties will arrange themselves in accessible sites so as to minimize the total energy of the system. On the other hand, isotopic exchanges of a single element between phases contribute little to the energy balance of the system and such isotopes will be evenly distributed so as to maximize the entropy of the system. For the needs of geochronology, both natural mass fractionation and instrumental mass bias are purely and simply eliminated by internal normalization against some arbitrary reference ratio (see box). As carbonates precipitate out from seawater, the $^{87}Sr/^{86}Sr$ ratio is exactly the same in calcite and in the seawater from which it precipitates;

as the mantle melts, ^{143}Nd/^{144}Nd is the same in the molten liquid as in the residue. There are, admittedly, processes where melting occurs in disequilibrium, but even if thermodynamic fractionation persisted at such high temperatures it would be corrected out by the internal standardization procedure (see box). In the rest of the discussion, we will therefore ignore mass-dependent fractionation.

Why radiogenic tracers are insensitive to phase changes and normalization values

The question often arises of how "are we so sure that these tracers are not upset by magmatic, metamorphic, or sedimentary processes just as oxygen and carbon isotopes are?" A common answer, albeit quite incorrect, is that the atomic weights of these tracers are heavy enough to make such fractionation negligible. Fractionation of the stable isotopes of strontium (A $\approx$ 88) and mercury ($\approx$ 200) are well established, both in nature and during mass-spectrometric analysis. The correct answer is that stable isotope fractionation is carefully separated from the fractionation induced by radiogenic decay by internal normalization to a reference isotope ratio. To measure isotopic ratios of interest for radiogenic isotopes, e.g. the ^{87}Sr/^{86}Sr ratio, both isotopic instrument bias and natural thermodynamic and kinetic fractionation are eliminated by standardization to an arbitrary stable isotope ratio (Fig. 4.2) using the theory developed for stable isotopes. For strontium, the universal choice is ^{88}Sr/^{86}Sr = 8.3752. From a linear approximation similar to (3.27), the magnitude f of the mass bias can be inferred from the measurement of this ratio:

$$\left(\frac{^{88}\text{Sr}}{^{86}\text{Sr}}\right)_{\text{meas}} = 8.3752\,(1 + 2f) \tag{4.8}$$

and the resulting value of f introduced into the similar expression for the ^{87}Sr/^{86}Sr ratio:

$$\left(\frac{^{87}\text{Sr}}{^{86}\text{Sr}}\right)_{\text{normalized}} = \left(\frac{^{87}\text{Sr}}{^{86}\text{Sr}}\right)_{\text{meas}} (1 + 1f) \tag{4.9}$$

$$= \left(\frac{^{87}\text{Sr}}{^{86}\text{Sr}}\right)_{\text{meas}} \times \left[1 + \frac{1}{2}\frac{\left(^{88}\text{Sr}/^{86}\text{Sr}\right)_{\text{meas}} - 8.3752}{8.3752}\right] \tag{4.10}$$

This completes the internal isotopic normalization which removes any isotope fractionation that is not due to radioactive decay. Note the values +2 and +1 for the mass differences between the masses at the numerator and denominator of the isotopic ratios.

For technical and historical reasons, the standardization of isotopic ratios of certain elements such as neodymium may refer to different isotopic values: the ^{143}Nd/^{144}Nd ratios taken from the literature must therefore be compared with utmost care. We saw in the previous chapter that with the delta notation, the problem of bias calibration among different laboratories or using different reference ratios disappears as soon as the isotope compositions are taken with respect to a common reference sample. For neodymium and hafnium, it is common practice to compare the sample isotopic compositions to that of the mean of

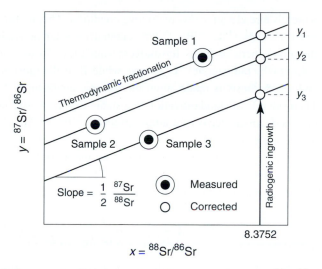

$$x = {}^{88}Sr/{}^{86}Sr$$

Figure 4.2   Principle of standardization of radiogenic isotope abundances (here $y = {}^{87}Sr/{}^{86}Sr$) relative to a reference ratio (here $x = {}^{88}Sr/{}^{86}Sr$). The various straight lines represent both natural and analytic thermodynamic isotopic fractionation depending on mass. The different ordinates between the straight lines represent the effect of radioactive accumulation of ${}^{87}Sr$, which varies from one sample to another. Standardization to $x = 8.3752$ eliminates thermodynamic fractionation leaving only radiogenic variability y_1, y_2, and y_3. Note the slope equal to $(87-86)/(88-86) \times {}^{87}Sr/{}^{88}Sr$.

chondritic meteorites, and to use a relative deviation notation, analogous to the δ notation in use for oxygen isotopes: $\varepsilon_{Nd}(T)$ is defined as:

$$\varepsilon_{Nd}(T) = \left[\frac{\left({}^{143}Nd/{}^{144}Nd\right)_{sample}(T)}{\left({}^{143}Nd/{}^{144}Nd\right)_{chondr}(T)} - 1 \right] \times 10\,000 \qquad (4.11)$$

which is the deviation in parts per 10 000 of the ${}^{143}Nd/{}^{144}Nd$ ratio in the sample relative to that of chondrites of the same age T. In a similar way, $\varepsilon_{Hf}(T)$ can be defined for the ${}^{176}Hf/{}^{177}Hf$ ratio.

Lead is an exception because it has only a single stable isotope, which rules out internal standardization and explains the intrinsically lower precision of measurement of some data (a few parts per 10 000) compared with that of Sr, Nd, or Hf ($10-30 \times 10^{-6}$).

In (4.2), let us divide P by the number P' of atoms of a stable isotope of the same element as the radioactive nuclide. As the system is closed, the number of stable nuclides P' remains constant, which we denote $P' = P'_0$. This gives:

$$\left(\frac{P}{P'}\right)_t = \left(\frac{P}{P'}\right)_0 e^{-\lambda t} \qquad (4.12)$$

The additional condition required to make the decay equation a chronometer is no longer to assume P_0 but rather to determine the isotope ratio $(P/P')_0$ when the system formed, which is already a far less restrictive condition. This method is employed for many short-lived nuclides (^{14}C, ^{10}Be, ^{210}Pb) created by solar or galactic radiation interacting with the atmosphere or rocks (cosmogenic nuclides). In the case of the carbon-14 clock, P refers to the radioactive ^{14}C isotope, while P' is the most abundant isotope ^{12}C of carbon, and a hypothesis is made about the isotopic abundance of ^{14}C in the upper atmosphere. The principle of standardization to a stable isotope is also utilized for radioactive nuclides derived from the decay of uranium isotopes (^{234}U, ^{230}Th, ^{231}Pa), but the equations are then a little more complex.

Equation (4.5) may also be divided by the number D' of atoms of a stable isotope of the same element as the radiogenic nuclide. As the system is closed, the number of stable nuclides remains constant and $D' = D'_0$. This yields:

$$\left(\frac{D}{D'}\right)_t = \left(\frac{D}{D'}\right)_0 + \left(\frac{P}{D'}\right)_t (e^{\lambda t} - 1) \tag{4.13}$$

Equation (4.13) is known as the isochron equation: in a plot of $x = P/D'$ and $y = D/D'$, a set of sub-systems of the same age T and the same initial isotope ratio $(D/D')_0$ will lie on a straight line of slope $e^{\lambda t} - 1$. The P/D' ratio is usually referred to, somewhat improperly, as the parent/daughter ratio. Writing this equation for two samples 1 and 2 that formed at the same time with the same isotope composition $(D/D')_0$ and subtracting them from one another, we obtain an expression for the time:

$$t = \frac{1}{\lambda} \ln \frac{(D/D')_2 - (D/D')_1}{(P/D')_2 - (P/D')_1} \tag{4.14}$$

Thus, for the system ^{87}Rb–^{87}Sr, P stands for ^{87}Rb, D for ^{87}Sr, and D' for ^{86}Sr, and we can write:

$$\left(\frac{^{87}\text{Sr}}{^{86}\text{Sr}}\right)_t = \left(\frac{^{87}\text{Sr}}{^{86}\text{Sr}}\right)_0 + \left(\frac{^{87}\text{Rb}}{^{86}\text{Sr}}\right)_t \left(e^{\lambda_{^{87}\text{Rb}} t} - 1\right) \tag{4.15}$$

This expression defines Nicolaysen's (1961) isochron. It has the familiar form of the equation of a straight line $y = y_0 + mx$, where $x = (^{87}\text{Rb}/^{86}\text{Sr})_t$, $y = (^{87}\text{Sr}/^{86}\text{Sr})_t$, with intercept $y_0 = (^{87}\text{Sr}/^{86}\text{Sr})_0$ and slope $m = e^{\lambda_{^{87}\text{Rb}} t} - 1$. Time t is derived from the expression:

$$t = \frac{1}{\lambda_{^{87}\text{Rb}}} \ln \frac{\left(^{87}\text{Sr}/^{86}\text{Sr}\right)_2 - \left(^{87}\text{Sr}/^{86}\text{Sr}\right)_1}{\left(^{87}\text{Rb}/^{86}\text{Sr}\right)_2 - \left(^{87}\text{Rb}/^{86}\text{Sr}\right)_1} \tag{4.16}$$

(all the ratios measured today).

For an isolated sample with a high parent/daughter ratio, i.e. $(P/D')_1 \ll (P/D')_2$ and for which the $(D/D')_1$ can be assumed, the age derived in this way is a *model age*.

Figure 4.3 shows two examples of isochron diagrams. A simple way of understanding isochrons is to appreciate that the existence of an alignment in the isochron diagram, in which relations between the quantities plotted change with time, cannot be fortuitous and depend on us to observe it today: if the samples form an alignment at the present time, an alignment must also have existed at each time since their formation. Since the D/D' ratio

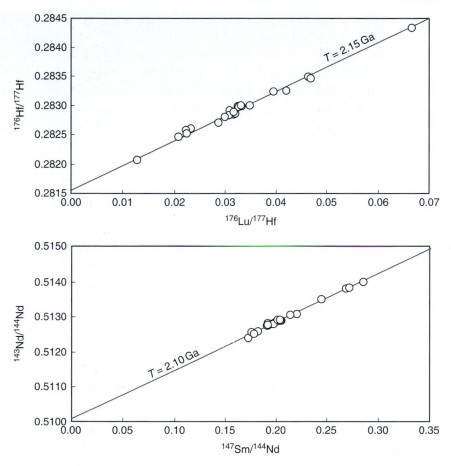

Figure 4.3   Two examples of isochrons for the ^{176}Lu $\rightarrow$ ^{176}Hf (top, λ^{-1} = 1.865 × 10^{-11} a^{-1}) and the ^{147}Sm $\rightarrow$ ^{143}Nd (bottom, λ^{-1} = 0.654 × 10^{-11} a^{-1}) systems on the same samples of Proterozoic basalts from West Africa (Abouchami *et al.*, 1990; Blichert-Toft *et al.*, 1999). Samples of the same age, and for which the range of isotopic compositions of the daughter element at the time of emplacement can be considered negligible with respect to the range caused by radiogenic ingrowth, will lie on a straight line. The slope of the isochron gives the age I of the lava flows and the intercept the mean isotopic ratio of the daughter element at that particular moment.

of any sample devoid of parent nuclide is unchanged, alignments simply rotate with time around the intercept.

Isochrons can alternatively be viewed as mixing lines between two end-members. The first end-member ($x = 0, y = (D/D')_0$ is the initial inventory of the daughter element, whereas the second end-member ($x = \infty, y = \infty, y/x = e^{\lambda t} - 1$) represents pure radiogenic ingrowth.

Let us now take a quick guided tour of dating methods, beginning with those based on the measurement of radioactive nuclides; then the systems with high parent/daughter ratios, in which the initial quantities of radiogenic nuclides can be neglected; and, finally, systems with low parent/daughter ratios, where the isochron method is applicable. In general, a clock can be applied to samples whose age does not exceed five times the radioactive

Table 4.1 Decay constants of the major radioactive systems: the daughter nuclide is shown when it is used for dating

System	$\lambda\ (y^{-1})$	System	$\lambda\ (y^{-1})$	System	$\lambda\ (y^{-1})$
$^{138}La-^{138}Ce$	2.24×10^{-12}	$^{40}K-^{40}Ca$	4.96×10^{-10}	^{26}Al	9.80×10^{-7}
$^{147}Sm-^{143}Nd$	6.54×10^{-12}	$^{235}U-^{207}Pb$	9.85×10^{-10}	^{36}Cl	2.30×10^{-6}
$^{87}Rb-^{87}Sr$	1.42×10^{-11}	$^{146}Sm-^{142}Nd$	6.73×10^{-9}	^{230}Th	9.20×10^{-6}
$^{187}Re-^{187}Os$	1.64×10^{-11}	^{244}Pu	8.66×10^{-9}	^{234}U	2.83×10^{-6}
$^{176}Lu-^{176}Hf$	1.865×10^{-11}	$^{182}Hf-^{182}W$	7.7×10^{-8}	^{231}Pa	2.11×10^{-5}
$^{232}Th-^{208}Pb$	4.95×10^{-11}	$^{129}I-^{129}Xe$	4.30×10^{-8}	^{14}C	1.21×10^{-4}
$^{40}K-^{40}Ar$	5.81×10^{-11}	$^{53}Mn-^{53}Cr$	1.87×10^{-8}	^{226}Ra	4.33×10^{-4}
$^{238}U-^{206}Pb$	1.55×10^{-10}	^{10}Be	4.62×10^{-7}	^{210}Pb	3.11×10^{-2}

period. Table 4.1 shows that the clocks spread over a wide range but certain age ranges are not well covered, especially that at around one million years.

Note that physical time elapsing in the real world is normally given in seconds (s), which is not a very helpful unit in the Earth sciences, while geological ages, through which we go back through time, are noted in anni (a), from the Latin *annus*. Derived units ky and ka (thousand years), My and Ma (million years), Gy and Ga (billion years) apply to physical time and time interval (or age), respectively. Appendix D shows the division of geological time into absolute ages. The geological time scale is the product of the work of literally thousands of scientists throughout the last century and cannot be credited to one particular work. An overview of the techniques commonly used for the determination of elemental concentrations and isotopic ratios is given in Appendix E.

4.1 Dating by radioactive nuclides

This group of methods relates essentially to nuclides produced by cosmic radiation, but we will see that the approach can be generalized to the descendants of uranium and thorium with methods based on the surpluses of these nuclides. Here we make an assumption about the initial isotopic composition of the element to which the radioactive nuclide belongs.

4.1.1 Carbon-14

This method of dating, which is certainly the most familiar to the general public, is not the oldest historically. However, it has revolutionized archeology and earned its inventor, Libby (see Arnold and Libby, 1949), the Nobel Prize for Chemistry in 1960. The Earth is subjected to bombardment from high-energy galactic cosmic rays, mostly protons and α particles, which react with the Earth's atmosphere. The interaction of these particles with nitrogen and oxygen produces secondary particles, mostly neutrons. In spite of a limited lifetime, neutrons do not have to overcome the Coulomb barrier of the nucleus and react more easily than charged particles of the same energy, such as protons. An important source of ^{14}C is the reaction between neutrons and nitrogen, which produces radioactive carbon-14 and a proton:

$$^{14}N + n \rightarrow {}^{14}C + p \tag{4.17}$$

The ^{14}C atom decays to ^{14}N by β^- emission with a decay constant λ_{14C} of 1.2×10^{-4} y^{-1}. Before disappearing, it mixes very quickly with the stable isotopes of carbon, the most important of which is ^{12}C. Equation 4.8 can now be written:

$$\left(\frac{^{14}C}{^{12}C}\right)_t = \left(\frac{^{14}C}{^{12}C}\right)_0 e^{-\lambda_{14C}t} \tag{4.18}$$

Plants exchange their carbon with the atmosphere, with which they are in isotopic equilibrium until they die. It can be seen that if the ratio $(^{14}C/^{12}C)_0$ in the atmosphere is constant and known, measurement of the $^{14}C/^{12}C$ ratio in wood or a fossil carbonate will date the death of the organism.

This approach is complicated by several effects. First, ever since the nineteenth century, the burning of coal and oil has released a large quantity of "dead" carbon devoid of ^{14}C into the atmosphere, thus complicating estimates of the $(^{14}C/^{12}C)_0$ ratio. To make things worse, above-ground nuclear explosions until the mid 1970s have contaminated the atmosphere with artificial ^{14}C, some of which has invaded the surface of the oceans. Finally, the variation in solar activity modulates galactic cosmic radiation received by the Earth and therefore changes the rate of production of ^{14}C in the atmosphere over very long periods. To overcome these difficulties, the ^{14}C scale has been calibrated for more recent ages by dendrochronology – a method based on counting the growth rings of very old trees – Californian bristle-cone pines or German oaks. For older ages, calibration is achieved by comparison with the thorium-230 method on corals.

There are various applications for the ^{14}C method; it can provide dates as old as 40 000 years ago. Measurement methods using a linear accelerator have pushed this limit a little further back, but above all they have reduced the quantities of material required for analysis to be conducted. The ^{14}C method has been used very successfully in archeology and Quaternary geology. It also has applications in dating groundwater sources and, as we will see later, deep water of the oceans.

4.1.2 Beryllium-10

The radioactive nuclide ^{10}Be is one of the pieces produced by spallation, the breaking of atmospheric ^{14}N and ^{16}O nuclei by cosmic rays. Metallic Be is rapidly oxidized to BeO, scavenged by atmospheric particles, and finally incorporated into soil and sediments by rain water and run off. The ^{10}Be decay constant is 4.62×10^{-7} y^{-1} and it is customary to normalize its abundance to that of the stable isotope ^9Be. Beryllium is an element similar to aluminum and is found in clay and soils. Beryllium-10 dating is much used in oceanography for measuring sedimentation rates or manganese-nodule growth rates. For samples

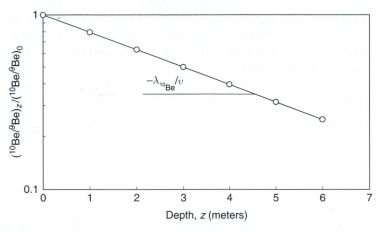

Figure 4.4 Measurement of sedimentation rate v by variation in relative abundance of the cosmogenic isotope ^{10}Be with depth. The isotopic ratio is standardized to its surface value (0). The scale is semi-logarithmic.

taken at depth z in a sediment core in which the sedimentation rate can be assumed to have remained constant, the equivalent to (4.18) can be written as:

$$\left(\frac{^{10}\text{Be}}{^{9}\text{Be}}\right)_t = \left(\frac{^{10}\text{Be}}{^{9}\text{Be}}\right)_0 \exp\left(-\lambda_{10_{\text{Be}}}\frac{z}{v}\right) \qquad (4.19)$$

where v is the sedimentation rate. On the condition that the ^{10}Be/^{9}Be ratio at the time of deposition does not vary locally over the time interval sampled by the core, the ^{10}Be/^{9}Be ratios measured in different samples should decrease exponentially with age. The logarithm of the ^{10}Be/^{9}Be ratio should therefore vary linearly with depth z (Fig. 4.4) in the core. The slope $-\lambda_{10_{\text{Be}}}/v$ is hence a measure of sedimentation rate.

This method can also be employed for measuring erosion rates. Neutrons produced by interaction of cosmic radiation with the atmosphere are stopped by the ground, where they cause spallation reactions, i.e. fragmentation of the atoms of the material. The production of ^{10}Be from silicon atoms in quartz is of particular interest. The number of ^{10}Be atoms produced in the ground per unit volume varies with depth z beneath the surface as $R_0 \, e^{-z/l}$, where R_0 is the surface production and l is the attenuation distance or mean penetration distance of cosmic particles into the rock. Over time t, thickness $dz = v dt$ of the rock has been eroded, where v is the erosion rate. The number dN of ^{10}Be atoms transiting between depth z and depth $z + dz$ is $-(vN(z + dz) - vN(z))$ per unit time, since the erosion rate is directed toward negative depths. Let us assume that the erosion rate is constant and that the ^{10}Be production rate remains constant over the time interval of interest. At equilibrium we can write that removal by erosion and radioactivity balances production by cosmic radiation:

$$-\left[vN(z + dz) - vN(z)\right] + R_0 e^{-z/l} dz - \lambda N(z) \, dz = 0 \qquad (4.20)$$

and a little mathematics give the expression:

$$N(z) = \frac{R_0}{\lambda + v/l}e^{-z/l} \qquad (4.21)$$

It can be seen, then, that if we know the rate of production R_0 and the attenuation distance l, the abundance $N(z = 0)$ of ^{10}Be at the surface provides a direct estimate of the rate v of erosion.

4.1.3 The thorium-230 excess method

This is a somewhat different case, as this nuclide, with a radioactive half-life of 75 000 years, is not created by radiation but by another parent nuclide whose half-life is long enough for its rate of production to be considered constant on time scales of less than one million years. It is one of the examples of clocks based on a chain of radioactive decay (Fig. 4.5) in which nuclides decay from one to another by α or β^- radioactive processes. The vast majority of intermediate nuclides have half-lives too short to act as useful clocks. There are only four radioactive chains. These chains, which have a long-lived, heavy nuclide as their parent, are those of ^{232}Th, ^{235}U, ^{238}U, and ^{237}Np. The first three end with three isotopes of lead, ^{208}Pb, ^{207}Pb, ^{206}Pb, and we will see that their relative abundances in modern lead are utilized as clocks. The fourth chain is that of neptunium-237, an extinct radiogenic nuclide ending with the single isotope bismuth-209. It is not detectable in natural products.

Uranium-238 decays to ^{234}Th, then to ^{234}U and, finally, to ^{230}Th, which is itself radioactive. The two intermediate nuclides, ^{234}Th and ^{234}U, are so short-lived that we can ignore them here. The change in the number $N_{230_{Th}}$ of ^{230}Th nuclides can therefore be written as the difference between production by the parent ^{238}U and radioactive decay:

$$\frac{dN_{230_{Th}}}{dt} = -\lambda_{230_{Th}} + \lambda_{238_U} \qquad (4.22)$$

The complete set of equations and their solutions were found by Bateman (1910). After a few hundreds of thousand years, the number of ^{230}Th nuclides reaches steady state and the two terms of the right-hand side cancel out. This state, where all activity levels are equal for all the nuclides in the chain, is known as secular equilibrium. The activity $[^{238}U]$ remains essentially constant during the time it takes to establish secular equilibrium (roughly $1/\lambda_{230_{Th}}$, or several $100,000$ years) and variation $d[^{238}U]/dt$ with time is therefore zero. Allowing for this property and multiplying (4.22) by λ_{230Th}, we obtain:

$$\frac{d\{[^{230}Th] - [^{238}U]\}}{dt} = -\lambda_{230_{Th}}\{[^{230}Th] - [^{238}U]\} \qquad (4.23)$$

in which the square brackets represent activity. The difference $[^{230}Th] - [^{238}U]$ is known as ^{230}Th excess (the term excess is taken by reference to the amount present at secular equilibrium), and is written $[^{230}Th]_{ex}$. Equation (4.23) integrates as:

$$[^{230}Th]_{ex} = [^{230}Th]_{ex,0}\, e^{-\lambda_{230_{Th}}t} \qquad (4.24)$$

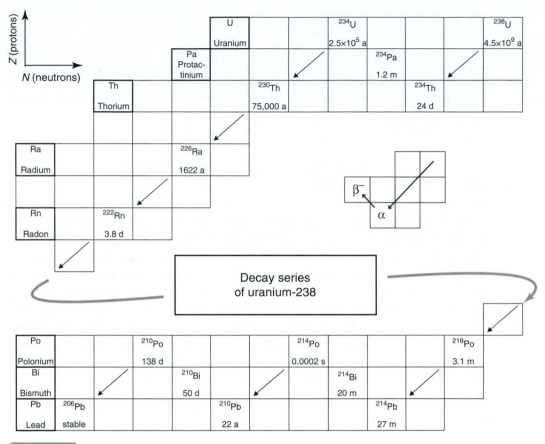

Figure 4.5 Decay series of ^{238}U. The decay period of unstable nuclides is shown. The number of neutrons is given on the *x*-axis and the atomic number on the *y*-axis. Jumps induced by α and β$^-$ decay are shown in the inset box. Units: year (a), day (d), minute (m), second (s).

where the subscript 0 denotes the initial time. If, as with ^{10}Be, we wish to measure a sedimentation rate, time *t* is replaced by the ratio between depth *z* and rate of sedimentation *v*, and the logarithm of $\left[^{230}\text{Th}\right]_{\text{ex}}$ is plotted as a function of *z* for several samples collected at different depths from a single core: the slope of the alignment indicates the $-\lambda_{230\text{Th}}/v$ ratio, and so *v* can be determined from it.

These concepts apply to excesses of several nuclides descending from ^{238}U, such as ^{234}Th, ^{234}U, ^{210}Pb, and from ^{235}U, such as ^{231}Pa. These methods are often said to be based on "uranium disequilibrium series," to indicate that the chronometric information lies with the deviation of a given parent/daughter pair from secular equilibrium. There are many fields of application, from oceanography to dating of Quaternary lacustrine or marine sediments. Lead-210 $\left(^{210}\text{Pb}\right)$, with a half-life $T^{1/2} = 23.3$ years, notably forms in the atmosphere and in rain water by decay of radon-222 (the offspring of ^{238}U, see Fig. 4.5), a gaseous nuclide given off constantly by the ocean, mantle, and crust. It is used for determining the rates of accumulation of ice or of sedimentation during the last century. It is also very helpful for studying pollution processes.

4.2 Systems with high parent/daughter ratios

There are many important clocks for which the quantities of daughter isotopes at time $t = 0$ can be ignored. The potassium–argon method and uranium–lead method on zircon are the best known, but the old uranium–helium method has recently been resurrected to address the chronology of erosion processes. Also worth mentioning in this category are the many clocks based on the descendants of uranium, such as $^{238}U-^{230}Th$, which have been amazingly successful at dating corals and cave deposits.

4.2.1 The potassium–argon method

This is the workhorse of geochronology. The method relies on the potassium-40 nucleus capturing a K-shell electron so that $^{40}K + e^- \rightarrow {}^{40}Ar$. The radioactive constant or probability of decay per unit time for this process is $\lambda_\varepsilon = 5.81 \times 10^{-11}$ y^{-1}. Potassium-40 also decays by an ordinary β^- process into ^{40}Ca (dual decay), for which the radioactive constant is $\lambda_\beta = 4.96 \times 10^{-10}$ y^{-1}. The proportion of ^{40}K atoms taking the ^{40}Ar pathway is equal to the relative probability $\lambda_\varepsilon /(\lambda_\varepsilon + \lambda_\beta)$ or 10.5%. Using N for the number of nuclides, equation (4.5) then becomes:

$$N_{40_{Ar}}(t) = N_{40_{Ar}}(0) + \frac{\lambda_\varepsilon}{\lambda_\varepsilon + \lambda_\beta} N_{40_K}(t) \left[e^{(\lambda_\varepsilon + \lambda_\beta)t} - 1 \right] \qquad (4.25)$$

with the sum of probabilities of decay, by one or other pathway, in the exponential, and the proportion of daughter nuclides that are atoms of ^{40}Ar in the factor term. The principle behind the method is that the $^{40}Ar_0$ term can be neglected relative to the second term. Although argon is present in notable quantities in the atmosphere (1%) and in the interstitial gases of rocks, this inert gas is not very soluble at ambient pressure in melts and minerals since it forms only weak van der Waals'-type bonds with mineral ions. At higher pressure, however, substantial amounts of argon may remain trapped in submarine glasses or metamorphic minerals (excess argon).

For analysis, argon is extracted from rocks by heating and melting in ultra-vacuum lines and analyzed with a mass spectrometer. The ultra-vacuum is necessary to prevent atmospheric contamination. Such contamination inevitably occurs anyway as minerals adsorb small quantities of atmospheric gases onto their surfaces or in grain fractures. To obviate this, we use the fact that atmospheric argon has several isotopes and that the atmospheric $^{40}Ar/^{36}Ar$ ratio is 296. We simply subtract 296 times the quantity of ^{36}Ar measured, from the ^{40}Ar measured in the sample, to obtain the radiogenic argon (Fig. 4.6). This can only be done with precision when radiogenic ^{40}Ar is not dominated by that originating in the atmosphere: measuring young ages for rocks with a low potassium content is therefore a technical feat of skill. It is thanks to the potassium–argon method that the scale of magnetic reversals can be calibrated, as it is the only method for dating the lava on which the paleomagnetic measurements are made. It will be seen when studying the thermal history

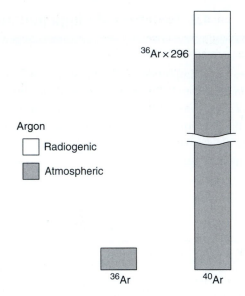

Figure 4.6 Correction of the atmospheric argon contribution to mass 40. The ^{40}Ar/^{36}Ar ratio in the atmosphere being invariably constant and equal to 296, radiogenic ^{40}Ar is inferred by the difference between total ^{40}Ar and 296 × ^{36}Ar.

of rocks in the chapter on diffusion, that this chronometer also provides a measure of post-orogenic cooling rates and rates of exhumation of mountain ranges. Applications in other fields find it difficult to compete with other methods, notably U–Pb on zircons.

An important and widely used variant of the standard ^{40}K–^{40}Ar method involves irradiating the sample in a nuclear reactor with fast neutrons (energy > 1 MeV) to produce the reaction ^{39}K + n ⇒ ^{39}Ar + p. It is here not ^{40}K, but the ^{39}Ar produced in the reaction that is measured. Accordingly, this is termed the ^{39}Ar–^{40}Ar method, which replaces the separate measurement of two isotopes of two different elements, ^{40}K and ^{40}Ar, with a precise measurement of the ^{40}Ar/^{39}Ar ratio of two isotopes of argon. A control sample of known age (monitor), irradiated at the same time, provides a measure of the yield of the reaction and thus allows the age of the samples to be determined by cross-calibration with the standard. This method, combined with progressive extraction of argon from the irradiated sample, has several advantages, described in more specialized texts (e.g. McDougall and Harrison, 1999), including the identification of possible argon losses after the formation of minerals.

4.2.2 Dating zircons by the uranium–lead method

This method is the Ferrari of geochronology for long geological time scales; it is precise and resistant against disturbances occurring after closure of the system, but difficult to implement. In recent years, its development has considerably benefited from the improved cleanliness of chemical extractions and *in situ* methods of analysis (secondary ion mass spectrometry and laser-ablation inductively coupled plasma mass spectrometry, ICP-MS). The advantage of the method lies in the radioactive (^{238}U and ^{235}U) and radiogenic nuclides (^{206}Pb and ^{207}Pb) being isotopes of the same elements: uranium for one and lead for the other. In the absence of any initial radiogenic lead, (4.13) applied to the systems ^{238}U–^{206}Pb ($\lambda_{238_U} = 0.155\ 125 \times 10^{-9}\ \text{y}^{-1}$) and ^{235}U–^{207}Pb ($\lambda_{235_U} = 0.984\ 85 \times 10^{-9}\ \text{y}^{-1}$) gives:

$$\left(\frac{^{206}\text{Pb}}{^{238}\text{U}}\right)_t = e^{\lambda_{238_U} t} - 1$$

$$\left(\frac{^{207}\text{Pb}}{^{235}\text{U}}\right)_t = e^{\lambda_{235_U} t} - 1 \tag{4.26}$$

This double clock is routinely applied to the radiogenic lead and uranium of an accessory, but common, zirconium silicate of granite and metamorphic rocks, zircon (ZrSiO_4). Uranium U^{4+} substitutes in large quantities for Zr^{4+}; but Pb^{2+}, which is of very different ionic radius (0.133 nm) and charge from Zr^{4+} (0.084 nm), is essentially excluded at equilibrium. As with atmospheric argon, there may nonetheless be contamination by lead at mineral surfaces or in grain fractures. Because of the presence in the atmosphere of tetraethyl lead, used until the 1990s as an anti-knocking agent in fuel, man-made pollution may also be significant. A very similar technique to that described for argon is employed, involving subtraction from the total lead content of the contaminated lead, whose isotopic composition is relatively well known by using a stable isotope ^{204}Pb. *In situ* isotopic analysis using modern ion probes also allows the zones for analysis to be selected so that contamination is almost completely eliminated.

The pair of equations in (4.26) defines the locus of points for which the ages indicated by both methods concur, the locus being traditionally called concordia proposed by Wetherill in 1956. This concordia flattens out toward older ages (Fig. 4.7), as ^{235}U decays much more rapidly than ^{238}U: natural uranium today contains only 0.7% ^{235}U compared with 8% ^{235}U three billion years ago. Although methods were developed to attempt to correct the effect of disagreement related to losses of lead after closure of the system, they are now of little value because of the improvement in techniques. The ratio of the x-axis to the y-axis is proportional to the isotope ratio ^{206}Pb/^{207}Pb:

$$\frac{\left(^{206}\text{Pb}/^{238}\text{U}\right)_t}{\left(^{207}\text{Pb}/^{235}\text{U}\right)_t} = \frac{1}{\left(^{207}\text{Pb}/^{206}\text{Pb}\right)_t} \times \frac{1}{\left(^{238}\text{U}/^{235}\text{U}\right)_t} \tag{4.27}$$

The second term on the right-hand side is constant and equal to 1/137.88. The denominators of the ratios plotted on the x- and y-axes are therefore proportional and their ratio ^{238}U/^{235}U is constant. It follows from the discussion of ratio behavior during mixing, presented earlier (Fig. 2.4), that mixtures of zircons or overgrowth will be reflected by alignments in the concordia plot: the intercepts of these alignments with the concordia will therefore a priori yield interesting ages.

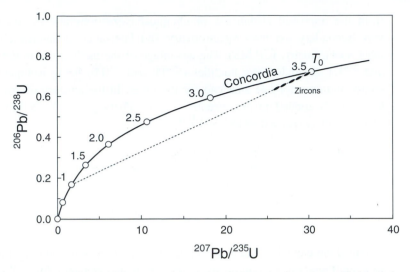

Figure 4.7 The concordia curve is the locus of points for which the *x*- and *y*-axes yield identical ages for both the $^{235}U-^{207}Pb$ and $^{238}U-^{206}Pb$ methods. The numbers 1, 2.0, 2.5, etc., on the curve correspond to geological ages of 1, 2.0, 2.5, etc., billions of years. Zircons extracted from a single sample, and shown here by the black ellipses, indicate an age T_0 of crystallization of 3.5 Ga, with 1.0 Ga old overgrowth. The dashed lined is a mixing line between the two zircon generations.

4.3 The isochron method

When minerals and rocks form, they already contain some of the radiogenic isotopes used for dating. The daughter nuclide may be present in large, yet unknown, concentrations. The isochron method was devised to provide an age, even when the amount of radiogenic isotope initially present in the system is not negligible with respect to that produced by radioactive decay after its formation. The key assumption is that the initial isotopic composition of the element to which the radiogenic nuclide belongs is unknown, but constant, in all the samples analyzed. Isotopic homogenization is assumed to be complete at $t = 0$, which may be the case where minerals crystallize from a magma or from seawater within a time interval that can be considered as very short compared to the age of the rocks. A large number of geochronological systems are used in this way: $^{87}Rb-^{87}Sr$, $^{147}Sm-^{143}Nd$, $^{176}Lu-^{176}Hf$, and $^{187}Re-^{187}Os$ are examples (note that chronological systems are denoted by hyphenating the parent and daughter isotopes in the order $P-D$). In addition to the chronological aspect, the variations in the isotopic abundances of radiogenic isotopes are useful in studying many geological processes. Let us take as an example the $^{147}Sm-^{143}Nd$ system, for which the stable reference isotope is usually ^{144}Nd, and in (4.13) replace P with ^{147}Sm, D with ^{143}Nd, and D' with ^{144}Nd. For a closed system, the isochron equation becomes:

$$\left(\frac{^{143}Nd}{^{144}Nd}\right)_t = \left(\frac{^{143}Nd}{^{144}Nd}\right)_0 + \left(\frac{^{147}Sm}{^{144}Nd}\right)_t \left(e^{\lambda_{147Sm}t} - 1\right) \qquad (4.28)$$

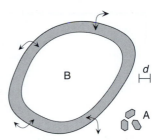

Figure 4.8 Closed system and open system. If we imagine that during a metamorphic perturbation an element has moved an average distance d, small systems (minerals A) will be "opened" much more intensely than large ones (large sample B), where only the outer skin will have exchanged with the ambient medium. The shaded area represents the material involved in exchange outside the system.

Samples, for example basalts extracted from a single source in the mantle, that formed at t with the same $(^{143}Nd/^{144}Nd)_0$ ratio but with different $^{147}Sm/^{144}Nd$ ratios, presumably by different degrees of melting, will form an alignment known as the isochron (Fig. 4.3), whose slope gives the age of formation. It is commonly held that, for the isochron to be valid, the rocks must be "co-genetic" and the initial isotopic homogeneity must be perfect. However, as a result of improved analytical techniques and instrumentation, there is always an achievable level of precision for which the assumption of initial isotopic homogeneity breaks down. A less conservative, but more realistic, statement is that the initial isotopic variability must remain negligible compared with that related to the accumulation of radiogenic nuclides: this explains why isochrons of old age, such as those obtained from meteorites or lunar rocks, "look" better than those from recent granites, or why isochrons of reasonable age can be obtained from samples that are not necessarily taken from a homogeneous medium, such as sedimentary sample suites.

Let us look at the problem of how the isochron withstands subsequent perturbations of the closed-system regime. Let us imagine, for example, a series of basalts formed from a single mantle source in Archean time, some 2.7 billion years ago, and subjected to intense thermal perturbation when, caught up in the formation of a mountain belt some 600 million years ago, they were transported to great depth and raised to temperatures of the order of 600–800 °C. Such metamorphism will create new minerals in basaltic rocks, probably an assemblage of pyroxene and garnet known as eclogite. Suppose that a mean distance d can be defined over which the elements Sm and Nd migrate in the course of metamorphic recrystallization, ranging probably from a few millimeters to a few centimeters (Fig. 4.8). If our sample is very much smaller than d, we can presume that the exchanges of Sm and Nd are complete and that the minerals, normally of very small size, will all have adopted the ambient Nd isotopic composition, i.e. that of the rock at this location. If the sample size is much greater than d, only the outer part over a depth d will have engaged in exchange with the exterior. The interior will have been disturbed, but the exchanges will have remained confined within the sample. If this open fringe represents only a small volume compared with the bulk of the sample, it will be considered that the sample has not been affected by the disturbance.

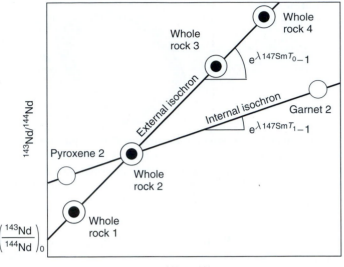

Figure 4.9 During a metamorphic event, large samples (whole rock) will remain virtually closed (see Fig. 4.8) and the "external" isochron will record the age T_0 of rock formation. However, the minerals of rock 2 (pyroxene, garnet) will have exchanged their isotopes and their $^{143}Nd/^{144}Nd$ isotope ratios will be in equilibrium with that of the mean ambient value, i.e. that of the rock. The "internal" isochron therefore gives the age T_1 of the disturbance.

   This pattern of exchange suggests the sampling strategy (Fig. 4.9). To determine the age of any metamorphic perturbation, the minerals of a rock are separated, here garnet and pyroxene, whose small size ensures that they will have been returned to isotopic equilibrium during metamorphism. The age indicated by what is termed the internal (or mineral) isochron will be that of the metamorphic event. Conversely, to obtain the age of formation of the rock, the largest possible samples must be taken so as to minimize the effect of exchanges with the exterior. The whole rock isochron will be obtained from small fractions of large samples that have been thoroughly ground and mixed.

   There are many methods of dating using isochrons. Historically, $^{87}Rb–^{87}Sr$ chronology has yielded many ages for the emplacement of granites, whereas the dating of white mica (muscovite) or black mica (biotite), whose high $^{87}Rb/^{86}Sr$ ratio ensures good age precision, have dominated metamorphic geochronology alongside $^{40}K–^{40}Ar$ ages. The mobility of rubidium, an alkali element, and of strontium, an alkaline-earth element, in metamorphic and hydrothermal fluids unfortunately often disturbs this chronometer. $^{147}Sm–^{143}Nd$ chronology is valuable for dating old basaltic rocks. Samarium and neodymium, two of the rare-earths, are much less mobile, but the method has also been known to go awry. Dating high-temperature and high-pressure metamorphism, especially with pyroxene–garnet internal isochrons, has yielded good results by this method, as has, more recently, the $^{176}Lu–^{176}Hf$ system. Dating of basaltic rocks and peridotites, but also of sulfide ore deposits and petroleum by the $^{187}Re–^{187}Os$ method, two highly siderophile elements, has also proved to be very successful.

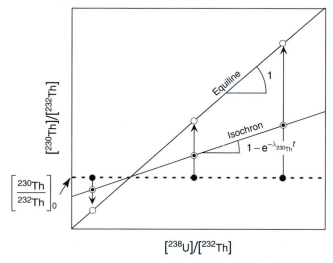

Figure 4.10 The U–Th isochron. Samples derived from the same sedimentary or magmatic reservoir initially have the same $[^{230}Th/^{232}Th]$ isotope ratio but different U/Th ratios. They line up first on a horizontal line and then migrate progressively toward the equiline, which is the stable state of secular equilibrium. Square brackets indicate activity.

A first special application of the isochron method is that of the ^{230}Th dating method, which is especially useful for recent volcanic rocks. If we re-write (4.24) by allowing for the definition of excess ^{230}Th and if we then divide both sides by the activity of the thorium-232 isotope (which decays slowly enough for this activity to be considered constant), we obtain:

$$\left(\frac{\left[^{230}Th\right]}{\left[^{232}Th\right]}\right)_t = \left(\frac{\left[^{230}Th\right]}{\left[^{232}Th\right]}\right)_0 e^{-\lambda_{230Th}t} + \left(\frac{\left[^{238}U\right]}{\left[^{232}Th\right]}\right)_t \left(1 - e^{-\lambda_{230Th}t}\right) \qquad (4.29)$$

The right-hand side is the sum of two terms, which we will meet many times: the first term expresses the demise of the initial condition, while the second term describes how fast the system is moving towards steady state. It can be seen from an isochron plot $x = \left[^{238}U\right] / \left[^{232}Th\right]$, $y = \left[^{230}Th\right] / \left[^{232}Th\right]$ (Fig. 4.10) that samples formed at $t = 0$ with the same $\left[^{230}Th\right] / \left[^{232}Th\right]_0$ ratio will lie on a horizontal line. With time the alignment pivots around the point of intersection with the line $y = x$, known as the equiline, until its slope becomes unity: at this point, the system is in secular equilibrium $\left(\left[^{230}Th\right] = \left[^{238}U\right]\right.$, see above). The slope of the alignment gives an age, at least as long as we are far from equilibrium. This method is often applied to minerals extracted from lavas to date their crystallization. It can only be applied for ages of less than 350,000 years.

A second special application is that of the lead–lead method. By utilizing the stable lead isotope of mass 204, the two equations of the isochron can be written:

$$\left(\frac{^{206}Pb}{^{204}Pb}\right)_t - \left(\frac{^{206}Pb}{^{204}Pb}\right)_0 = \left(\frac{^{238}U}{^{204}Pb}\right)_t \left(e^{\lambda_{238U}t} - 1\right) \qquad (4.30)$$

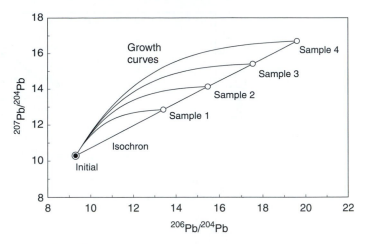

Figure 4.11 ^{206}Pb/^{204}Pb and ^{207}Pb/^{204}Pb ratios of samples formed with the same initial isotopic composition of lead but different U/Pb ratios evolve along growth curves so that the isotopic ratios remain on an isochron with a slope indicative of the age of formation.

$$\left(\frac{^{207}\text{Pb}}{^{204}\text{Pb}}\right)_t - \left(\frac{^{207}\text{Pb}}{^{204}\text{Pb}}\right)_0 = \left(\frac{^{235}\text{U}}{^{204}\text{Pb}}\right)_t \left(e^{\lambda_{235}\text{U}^t} - 1\right) \tag{4.31}$$

By dividing (4.31) by (4.30), we obtain:

$$\frac{\left(^{207}\text{Pb}/^{204}\text{Pb}\right)_t - \left(^{207}\text{Pb}/^{204}\text{Pb}\right)_0}{\left(^{206}\text{Pb}/^{204}\text{Pb}\right)_t - \left(^{206}\text{Pb}/^{204}\text{Pb}\right)_0} = \left(\frac{^{235}\text{U}}{^{238}\text{U}}\right)_t \frac{e^{\lambda_{235}\text{U}^t} - 1}{e^{\lambda_{238}\text{U}^t} - 1} \tag{4.32}$$

Because the present ratio ^{235}U/^{238}U is a constant equal to 1/137.88, (4.32) can be recast as $(y - y_0)/(x - x_0) = m$. In a plot $x = \left(^{206}\text{Pb}/^{204}\text{Pb}\right)_t$, $y = \left(^{207}\text{Pb}/^{204}\text{Pb}\right)_t$ (Fig. 4.11), (4.32) describes an isochron straight line going through the point of coordinates (x_0, y_0) and with slope m. The advantage of this method is that it requires only the isotope ratios of lead to be determined and not the concentrations, in particular that of uranium which is commonly strongly affected by water circulation in the water table and by weathering. For this reason, this method was commonly used for dating all sorts of sedimentary and magmatic rocks until superseded by zircon geochronology. It will be seen that this was the first method ever to yield the age of the Solar System and it is still widely used to date meteorites and planetary samples.

4.4 Radiogenic tracers

The property that phase separation, such as melting and crystallization, fractionates parent/daughter ratios has received enormous attention and created the fertile concept of radiogenic tracers. We have previously discussed the point that phase change does not fractionate the normalized isotopic ratios themselves, a common source of confusion.

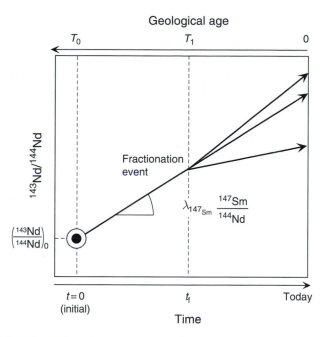

Figure 4.12 Evolution of the ^{143}Nd/^{144}Nd isotopic ratio by radiogenic decay of ^{147}Sm. This ratio changes almost linearly with time t and the slope of the evolution curve indicates the ^{147}Sm/^{144}Nd, and therefore the Sm/Nd ratio, of the system. Magmatic, sedimentary, or metamorphic events, here represented as occurring at T_1, may fractionate the Sm/Nd ratio of various sub-systems.

The initial isotopic ratios $(D/D')_0$ obtained as intercepts of isochrons were once a by-product of geochronology. These ratios, which describe the values of ^{87}Sr/^{86}Sr, ^{143}Nd/^{144}Nd, ^{176}Hf/^{177}Hf, etc., at the time the rocks formed, have become, together with the isotopic ratios measured in modern samples, a prime source of geochemical information about parent/daughter ratios in various geological systems. If (4.28) is considered not as an isochron equation but as expressing a change in the ^{143}Nd/^{144}Nd ratio versus time (Fig. 4.12), we have a means of evaluating the mean Sm/Nd ratio of the system. Let us recast this equation as:

$$\left(\frac{^{143}\text{Nd}}{^{144}\text{Nd}}\right)_t = \left(\frac{^{143}\text{Nd}}{^{144}\text{Nd}}\right)_0 + \left(\frac{^{147}\text{Sm}}{^{144}\text{Nd}}\right)_0 \left(1 - e^{-\lambda_{147\text{Sm}}t}\right) \qquad (4.33)$$

The slope of the evolution curve (the so-called growth curve) in Fig. 4.12 is simply obtained by taking the derivative of $1 - e^{-\lambda_{147\text{Sm}}t}$ with respect to t. Because of the very slow decay of ^{147}Sm, the approximation $e^{-\lambda_{147\text{Sm}}t} \approx 1 - \lambda_{147\text{Sm}}t$ holds. This slope is very nearly equal to $\lambda_{147\text{Sm}}(^{147}\text{Sm}/^{144}\text{Nd})_0$. Upon division by the appropriate isotopic abundance factors (here 0.15/0.24) and multiplication by the ratio of atomic masses of the two elements (150.4/144.2), the ^{147}Sm/^{144}Nd ratio can be converted into a Sm/Nd weight ratio.

Let us first use the chronometric equation to infer the mean "time-integrated" parent/daughter ratio that apparently prevailed during the entire history of a particular sample, from the Earth formed to the sample as part of the modern world. The initial ^{143}Nd/^{144}Nd ratio of the Solar System is known, typically from meteorite work. The mean $^{147}\text{Sm}/^{144}\text{Nd}$ ratio prevailing as the system evolved can be deduced from (4.33):

$$\left(\frac{^{147}\text{Sm}}{^{144}\text{Nd}}\right)_0 \approx \frac{1}{\lambda_{147\text{Sm}}t}\left[\left(\frac{^{143}\text{Nd}}{^{144}\text{Nd}}\right)_{\text{today}} - \left(\frac{^{143}\text{Nd}}{^{144}\text{Nd}}\right)_{\text{initial}}\right] \qquad (4.34)$$

Let us first deduce the time-integrated $^{147}\text{Sm}/^{144}\text{Nd}$ ratio of the source of different samples from their $^{143}\text{Nd}/^{144}\text{Nd}$ values. Let us pick up a basalt from the East Pacific Rise (0.5131) and a clay mud collected from the mouth of the Amazon (0.5108). The basalt source is the upper mantle, whereas the clay is derived from erosion of South American continental crust. Equation (4.34) can therefore be used to calculate the mean $^{147}\text{Sm}/^{144}\text{Nd}$ ratio of the upper mantle and that of the continental crust from the beginning of the Earth's history. Of course, these are virtual ratios, as the crust and mantle have complex histories that cannot be fully captured by a single parameter, but they give us useful information about the nature of the fractionation processes that have affected these geological units. Meteorite studies tell us that at the time the Solar System formed 4.56 billion years ago, the $^{143}\text{Nd}/^{144}\text{Nd}$ ratio was 0.5067. The mean $^{147}\text{Sm}/^{144}\text{Nd}$ ratio of the mantle in which the neodymium evolved before passing into the basalt was therefore:

$$\left(\frac{^{147}\text{Sm}}{^{144}\text{Nd}}\right)_{\text{mantle}} = \frac{0.5131 - 0.5067}{(6.54 \times 10^{-12}) \times (4.56 \times 10^9)} = 0.215$$

For the continent that was the source of the clay, this calculation gives a $^{147}\text{Sm}/^{144}\text{Nd}$ ratio of 0.138.

Similar calculations can be made for the $^{87}\text{Rb}-^{87}\text{Sr}$ and U–Pb systems. Let us assume that the $^{87}\text{Sr}/^{86}\text{Sr}$ ratio is 0.7023 for the basalt and 0.7140 for the clay, i.e. that this increases when $^{143}\text{Nd}/^{144}\text{Nd}$ decreases. Using equations similar to those we just derived for Nd and a chondritic initial $^{87}\text{Sr}/^{86}\text{Sr}$ ratio of 0.6992, the $^{87}\text{Rb}/^{86}\text{Sr}$ ratio turns out to be 0.046 for the upper-mantle and 0.22 for the continental-crust source of the clay. For lead, the the linear approximation would have to be abandoned, which raises no serious problem.

The Sm/Nd, Rb/Sr, and U/Pb ratios therefore differ between the continental crust and the upper mantle, so that the more incompatible element of each pair (Nd, Rb, and U) is more concentrated in the crust than the corresponding more compatible element (Sm, Sr, and Pb). A process capable of fractionating these ratios as the continental crust forms must therefore be imagined, for example, melting followed by selective extraction of magmas.

By plotting on an isochron diagram the present-day $^{143}\text{Nd}/^{144}\text{Nd}$ and $^{147}\text{Sm}/^{144}\text{Nd}$ ratios of any rock sample of continental crust and those of the average upper mantle (the mid-ocean ridge basalt source, see Chapter 12), a theoretical age T_{Nd} can be defined at which this particular piece of continental crust could have separated from the upper mantle. This age, usually referred to as the Nd model age of this particular crustal sample, is obtained by writing (4.34) once for the rock and once for the upper mantle, and by eliminating the $^{143}\text{Nd}/^{144}\text{Nd}_0$ ratio between the two expressions:

Table 4.2 Solid/melt fractionation of the parent and daughter nuclides of common systems and its effect on the long-term evolution of the isotope composition of the daughter element

Parent–Daughter	Parent/daughter ratio*		Daughter element*	
	In melt	In residue	In melt	In residue
^{87}Rb–^{87}Sr	Higher	Lower	More radiogenic	Less radiogenic
^{147}Sm–^{143}Nd	Lower	Higher	Less radiogenic	More radiogenic
^{176}Lu–^{176}Hf	Lower	Higher	Less radiogenic	More radiogenic
^{187}Re–^{187}Os	Higher	Lower	More radiogenic	Less radiogenic
^{232}Th–^{208}Pb	Higher	Lower	More radiogenic	Less radiogenic
^{238}U–^{206}Pb	Higher	Lower	More radiogenic	Less radiogenic

*With respect to source material

$$T_{Nd} = \frac{1}{\lambda_{147_{Sm}}} \times \frac{\left(^{143}\text{Nd}/^{144}\text{Nd}\right)_{rock} - \left(^{143}\text{Nd}/^{144}\text{Nd}\right)_{mantle}}{\left(^{147}\text{Sm}/^{144}\text{Nd}\right)_{rock} - \left(^{147}\text{Sm}/^{144}\text{Nd}\right)_{mantle}} \tag{4.35}$$

If our Amazon mud has a ^{147}Sm/^{144}Nd ratio of 0.120, a value that is quite representative of the continental crust in general, the model age for local separation of the crust and the upper mantle is therefore:

$$T_{Nd} = \frac{1}{0.654 \times 10^{-12}} \times \frac{0.5108 - 0.5131}{0.120 - 0.215} = 3.7 \times 10^9 \text{ years}$$

Radiogenic isotopes are most commonly utilized, then, for tracing sources, which we will examine in detail later. The principle is that when rock melts, the parent/daughter ratios are affected in a way that depends on the residual mineral assemblage, while the isotopic ratios remain unaffected. Thus, for mantle melting, Sm/Nd and Lu/Hf ratios are higher in the melt residue than in the liquid, while the contrary is true of Rb/Sr, Re/Os, and U–Th/Pb (Table 4.2).

4.5 Helium isotopes

Radiogenic helium ^{4}He is produced by the decay of the two natural isotopes of uranium (^{238}U and ^{235}U) and of the only long-lived isotope of thorium (^{232}Th). We have seen earlier that these decay processes are in fact the start of a chain of events. For example, in order to pass from ^{238}U to its distant descendant ^{206}Pb, the initial nuclide must lose (238–206)/4 = 8 α particles, which, by capture of the rock matrix electrons torn off during particle expulsion, become that number of ^{4}He atoms. The only stable reference isotope is ^{3}He. The growth equation of the ^{4}He/^{3}He isotope ratio is therefore:

$$\left(\frac{^4\text{He}}{^3\text{He}}\right)_t = \left(\frac{^4\text{He}}{^3\text{He}}\right)_0 + 8\left(\frac{^{238}\text{U}}{^3\text{He}}\right)_0\left(1 - e^{-\lambda_{238\text{U}}t}\right)$$

$$+ 7\left(\frac{^{235}\text{U}}{^3\text{He}}\right)_0\left(1 - e^{-\lambda_{235\text{U}}t}\right) + 6\left(\frac{^{232}\text{Th}}{^3\text{He}}\right)_0\left(1 - e^{-\lambda_{232\text{Th}}t}\right)$$

$$(4.36)$$

which can be rearranged as:

$$\left(\frac{^4\text{He}}{^3\text{He}}\right)_t = \left(\frac{^4\text{He}}{^3\text{He}}\right)_0 + \left(\frac{^{238}\text{U}}{^3\text{He}}\right)_0\left(1 - e^{\lambda_{238\text{U}}t}\right)$$

$$\times \left[8 + 7\left(\frac{^{235}\text{U}}{^{238}\text{U}}\right)_0\frac{1 - e^{-\lambda_{235\text{U}}t}}{1 - e^{\lambda_{238\text{U}}t}} + 6\left(\frac{^{232}\text{Th}}{^{238}\text{U}}\right)_0\frac{1 - e^{-\lambda_{232\text{Th}}t}}{1 - e^{\lambda_{238\text{U}}t}}\right]$$

$$(4.37)$$

At a given point in time $\left(^{235}\text{U}/^{238}\text{U}\right)_0$ is a constant, while $\left(^{232}\text{Th}/^{238}\text{U}\right)_0$ is indistinguishable from the nearly invariant Th/U ratio. The last expression therefore shows that ^{4}He/^{3}He essentially reflects the evolution of the ratio of the incompatible refractory isotope ^{238}U to the volatile stable isotope ^{3}He in the system. The isotope geochemistry of helium is an incomparable tool for investigating the outgassing of the mantle and also to model water circulation in aquifers. Note that, by force of habit, many geochemists continue to use the inverse ^{3}He/^{4}He ratios standardized to the value for atmospheric helium (1.4×10^{-6}) instead of the ^{4}He/^{3}He ratios.

Exercises

1. What is the proportion of radiogenic ^{40}Ar* in the total ^{40}Ar of a sample with a ^{40}Ar/^{36}Ar ratio of (1) 50 000 (2) 2000 (3) 300? If the total ^{40}Ar content is known to within 1%, what do you expect for the precision on the concentration of radiogenic ^{40}Ar* in each case? Assume that atmospheric argon has a ^{40}Ar/^{36}Ar ratio of ≈ 296.
2. Calculate the age of a basalt containing 1.7 wt% K_2O and $6.0\ 10^{-11}$ mol g^{-1} ^{40}Ar. Assume that the atomic proportion of ^{40}K in natural potassium is 0.0117 %.
3. Two samples are irradiated with fast neutrons in the same vial. The K–Ar age of one of them (the monitor) is known. It is observed that some of the ^{39}K of the samples is transformed into ^{39}Ar. Using the monitor to assess the yield of the nuclear reaction, devise a method to infer the K–Ar age of the unknown sample from the age of the monitor and the ^{40}Ar*/^{39}Ar ratios of the sample and the monitor (* labels radiogenic argon).
4. A famous Rb–Sr isochron work: the Baltimore gneiss (Wetherill *et al.*, 1968). Draw the whole-rock (WR) ^{87}Sr/^{86}Sr vs. ^{87}Rb/^{86}Sr and the biotite–whole-rock isochrons for

Table 4.3 Rb–Sr isotopic data for the Baltimore gneiss, Maryland (Wetherill *et al.*, 1968)

Sample		$^{87}Rb/^{86}Sr$	$^{87}Sr/^{86}Sr$
B105	WR	2.244	0.738
B20C	WR	3.642	0.7612
B20	WR	3.628	0.7573
	biotite	116.4	1.2146
B41	WR	6.59	0.7992
	biotite	289.7	1.969
B4	WR	0.2313	0.7074

Table 4.4 Sm–Nd isotopic data for Apollo 17 basalt 75075 (Lugmair *et al.*, 1975)

	$^{147}Sm/^{144}Nd$	$^{143}Sm/^{144}Nd$
Plagioclase	0.1942	0.51300
Ilmenite	0.2416	0.51417
Whole-rock	0.2566	0.51454
Pyroxene	0.2930	0.51542

the following samples (Table 4.3). Infer the emplacement age and the perturbation age of these rocks.

5. The Sm–Nd dating of Apollo 17 basalt 75075 (Lugmair *et al.*, 1975). Plot the results of Table 4.4 in a $^{143}Sm/^{144}Nd$ vs. $^{147}Sm/^{144}Nd$ isochron diagram, draw the internal (mineral) isochron and determine the eruption age of this basalt ($\lambda_{147_{Sm}} = 0.654 \times 10^{-11}$ a^{-1}).

6. ^{176}Lu–^{176}Hf dating of Birimian basalts from Burkina Faso (Blichert-Toft *et al.*, 1999). Plot the results of Table 4.5 in a $^{176}Hf/^{177}Hf$ vs. $^{176}Lu/^{177}Hf$ isochron diagram, draw the external (whole-rock) isochron and determine the emplacement age of these basalts ($\lambda_{176_{Lu}} = 1.865 \times 10^{-11}$ a^{-1}). Do you think these basalts were erupted with the same $^{176}Hf/^{177}Hf$ initial value? If not, why do you think the isochron looks so good?

7. Write the master equation for the ^{187}Re–^{187}Os isochron using ^{188}Os as the stable reference isotope.

8. Introducing ^{40}K–^{40}Ca dating: the Pikes Peak granite (Wyoming). What are the respective probabilities that ^{40}K decays into ^{40}Ca and ^{40}Ar? What is the isochron equation of the ^{40}K–^{40}Ca system if one uses the stable isotope ^{42}Ca as the reference? Draw the internal (mineral) isochron for the PP 76-2 sample (Table 4.6). What is the age of the last homogenization of Ca in the Pikes Peak granite?

9. Minerals extracted from a basaltic andesite at the Soufrière St. Vincent in the Antilles give the activity ratios listed in Table 4.7. Plot the Th–U isochron diagram and calculate the age of the last Th isotopic homogenization of these minerals.

Table 4.5 Lu–Hf isotopic data for basalts from Burkina Faso

Sample	$^{176}Lu/^{177}Hf$	$^{176}Hf/^{177}Hf$
B44	0.0325	0.282976
B47B	0.0331	0.282992
B46	0.0209	0.282461
BN75	0.0224	0.282515
BN11A	0.0352	0.283011
BN208A	0.0667	0.284325
BN209A	0.0848	0.285037
BN263A	0.0469	0.283458
BN280B	0.0652	0.284246

Table 4.6 K–Ca isotopic data for Pikes Peak sample PP 76-2 (Marshall and DePaolo, 1982)

	$^{40}K/^{42}Ca$	$^{40}Ca/^{42}Ca$
Whole-rock	0.1224	151.109
Plagioclase	0.0123	151.040
K-feldspar	0.800	151.577
Biotite	1.294	151.941

Table 4.7 U–Th activity ratios of minerals in a basaltic andesite from the Soufrière St. Vincent in the Antilles (Heath *et al.*, 1998)

	$\left[^{238}U\right]/\left[^{232}Th\right]$	$\left[^{230}Th\right]/\left[^{232}Th\right]$
Whole-rock	1.253	1.496
Olivine	1.089	1.183
Clinopyroxene	1.145	1.259
Plagioclase	1.136	1.244
Groundmass	1.244	1.450
Magnetic	1.230	1.458

10. Determination of the growth rate of a ferromanganese nodule from the excess of ^{230}Th in different layers below the nodule surface. Plot the excesses listed in Table 4.8 in a semi-log diagram. Find the growth rate from the slope of the alignment.

11. Dating the oldest terrestrial zircons (Wilde *et al.*, 2001). Table 4.9 lists the $^{206}Pb/^{238}U$ and $^{207}Pb/^{235}U$ ratios of four zircons from the Jack Hills conglomerate (Australia).

Table 4.8 $^{230}Th_{ex}$ (excess) at different depths in a ferromanganese nodule

Depth z (mm)	$^{230}Th_{ex}$
0.0	1242
0.2	788
0.4	488
0.6	285
0.8	180
1.0	119

Table 4.9 U–Pb isotopic data for four old zircons from Jack Hills (Australia)

	$^{206}Pb/^{238}U$	$^{207}Pb/^{235}U$
1	0.928	69.5
2	0.919	68.2
3	0.965	71.9
4	0.968	74.6

Table 4.10 U–Pb isotopic data for basaltic samples from Mauna Kea, Hawaii

Sample	Pb (ppm)	U (ppm)	$^{206}Pb/^{204}Pb$	$^{207}Pb/^{204}Pb$	$^{208}Pb/^{204}Pb$
SR0137-5.98	1.0	0.285	18.43	15.48	37.97
SR0157-6.25	0.593	0.192	18.44	15.48	38.00
SR0346-5.60	1.525	0.211	18.52	15.48	38.10
SR0664-5.10	0.765	0.415	18.55	15.49	38.14
SR0930-15.85	0.521	0.212	18.51	15.49	38.13
SR0967-2.75	0.695	0.319	18.49	15.48	38.11
Canyon Diablo			9.3066	10.293	29.475

Plot the concordia between 4.2 and 4.5 Ga and plot the Jack Hills data on the same diagram. What is the probable age of these zircons?

12. A rather difficult one! In the mid 1960s, it was proposed that an independent age of the Earth could be derived from the Pb isotope compositions of modern basalts, but this idea later proved to be incorrect. Plot $x = {}^{207}Pb^*/{}^{235}U$ and $y = {}^{206}Pb^*/{}^{238}U$ of basaltic samples from Mauna Kea, Hawaii, in the concordia diagram. The Pb isotope compositions and the U and Pb concentrations of the samples are provided in Table 4.10. In this context, the * superscript denotes the radiogenic isotopes accumulated since the Earth formed. Suppose that at that time, the Pb isotope composition of the Earth was that

Table 4.11 Pb isotopic data for five meteorites (Patterson, 1956)

Meteorite sample	$^{206}Pb/^{204}Pb$	$^{207}Pb/^{204}Pb$
Nuevo Laredo	50.28	34.86
Forest City	19.27	15.95
Modoc	19.48	15.76
Henbury	9.55	10.38
Canyon Diablo	9.46	10.34

Table 4.12 Sr isotopic data for six samples of modern shales

Sample	$^{87}Rb/^{86}Sr$	$^{87}Sr/^{86}Sr$
a	6.2	0.7121
b	2.7	0.7102
c	0.5	0.7083
d	52.5	0.7111
e	13.1	0.7081
f	21.4	0.7096

of the iron sulfide of the Canyon Diablo meteorite. We can safely assume a constant molar weight for Pb (207.2) and U (238.0) and the modern $^{238}U/^{235}U$ ratio is 137.8. What is the range of $^{207}Pb^*/^{206}Pb^*$ ratios of the mantle source of modern basalts (see Section 11.2)? Why is the age of the upper intercept of the concordia close to the age of the Solar System? Discuss where the pitfall is. (Hint: from the Pb concentrations, calculate the moles of initial ^{204}Pb per mass unit present in each sample and use the Canyon Diablo values to derive the same parameter for the other isotopes).

13. Pat's legacy: the age of the Solar System (Patterson, 1956). Lead isotopes were analyzed in the five meteorites listed in Table 4.11. Plot the $^{207}Pb/^{204}Pb$ vs. $^{206}Pb/^{204}Pb$ ratios. By trial and error, find Clair Patterson's estimate for the age of the last Pb isotopic homogenization of the Solar System as a whole.

14. Isochrons improve with age. Table 4.12 lists the Sr isotopic data for six samples of modern shales. Using your favorite spreadsheet or any convenient software, calculate the $^{87}Sr/^{86}Sr$ ratios of each sample 100 My, 2700 My, and 4560 My into the future. Plot the isochrons for today and for each of these future dates. Use your software to calculate the slope and the correlation coefficients and discuss the results.

15. Isochron or mixing line? A series of modern lavas are speculated to have formed by mixing (hybridization) of mingling basaltic and rhyolitic magmas. Table 4.13 lists the Nd isotopic data of two end-member magmas. Calculate the $^{147}Sm/^{144}Nd$ and $^{143}Nd/^{144}Nd$ ratios of hybrid rocks formed by 20–80, 40–60, 60–40, and 80–20 % of

Table 4.13 Nd isotopic data of the basaltic and granitic end-members			
	[Nd] (ppm)	$^{147}Sm/^{144}Nd$	$^{143}Nd/^{144}Nd$
Rhyolite	40	0.11	0.5115
Basalt	15	0.17	0.5128

basalt–rhyolite mixtures. Plot the mixing line with its end-members in the isochron diagram and "age" the samples by 1 Gy and 2 Gy. Calculate the apparent ages of the alignment for 0, 2, and 3 Gy into the future. Devise a strategy to decide whether an alignment in the isochron diagram is a mixing line or a true isochron.

16. Sketch out the evolution of the Bulk Silicate Earth ($^{147}Sm/^{144}Nd = 0.1967$ and $^{143}Nd/^{144}Nd = 0.512638$ today) in a $^{143}Nd/^{144}Nd$ vs. t evolution diagram between 2 Ga and the present and the evolution of a crustal rock formed at 2 Ga from the primitive mantle with a $^{147}Sm/^{144}Nd$ ratio of 0.11. What is the modern ε_{Nd} of this rock?

17. Compare the evolution diagram $^{143}Nd/^{144}Nd$ vs. t and the isochron diagram $^{143}Nd/^{144}Nd$ vs. $^{147}Sm/^{144}Nd$. Plot the points representing the modern depleted mantle ($^{147}Sm/^{144}Nd = 0.222$ and $^{143}Nd/^{144}Nd = 0.51312$) and a crustal sample with $^{147}Sm/^{144}Nd = 0.12$ and $^{143}Nd/^{144}Nd = 0.5111$. First, find graphically in each diagram the model age at which the protolith of this crustal sample was extracted from the depleted mantle. Second, derive a common expression that will let you calculate this model age with precision.

References

Abouchami, W., Boher, M., Michard, A., and Albarède, F. (1990) A major 2.1 Ga old event of mafic magmatism in West Africa: an early stage of crustal accretion. *J. Geophys. Res.*, **95**, 17 605–17 629.

♠ Arnold, J.R. and Libby, W. F. (1949) Age determination by radiocarbon content: Checks with samples of known ages. *Science*, **110**, 678–680.

♠ Bateman, H. (1910) Solution of a system of differential equations occurring in the theory of radioactive transformations. *Proc. Cambridge Phil. Soc.*, **15**, 423–427.

Blichert-Toft, J., Albarède, Rosing, M., Frei, R., and Bridgwater, D. (1999) The Nd and Hf isotopic evolution of the mantle through the Archean. Results from the Isua supracrustals, West Greenland, and from the Birimian terranes of West Africa. *Geochim. Cosmochim. Acta*, **63**, 3901–3914.

Heath, E., Turner, S. P., Macdonald, R., Hawkesworth, C. J., and van Calsteren, P. (1998) Long magma residence times at an island arc volcano (Soufriére, St. Vincent) in the Lesser Antilles: evidence from ^{238}U–^{230}Th isochron dating. *Earth Planet. Sci. Letters*, **160**, 49–63.

Lugmair, G. W., Scheinin, N. B., and Marti, K. (1975) Sm–Nd age and history of Apollo 17 basalt 75075: Evidence for early differentiation of the lunar exterior. *Proc. Lunar Sci. Conf.*, **6**, 1419–1429.

Marshall, B. D. and Depaolo, D. J. (1982) Precise age determinations and petrogenetic studies using the K–Ca method. *Geochim. Cosmochim. Acta*, **46**, 2537–2545.

McDougall, I. and Harrison, T. M. (1999) *Geochronology and Thermochronology by the ^{40}Ar–^{39}Ar method*. Oxford: Oxford University Press.

♠ Nicolaysen, L. (1961) Graphic interpretation of discordant age measurements on metamorphic rocks. *Ann. N.Y. Acad. Sci.*, **91**, 198–206.

♠ Patterson, C. (1956) Age of meteorites and the Earth. *Geochim. Cosmochim. Acta*, **10**, 230–237.

♠ Wetherill, G. W. (1956) Discordant uranium–lead ages. *Trans. Amer. Geophys. Union*, **37**, 320–326.

♠ Wetherill, G. W., Davis, G. L., and Lee-Hu, C. (1968) Rb-Sr measurements on whole rocks and separated minerals from the Baltimore gneiss, Maryland. *Geol. Soc. Amer. Bull.*, **79**, 757–762.

Wilde, S. A., Valley, J. W., Peck, W. H., and Graham, C. M. (2001) Evidence from detritical zircons for the existence of continental crust and oceans on the Earth 4.4 Gyr ago. *Nature*, **409**, 175–178.

5 Element transport

The theory of element transport is a way of representing the spatial changes in geochemical properties in various contexts, such as movement in the ocean or mantle, the migration of geological fluids or magmatic liquids within a rock matrix, or the attainment of chemical and isotopic equilibrium among minerals within the same rock, etc. It is, in fact, a set of rather complex theories involving some heavy-going mathematics, which we can only touch lightly upon in this book.

The essential concepts forming the core of this theory are those of conservation, flux, sources, and sinks. A conservative property is additive and can only be altered by addition or subtraction at the system boundaries or by the presence of sources and sinks. Mass or number of moles are conservative properties; concentration is not: if a mole of salt is added to a solution already containing one mole, the resulting solution will contain two moles, regardless of how the salt is added. In contrast, two solutions of one mole per liter combine to form two liters of a solution at one mole per liter. A flux is a quantity of something (mass, moles, energy, etc.) crossing a unit surface per unit time. The most familiar of these fluxes is volume flow, which is quite simply the velocity v (in cubic meters per square meter per second). Mass flux is the mass content of volume flow, i.e. ρv, where ρ is the density of the medium. The flux of an element i is $\rho v C^i$, where C^i is its local concentration in kilograms (or moles) per kilogram. If the flux of a compound or an element changes suddenly at one point, then a source or sink of this compound or element is present: generally, a chemical reaction or radioactive process is responsible for this.

Transport from one point to another is either by advection or by diffusion. Think of fish in a river. The general motion of water ensures advective transport of the fish; the movement of the fish relative to the water is diffusive transport. Generally, advection is transport of elements at the speed of the ambient medium, and diffusion of an element is its transport relative to the ambient medium.

The principles from which the transportation equations derive are difficult to formulate in mathematical terms, but they are easy enough to understand:

- Conservation of total matter, which produces the continuity equation.
- Conservation of element i

$$\frac{\text{variation of mass of } i}{\text{per unit of time}} = \sum \text{flux of } i \text{ across the surface}$$
$$+ \sum \text{sources and sinks} \tag{5.1}$$

In just the same way as we can count our fish from a bridge, from a boat drifting with the current, or from a motor boat, the terms of this equation can be expressed relative to a fixed, or moving, arbitrary reference point.

5.1 Advection

Advection is bulk transport, which is easier to understand in one dimension. Let us consider a medium of density ρ moving at speed v. If, at a point x and over an area A, we consider a slice of matter of thickness l, the balance of variation of mass of i per unit time in this slice is equal to the difference between the incoming and outgoing fluxes:

$$A\, l\rho \frac{\mathrm{d}C^i(x)}{\mathrm{d}t} = A\rho v C^i(x) - A\rho v C^i(x+l) \tag{5.2}$$

or, if l tends toward 0:

$$\frac{\partial C^i}{\partial t} = -v \frac{\partial C^i}{\partial x} \tag{5.3}$$

where the partial derivatives are used to show relative to which variable derivation is applied. If we imagine an observer who, like the boat drifting on the river, moves with the speed v of the medium, the apparent speed of the observer relative to the medium cancels out and the right-hand side disappears. Again, we find the condition that local concentration is invariant: this is known as the Lagrangian representation of the variation. This amazingly simple principle is the one applied in the study of convection of the mantle or the ocean, and in the dispersion of what are known as passive tracers, i.e. tracers not involved in reactions (Fig. 5.1): if we know the field of velocities, we can determine the motion of any point of the medium between two instants and therefore track the position of an atom of element i over time.

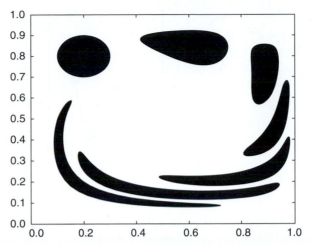

Figure 5.1 Advective dispersion of passive tracers in a velocity field of the x (abscissa) y (ordinate) plane. This diagram shows the evolution of an initial mass of tracers injected at the top left of the system at time t and which follows the clockwise motion of the material (advection). This may represent the evolution of a block of oceanic crust caught up in mantle convection or the dispersion of pollutants released into the ocean.

5.2 Diffusion

Diffusive transfer is transfer over short distances caused by the thermal agitation of atoms or the turbulence of the medium. Let us consider the one-dimensional migration of an element i along the x-axis. Let P be the probability of an atom jumping per unit time from its original site and suppose that it can only jump from one site to an adjacent site at a distance l. A site loses atoms with the same probability to the site to its left as to the site to its right, but only gains half of the atoms lost from each adjacent site. The balance can be written as:

$$\frac{\partial C^i(x)}{\partial t} = \frac{P}{2}C^i(x-l) + \frac{P}{2}C^i(x+l) - PC^i(x) \tag{5.4}$$

By expanding $C^i(x+l)$ and $C^i(x-l)$ to the second order, we obtain:

$$C^i(x+l) = C^i(x) + l\frac{\partial C^i(x)}{\partial x} + \frac{l^2}{2}\frac{\partial^2 C^i(x)}{\partial x^2} + \mathcal{O}(l^3) \tag{5.5}$$

$$C^i(x-l) = C^i(x) - l\frac{\partial C^i(x)}{\partial x} + \frac{l^2}{2}\frac{\partial^2 C^i(x)}{\partial x^2} - \mathcal{O}(l^3) \tag{5.6}$$

Adding these two equations, ignoring terms of order higher than two, and transferring the result into (5.4), gives:

$$\frac{\partial C^i}{\partial t} = \frac{Pl^2}{2}\frac{\partial^2 C^i}{\partial x^2} = D^i\frac{\partial^2 C^i}{\partial x^2} \tag{5.7}$$

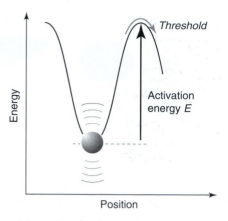

Figure 5.2 The activation energy threshold of diffusion. The maxima represent intermediate energy barriers that the thermally activated atoms need to pass in order to reach the next stable lattice positions represented by the minima.

where the diffusion coefficient or diffusivity $D^i = Pl^2/2$ is expressed in square meters per second. The probability P is proportional to (Fig. 5.2): (a) the number of jump attempts per unit time, i.e. the frequency of vibrations ν of the atom around its resting position; and (b) the proportion of atoms with enough energy to cross the energy threshold E^i, known as the activation energy, dependent on the nature of the atom and the atomic configuration of the site. From Boltzmann's law, this proportion is equal to $\exp(-E^i/RT)$ and the temperature dependence of the diffusion coefficient is therefore expressed:

$$D^i = D_0^i \exp\left(-\frac{E^i}{RT}\right) \tag{5.8}$$

where D_0^i contains the vibrational term. If we introduce the logarithm of D^i as a function of the inverse $1/T$ of absolute temperature (Arrhenius plot), we obtain a straight line of slope $-E^i/R$. A few values of the diffusion coefficient in water, magmatic liquids, and minerals at magmatic temperatures are shown in the Arrhenius plot in Fig. 5.3. The tendency of the Arrhenius lines for different elements to fan out from a common point at high temperature reflects an inverse relationship between the activation energy and the pre-exponential factor and is known as the Meyer–Neldel, or Winchell, "compensation" law.

The physical meaning of the diffusion coefficient can also be understood in a different way. Equation (5.7) can be re-written as follows (quite analogous to (5.1)):

$$\frac{\partial C^i}{\partial t} = -\frac{\partial J^i}{\partial x} \tag{5.9}$$

where the diffusive flux J^i of species i is defined as:

$$J^i = -D^i \frac{\partial C^i}{\partial x} \tag{5.10}$$

This shows that the transport of i by diffusion is proportional to the concentration gradient $\partial C^i/\partial x$ and that the diffusivity D^i is the constant of proportionality. The minus sign

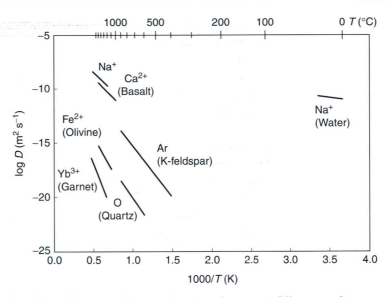

Figure 5.3 Arrhenius plot of the diffusion coefficients of various elements in different media. Semi-logarithmic graph. Notice temperatures are in K at the bottom and °C at the top.

ensures that the flux is directed toward the part of the system with a lower i concentration. Equation (5.10) is Fick's law.

For example, we can calculate the diagenetic diffusive flux of manganese ($i = $ Mn) in sedimentary pore water to ocean bottom water. Let us assume that, because of its reducing character, interstitial water 10 cm beneath the surface has a manganese concentration of $10 \, \mu\text{mol} \, l^{-1}$. Seawater, being highly oxidized, has a manganese concentration that is virtually nil. The diffusivity of manganese in seawater is $D^{\text{Mn}} = 0.7 \times 10^{-9} \, \text{m}^2 \, \text{s}^{-1}$. The local diffusive flux of manganese to the ocean in absolute terms is therefore:

$$J^{\text{Mn}} = 0.7 \times 10^{-9} \frac{10 \times 10^{-6} \times 10^{-3} - 0}{0.1} = 0.7 \times 10^{-16} \, \text{mol} \, \text{m}^{-2} \, \text{s}^{-1}$$

Clearly, as we considered the probability of a forward or backward jump as equal (Brownian motion), the mean displacement of an atom is zero. Nonetheless, we can calculate a measure of the dispersion of a population of atoms initially in position x through another simple statistical concept, the standard deviation of displacement, i.e. the mean quadratic displacement. During time t, the number of jumps by an atom is equal to Pt and the quadratic distance traveled by an atom is Ptl^2, or $2D^i t$. The mean quadratic deviation, i.e. the distance δ over which two-thirds of the atoms will be found on either side of their initial position, is therefore $\delta = \sqrt{2D^i t}$. This distance is a measure of the migration characteristic of the atoms under the effect of diffusion and is generally short; it is typically a few centimeters per year for an ion in seawater, meters per million years for the Fe–Mg exchange of olivine in the mantle, or a micrometer per million years for oxygen isotope exchange in metamorphic quartz.

Diffusion is not therefore a process for long-distance transport, but it usually does allow minerals to achieve equilibrium with each other or with solutions percolating through the rock. The time required for a spherical mineral of radius a to lose a fraction F of an element it initially contained complies fairly well with the equation:

$$F = 1 - 6\sqrt{\tau/\pi} + 3\tau \qquad (5.11)$$

where $\tau = D^i t/a^2$ is a scalar proportional to time known as dimensionless time. This equation is widely utilized for calculating diffusion coefficients; for example, for rare gases, by evaluating the rate of degassing of a particular mineral at a certain temperature as a function of time, or for soluble compounds, by immersing them in a liquid.

When advection and diffusion are both operative at the same time, their relative importance can be evaluated with the Peclet number Pe, with $Pe = vd/D^i$, where d is a characteristic dimension of the system (e.g. grain size or conduit diameter). Diffusion predominates when $Pe < 1$, and advection for $Pe > 1$.

5.2.1 Closure temperature: chronometers, thermometers, and barometers

Diffusion theory can be applied to calculate the resistance of radioactive chronometers to thermal disturbances or for evaluating the equilibration temperature of a mineral assemblage (thermobarometry). A number of important applications in geochronology rely on the concept of closure temperature defined for radiogenic isotopes by Dodson (1973). The high activation energies of diffusion in solids (Eq. (5.8)) cause diffusion coefficients to vary closely with temperature. The transition from open system to closed system therefore occurs within a narrow range of temperature (Fig. 5.4). Above a certain temperature T_c and therefore for a diffusion coefficient greater than D_c^i, where:

$$D_c^i = D_0^i \exp\left(-\frac{E^i}{RT_c}\right) \qquad (5.12)$$

the system can be considered open. The radiogenic isotope will only begin to accumulate below T_c, the temperature at which the stop-watch is started, with the transition occurring within a few tens of degrees. Let us see how to evaluate T_c for a given isotope in a given mineral. From an equation such as (5.11), we see that the critical parameter of the loss process is dimensionless time $\tau = D^i t/a^2$. Let us decide that the system is open for $\tau > \tau_c$ and closed for $\tau < \tau_c$, where τ_c is a constant factor dependent on geometry (0.02 for a sphere and 0.12 for a sheet). If the significance of the radius a of the mineral is clearly perceived, what is the time characteristic to be introduced in τ? Dodson suggests that a characteristic time is the time $t = \theta$ required for the diffusion coefficient to reduce by a factor e, and is equal to:

$$\frac{1}{\theta} = -\frac{d \ln D^i}{dt} = \frac{E^i}{R}\frac{d(1/T)}{dt} = -\frac{E^i}{RT^2}\frac{dT}{dt} \qquad (5.13)$$

This time can be estimated from experimental data on diffusion coefficients and even approximate knowledge of the local rate dT/dt of cooling after formation of the minerals under analysis. Let us define the dimensionless time τ_0 as:

$$\tau_0 = \frac{D_0^i \theta}{a^2} \qquad (5.14)$$

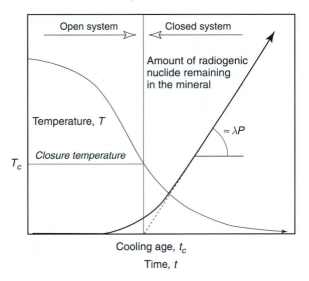

Figure 5.4   Cooling age and closure temperature of a chronometric system with decay constant λ. The cooling
time and the corresponding cooling age t_c are defined by linearly extrapolating the ingrowth of
radiogenic nuclides at zero, while the closure temperature T_c is the temperature of the mineral at
this time. The system is open to the loss of the daughter isotope for $T > T_c$ and closed for $T < T_c$.
The slope of the daughter-isotope evolution curve is simply the activity λP.

Dividing τ_c by τ_0 gives:

$$\frac{\tau_c}{\tau_0} = \exp\left(-\frac{E^i}{RT_c}\right) \tag{5.15}$$

or

$$T_c = \frac{E^i}{R\ln(\tau_0/\tau_c)} = \frac{E^i}{R\ln\left(D_0^i\theta/a^2\tau_c\right)} \tag{5.16}$$

We now express θ using (5.13) for $T = T_c$:

$$T_c = \frac{E^i}{R\ln\left[D_0^i RT_c^2/E_i a^2\tau_c\,(-\mathrm{d}T/\mathrm{d}t)\right]} \tag{5.17}$$

which is Dodson's equation.

We can now evaluate T_c by trial and error and iteration. For example, let us consider the
closure of potassium feldspar crystals, assumed for simplicity to be spherical with a radius
of 2 mm, to the diffusion of ^{40}Ar produced by decay of ^{40}K. We assume that argon diffusion is described by the following parameters: $D_0 = 1.4 \times 10^{-6}$ m^2 s^{-1}, $E^i = 180\,000$ J
mol^{-1}, $a = 0.002$ m, $\mathrm{d}T/\mathrm{d}t = 5$ K My^{-1}. We would normally have to find the closure
temperature by trial and error and we would converge to $T_c = 615$ K. Upon inserting
$T = 342\,°\mathrm{C} = 615$ K into (5.13), we obtain $\theta = 1.10 \times 10^{14}$ K, which corresponds to
$\tau_0 = 3.85 \times 10^{13}$ (Eq. (5.14)). Inserting these values into (5.17), we can check that $T = T_c$.
We conclude that crystals of potassium feldspar with a radius smaller than 2 mm start los-
ing their radiogenic argon when heated above 342 °C, or, equivalently, that they start being

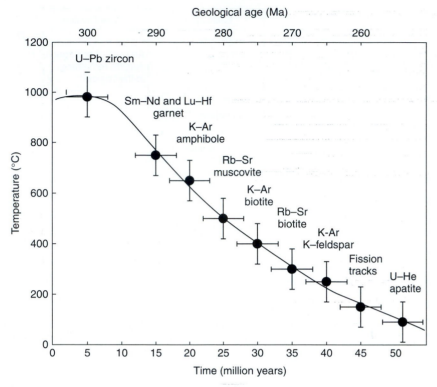

Hypothetical changes in temperature during the uplift and erosion of land after an orogenic period. The example is for a rock whose zircons crystallized 300 million years ago. The different closure temperatures of the various clocks and the ages define a temperature–time curve.

retentive for argon as soon as ambient temperature drops below 342 °C. Closure temperature depends mostly on grain size. Error bars on closure temperatures are of the order of 50–80 °C.

Closure temperatures of different chronometers for specific minerals can be calculated from diffusion experiments. Figure 5.5 shows a hypothetical curve constructed to show cooling after the formation of mountain ranges in a given area. If equilibration pressure is known, all of this information forms the pressure–temperature–time (PTt) path of exhumation of the orogen, an essential tool in modern tectonics. This same tool is also much used in petroleum exploration to determine the thermal evolution of sedimentary basins and the probability of temperatures suitable for oil formation being attained locally.

The concept of closure is not restricted to chronometers but applies to most assemblages of co-existing phases and begs the question of what mineral thermometers and barometers (barothermometry) actually record. Let us consider a rock made of four different minerals, e.g. a granite or a schist, with different closure temperatures (Fig. 5.6) and soaked in a percolating hydrothermal or metamorphic fluid. At high temperatures, all the minerals exchange material among themselves and with the fluid in order to keep the overall energy of the system to a minimum (equilibration). When temperature drops below the closure temperature of any particular mineral, exchange between this mineral and the rest

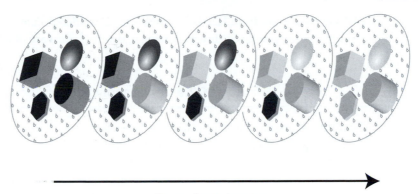

Decreasing temperature

Figure 5.6   Evolution of chemical exchanges between the minerals co-existing in a rock upon cooling. A fluid interstitial phase is assumed to be present. At high temperatures (dark tones), all minerals actively exchange material. When temperature drops below the closure temperature of one mineral phase, exchanges between this mineral and its environment comes to a halt (light tones). Note the strong differences between this representation of mineral compositions and that entailed by equilibrium between the co-existing mineral phases.

of the rock and its interstitial fluid comes to an end. The energy is no longer at its lowest point but, because diffusion is thermally activated, the cost of reaching the true minimum is higher than the energy available in the system: the system is caught in a sequence of local minima, a series of states called metastable. Each mineral in turn withdraws from the energy fair until the last mineral itself stops equilibrating with the fluid. This is Nature's mineral performance of Haydn's Farewell Symphony in which each artist, at the conclusion of his part, blows the candlelight set on his podium. A rock can rarely be regarded as a true mineral assemblage unless all the mineral phases reach their closure temperature in a very short time interval. This perception is of course true for the exchange of both elements and isotopes. Thermometers and barometers follow the same rules as chronometers: they record the temperature and the pressure at the time of the last equilibration between the phases involved in the exchange. A temperature of 700 °C obtained between a garnet–clinopyroxene pair in an eclogite only measures the lowest temperature prevailing when these two minerals last exchanged. More informative observations are those derived by *in situ* analysis, when the adjacent rims of both minerals are measured. Still, uncertainties remain about what the numbers actually mean. The same holds true for oxygen isotope thermometry which, for these reasons, fell into near oblivion more than three decades ago.

Minerals with very low diffusivity are therefore in very high demand because their closure temperature is high. We have seen that zircon is an excellent chronometer for the U–Pb systems, which it owes to a closure temperature for Pb diffusion in excess of 900 °C (Cherniak *et al.*, 2001). Zircon crystallizes early in granites. It nevertheless preserves the oxygen isotope composition of its parent magma throughout the severe disturbances imposed by the intense percolation of fluids at temperatures of 400–600 °C recorded in major minerals such as quartz, feldspars and micas. This property has found an important application in demonstrating that the source of early Archean granites, in which the

>4.1 Gy old zircons from Jack Hills (Australia) crystallized, contained rocks modified by low-temperature processes, thereby hinting at the early presence of a hydrosphere.

5.2.2 Other applications

The diffusion equation, with or without an advection term, is involved in the description of many other geochemical processes. A very important case is that of early diagenesis, a set of reactions occurring a few meters beneath the sea floor whereby mud is transformed into sediment. An observer in a fixed position at the water–sediment interface "watches" the sediment sink at the rate of sedimentation and must describe the chemical processes in an advective context. The diffusion of elements in interstitial water and the reactions between that water and the sedimented particles control the nature of mineralogical transformations and the flux back into seawater or diagenetic liquids as the sediments become compacted. Of particular importance are the reduction reactions of nitrates to nitrogen and of sulfates to sulfides by which many micro-organisms oxidize sedimented organic matter. These processes lie behind the dissolution of manganese nodules and the formation of sedimentary pyrite (fool's gold), in particular. They will be discussed in more detail in Chapter 10.

Diffusion is also involved in the growth of minerals from solutions or magmas: an olivine crystal that grows from a basalt is surrounded by a boundary layer depleted in the essential components of this mineral, notably Mg. This layer can often be observed under the microscope. The diffusion of these components allows re-supply and, consequently, the continued growth of the mineral. In the mantle, homogenization of compositions at distances greater than about ten meters is thought to be achieved by mechanical stirring caused by convection. Conversely, diffusion ensures homogenization over short distances, in the order of a few centimeters.

5.3 Chromatography

Imagine, on a hot summer day, a parade stretching along a long avenue lined with cafes and shady terraces. It is likely that some of the participants from time to time will succumb to the temptation of enjoying a few minutes of refreshment before joining the parade again, and that some will even take a liking to these halts. We can guess what our parade will look like after a few hours: the head of the procession will be reduced to the virtuous and sober participants; while behind, the atmosphere will be enlivened by all those who have made frequent stop-offs at the terraces. This type of mechanism is the essence of chromatographic separation.

In analytical chemistry, chromatography is a technique of element separation that involves the percolation of a liquid (the eluent or mobile phase) through a porous matrix (the stationary phase composed, for example, of ion-exchange resin) and exchange elements with this matrix. The variable affinity of the stationary phase for the elements lets each of them percolate through the matrix with a different velocity, which eventually ensures their separation. In rocks, similar processes take place when geological fluids move through the pores of a sediment during diagenesis or through rock layers during metamorphism: because of the large volumes of fluid involved and because of the broad range of reactivity from one element to another, considerable chemical separation in the fluid and strong mineralogical and geochemical modifications of the rock matrix are commonplace.

In a two-phase flow, such as water percolating along an aquifer through a porous soil or a magmatic liquid in a molten rock matrix, or of particles sedimenting in the ocean, the processes are both more complicated and more diverse. The constitutive equations of chromatography are difficult to establish because, besides continuity and conservation equations for each species in each phase (e.g. solid and liquid), we have to add a condition describing transfers between phases, such as phase-change kinetics (melting, dissolution, or crystallization) and chemical fractionation. The concentration of an element i in the interstitial mobile phase (liquid) can generally be described by a balance statement of the type:

$$\frac{dC_{\text{liquid}}^{i}}{dt} = \text{diffusion} + \text{advection} + \text{phase changes} \qquad (5.18)$$

To illustrate the concepts in the simplest way, let us ignore the diffusive term and the term related to transfer during phase change. The advective effect is then dominant. Let us denote v_{liquid} the velocity at which the liquid moves relative to the solid matrix and v^{i} the mean velocity at which element i moves ($v^{i} < v_{\text{liquid}}$). The theory of chromatography is quite heavy-going, but we can try to find a way around the worst difficulties by writing that the material flux of i moving with the velocity v_{liquid} must be the sum of the fluxes of i moving with the element velocity $v_{\text{liquid}}C^{i}$ in both the solid and in the liquid:

$$\varphi v_{\text{liquid}} C_{\text{liquid}}^{i} \approx \varphi v^{i} C_{\text{liquid}}^{i} + (1 - \varphi)\, v^{i}\, C_{\text{solid}}^{i} \qquad (5.19)$$

(note that the fluxes have to be weighted by the porosity φ for the liquid and by the fraction $1 - \varphi$ for the matrix). This equation can be rearranged as:

$$v^{i} \approx \frac{\varphi}{\varphi + (1 - \varphi)\, C_{\text{solid}}^{i}/C_{\text{liquid}}^{i}}\, v_{\text{liquid}} \qquad (5.20)$$

Let us define in the usual way the partition coefficient $C_{\text{solid}}^{i}/C_{\text{liquid}}^{i}$ as D^{i}. Ions that migrate with the water of an aquifer without interacting with the impregnated rock ($D^{i} \approx 0$) will therefore move faster than the ions that readily exchange between rock and water ($D^{i} \gg 0$) and spend a large part of their time of transit in the solid rock. For instance, chlorine ions have no affinity for minerals and, like a dying substance, move at the same speed as the water. Ions such as Zn^{2+} or Pb^{2+} are readily absorbed by the surface of the minerals. They are very much delayed by this reactivity. This is the principle behind the purification of natural water in the ground. The containment within nuclear-waste repositories of radioactive actinides, which normally have mineral/water $D^{i} \gg 0$, also relies heavily on this property.

Let us now consider the situation where the ratio $C_{\text{solid}}^{i}/C_{\text{liquid}}^{i}$ and, therefore, the velocity of i in the interstitial mobile phase, depends on its concentration. One more time, we will draw on a real-life analogy now taken from the experience of driving on highways at rush hour (Fig. 5.7). We assume that the highway has only one lane and that passing is not

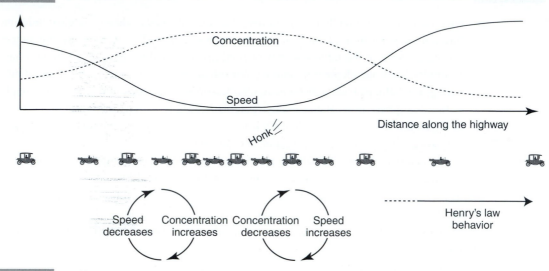

Figure 5.7 Traffic traveling waves. As long as vehicles are few, their speed is independent of their concentration on the lane. This is Henry's law regime. When more cars join the traffic, car spacing decreases and drivers reduce their speed which in turn increases concentration until cars pack to a jam. Ahead, traffic lightens up and cars are allowed to move faster again. Such alternating zones of light and dense traffic move up the highway. Chromatographic waves are produced by concentration-dependent velocities.

permitted. Concentration is the number of cars per km and flux is the product of concentration by speed. When traffic is light, car spacing exceeds the safety distance between cars and the drivers presumably adjust to a speed close to the speed limit. The speed is independent of the concentration of cars: this is Henry's law regime. When the density of traffic increases, each driver must reduce the car speed to avoid crashing into the vehicle ahead: velocity now decreases as car concentration increases. New cars merging into the highway make spacing shorter and shorter and progressively bring traffic to a halt. After a while, as the cars ahead of the jam move away, the traffic lightens up thereby increasing the distance and therefore the speed. Under conditions of heavy traffic, the alternating zones of light and dense traffic move up the highway as traveling waves. This is mechanical chromatography!

Let us now go back to our moving fluids and the elements they transport and see how "traffic waves" occur in nature. Suppose that D^i decreases when the concentration in i increases. In this case, the greater the concentration of the element, the less reactive it will be. By applying (5.20), it can be seen that the ions of highly concentrated zones will catch up with those of less concentrated zones (Fig. 5.8), thus forming what are known as metasomatic fronts or even more complicated geometric patterns. The description of magma migration is far more complex still because, contrary to the groundwater table where the water itself reacts very little with the rock matrix, all the major elements of the magma exchange intensely with the solid residue. In other words, incompatible elements cannot scurry ahead of the melt, which is of course made of compatible major elements.

Chromatography theory applies to other cases of two-phase transport. An important concept in oceanography is the entrainment – or scavenging – of elements by particles such as dead organisms, waste matter, or atmospheric dust. In this case, the

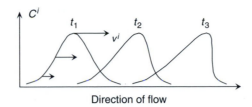

Direction of flow

Figure 5.8 Principle of chromatographic fractionation. Different ions propagate along the column at different speeds. Each curve represents the concentration profile of the same element at different times ($t_1 < t_2 < t_3$). If the speed of propagation of an element i in the porous medium depends on its concentration c^i in the liquid and partition coefficient D^i, concentration fronts rise and collapse.

particles take the place of the liquid occupying porosity φ of the general theory, while seawater plays the role of the rock matrix. What conditions reactivity is the seawater–particle partition coefficient D^i of elements between seawater and particulate matter (note the direction of exchange, which gives reactivity a counter-intuitive character). Elements that are not reactive with particles ($D^i \approx \infty$), such as sodium and chlorine, remain motionless and are not affected by particle flow. Elements subjected to intense entrainment by solids ($D^i \approx 0$), such as lead, cadmium, and rare-earths, move down the water column almost as quickly as the particles and are soon evacuated from the system.

5.4 Reaction rates

Let us consider the reaction by which marine sulfates are reduced at the water–sediment interface by accumulated organic matter, which we will represent by the very simple chemical formula CH_2O (the precise formula is immaterial to the explanation). This reaction produces hydrogen sulfide, which will precipitate as pyrite FeS_2 with the ambient iron, and a bicarbonate ion from the reaction:

$$2CH_2O + SO_4^{2-} \rightarrow H_2S + 2HCO_3^- \tag{5.21}$$

By indicating the molar concentrations of the species with square brackets, the conservation of matter during the reaction imposes the following conditions of evolution (see Appendix C):

$$-\frac{d\,[CH_2O]}{2} = -\frac{d\left[SO_4^{2-}\right]}{1} = +\frac{d\,[H_2S]}{1} = +\frac{d\left[HCO_3^-\right]}{2} = d\xi \tag{5.22}$$

where ξ is the degree of advancement of the reaction. The quantity $d\xi/dt$ is the reaction rate. This depends on the probability that the reactants will collide and, therefore, on their concentration; and, as with a diffusion coefficient, on the frequency ν of vibrations and on

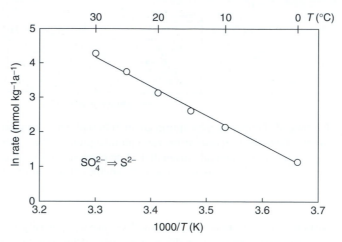

Figure 5.9   An example of reaction kinetics: sulfates being reduced to sulfides in diagenetic solutions. A linear relation is found in the semi-logarithmic Arrhenius plot (data from Westrich and Berner, 1984).

the potential barrier ΔE to be crossed so as to "activate" the reactants. This is therefore written:

$$\frac{d\xi}{dt} = k \left[CH_2O\right]^2 \left[SO_4^{2-}\right] \tag{5.23}$$

where it can be seen that the probabilities are multiplicative properties, with

$$k = k_0 \nu \exp\left(-\frac{\Delta E}{RT}\right) \tag{5.24}$$

where k_0 is a constant that includes many parameters, notably the specific surface area of the reactants (a finely divided mineral is more reactive than a single large crystal). We recognize a Boltzmann law, which can be represented in an Arrhenius plot $\ln k = A + B/T$ (Fig. 5.9). A large number of kinetic constants were determined in the 1980s and 1990s. At high temperatures, reaction kinetics is normally fast but, in the context of diagenesis and weathering, dissolution is slow. With the advent of fast computers, the calculation of kinetic constants using ab initio methods has made very substantial progress and kinetic modeling of reaction paths has become a quantitative field to the point that reaction mechanisms can be tested very accurately.

The last equation has particularly important consequences for kinetic isotope effects (see Section 3.1). First, let us go back to Fig. 3.2 and observe that a bond involving a light isotope is higher up on the energy scale and therefore easier to break than the same bond involving a heavy isotope. For example, proceeding along the reaction path therefore requires less energy for an OH bond than for an OD bond (Fig. 5.10). Second, the frequency term ν varies with $1/\sqrt{\mu}$, where μ is the reduced mass of the oscillator: the vibrational frequencies of bonds involving light isotopes therefore are higher than those of bonds involving heavy isotopes. Light isotopes will go through more frequent attempts to engage into the reaction pathway than heavy isotopes. Species of elements such as C, S, and N have very different coordinations under different redox conditions, in particular those prevailing during biological reactions. The term "kinetic isotope effect" is used to

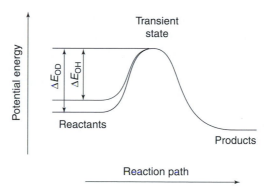

Figure 5.10 The kinetic isotope effect: a bond involving a light isotope is higher up on the energy scale than the same bond involving a heavy isotope. Breaking the bond and proceeding along the reaction path therefore requires less energy for an OH bond than for an OD bond.

reflect that changes in coordination are usually slow processes that rarely reach equilibrium and that isotope fractionation away from equilibrium therefore is the rule whenever organisms such as bacteria are involved in biogeochemical processes.

5.5 Adsorption

Capture of atoms, ions, and molecules by surfaces is known as adsorption. This process is critical to the understanding of geochemical cycles because it accounts for why the bulk of many elements carried by rivers do not contribute to the inorganic salts dissolved in seawater as much as expected: upon mixing of freshwater and seawater in estuaries, iron hydroxides precipitate and entrain, adsorbed on their surface, most of the transition elements, the rare-earth elements, and high-field-strength elements dissolved in the run-off. This is also a process at work in carbonated springs: decompression removes CO_2 and increases the pH; iron hydroxides precipitate and "clean up" the spring water from many of its constituents. Likewise, the scavenging of shallow waters by wind-blown particles carried from the deserts and by organic material depletes the surface ocean in the same elements.

Adsorption processes are conventionally separated by the energy binding the element to the surface. Adsorption with strong binding energies between ions and charged substrate is referred to as chemisorption. Weak binding energies resulting in particular from van der Waals forces characterize physisorption. Although the physical forces underlying these two processes are different, they may be simply described by similar phenomenological laws relating pressure in the gas or concentration in the liquid and the fraction θ of active sites on the solid surface filled by the adsorbed element and known as the coverage of the surface.

As an illustration, let us suppose that adsorption of species i takes place as a single layer on the surface of a solid phase in a gas at pressure P_i and that temperature is constant during the process. Let σ be the concentration of active sites capable of hosting the adsorbing species per unit area of the solid surface. To the first order, the rate of desorption R_{des} is proportional to the proportion of the occupied sites:

$$R_{des} = k_{des}\theta\sigma \tag{5.25}$$

where k_{des} is the rate constant. Likewise, the rate of adsorption R_{ads} is proportional to the proportion of free sites and to the pressure of species i:

$$R_{ads} = k_{ads}(1 - \theta)\sigma P_i \tag{5.26}$$

At equilibrium, these two rates are equal and we get:

$$\frac{1}{\theta} = 1 + \frac{k_{des}}{k_{ads}P_i} \tag{5.27}$$

which is the famous Langmuir "isotherm," a name by which it is implied that both the rate constants and pressure are temperature dependent. The parameter θ is the ratio of the volume V of adsorbed material to the volume V_0 corresponding to a continuous single layer and therefore:

$$\frac{1}{V} = \frac{1}{V_0} + \frac{k_{des}}{k_{ads}V_0 P_i} \tag{5.28}$$

By plotting the inverse of the volume V of species adsorbed as a function of $1/P_i$, the zero intercept gives the volume V_0 of the monolayer. Dividing V_0 by the footprint (cross-section) of the adsorbed species gives us the surface area of the solid. For adsorption in liquids, pressures should be replaced by concentrations. An equivalent isotherm exists that accounts for multilayer coverage and is known by the initials BET of its discoverers (Brunauer, Emmett, and Teller). Isotherms are routinely used to determine the "surface area" of various solids of geochemical interest: if for crystalline silicates, such as quartz or feldspars, the adsorption recovers values similar to those inferred from their geometry, the enormous surface area obtained for iron hydroxides ($600 \text{ m}^2 \text{ g}^{-1}$) attests to the fact that the surface of these solids is made of a myriad of nanoscale embayments and protrusions, which make them the prime scavengers of the sea.

Exercises

1. Which of these properties are conservative: total energy, kinetic energy, temperature, velocity, moles of iron content, moles of ferric iron, concentration of iron, pH, alkalinity?
2. At a given locality at the bottom of the ocean, the rate of sedimentation is 10 mm ky^{-1}. The density of the surface sediment is 2.0 g cm^{-3} and its phosphorus content is 0.65 wt%. What is the sedimentary (advection) flux of phosphorus in $\text{kg P m}^{-2} \text{ My}^{-1}$?
3. Explain why, during exhumation of mountain ranges, minerals should be treated as "Lagrangian" bodies.

4. During crystal growth from their melt, elements partition across the interface with, for element i, $C_{min}^i = K_{min/liq}^i C_{liq}^i$. The position of the mineral–melt interface is set at $x = 0$, the growth rate V is constant, and the density change during crystallization is neglected. We will assume that diffusion in the solid ($x < 0$) is slow enough to be neglected and that the diffusion coefficient of i in the melt ($x > 0$) is D_{liq}^i. Write the advection fluxes on each side of the interface, the diffusion flux in the liquid, and the overall transport balance of element i during crystallization. This situation is reminiscent of sedimentation. What is the major difference?

5. Experiments by van Orman *et al.* (2001) determined the parameters for Nd diffusion in pyrope crystals. The pre-exponential term is $D_0^{Nd} = 10^{-9.2}$ m^2 s^{-1} and the activation energy $E^{Nd} = 300$ kJ mol^{-1}. It is determined that the core of a spherical garnet of 1 cm radius contains about 1 ppm Nd. Concentration decreases in the depleted rim to a zero value at the surface and this is thought to be the result of a metamorphic pulse with a steady temperature of 800 °C. The thickness of the depleted rim (diffusion boundary layer) is 0.1 micron and we assume that concentration decreases linearly between the core and the surface. The specific weight of garnet is 3.5 g cm^{-3}. Calculate the total flux of Nd in kg s^{-1} out of the garnet during the metamorphic pulse.

6. Helium is lost by diffusion from an apatite crystal assumed to be spherical with a radius a of 100 microns (μm) during a short episode of reheating ($\Delta t = 200\,000$ years at 100 °C). The parameters for He diffusion in apatite are $D_0^{He} = 0.064$ m^2 s^{-1} and the activation energy $E^{He} = 148$ kJ mol^{-1}. Calculate the parameter $\tau = D^{He}\Delta t/a^2$ and the fraction of He lost at the end of the thermal disturbance.

7. Using the parameters given in the previous exercise, what is the closure temperature of He diffusion in apatite for a cooling rate of 10 K Ma^{-1} ? Hint: use a guess for the temperature to calculate θ then T_c, and proceed by successive refinements.

8. Cygan and Lasaga (1985) determined that the parameters for Mg diffusion in pyrope garnet are $D_0^{Mg} = 9.8 \times 10^{-9}$ m^2 s^{-1} and the activation energy $E^{Mg} = 239$ kJ mol^{-1}. If the cooling rate is 10 K Ma^{-1}, what is the closure temperature of Mg exchange between a spherical garnet of 1 cm radius and its neighbors?

9. A very large number of different ^{39}Ar–^{40}Ar analyses of K-feldspar in granites and gneisses showed that the Ar diffusion parameters are $E^{Ar} = 190$ kJ mol^{-1} and $D_0^{Ar}/a^2 = 10^5$ s^{-1}, where a is the "effective" diffusion radius (McDougall and Harrison, 1999). Calculate the closure temperature for Ar diffusion in feldspar for a cooling rate of 20 K Ma^{-1}. The value of D_0^{Ar} was also determined on homogeneous gem-quality material; by comparison with the previous data, it was shown that the "effective" diffusion radius a is only 6 μm. Discuss the implications of these observations for chronology.

10. Discuss the significance of the apparent temperatures given by (a) fractionation of oxygen isotopes between minerals and (b) Fe–Mg fractionation between co-existing clinopyroxene and orthopyroxene.

11. Groundwater percolates through the sedimentary basement at a velocity of 10 m a^{-1}. Rock volumic porosity is one percent and we neglect the difference in density between water and sediment. Let D^i be the partition coefficient of element i between the matrix

Table 5.1 Adsorption data of N_2 on goethite. Pressures refer to the nitrogen gas	
$P\ (10^{-12}\mathrm{Pa})$	$V\ (10^{-6}\mathrm{m}^3\ \mathrm{g}^{-1})$
251	8.5
176	7.9
128	7.2
97	6.6
70	6.0
45	4.9
30	3.9
20.7	3.1

and groundwater. How far will groundwater movement take a contamination in $10\,000$ years for the following elements: Cl ($D^{Cl} = 0$), I ($D^{I} = 0.01$), Sr ($D^{Sr} = 0.2$), U ($D^{U} = 100$), Th ($D^{Th} = 10^8$)?

12. Basaltic melt percolates through a matrix of molten peridotite. Let us assume that the degree of melting ($F = 2$ percent) does not change significantly over the distance of interest. Discuss the relative velocity of Ni ($D^{Ni} = 10$), Yb ($D^{U} = 1$), and Ba ($D^{Ba} = 0.0001$). Compare the velocity of incompatible elements (e.g. Ba) and major elements (e.g. Si, Al) and discuss which core assumption of the chromatography theory breaks down for the latter group.

13. Calculate the surface area in $\mathrm{m}^2\ \mathrm{g}^{-1}$ of goethite, FeO(OH), at ambient temperature using the volumes V of N_2 (in $\mathrm{m}^3\ \mathrm{g}^{-1}$) adsorbed under the pressures P of this gas as listed in Table 5.1. Use a Langmuir isotherm and a cross-section of $1.64 \times 10^{-18}\ \mathrm{m}^2$ for the N_2 molecule. One mole of gas under standard conditions is $0.0224\ \mathrm{m}^3$. Assume that the shape of the crystals is roughly spherical. The specific gravity of goethite is $3.6\ \mathrm{g\ cm}^{-3}$ and the mean diameter of mineral grains is $100\ \mu\mathrm{m}$. Compare the actual surface area of 1 g of powder with the surface area derived from the adsorption data.

References

Cherniak, D. J., Lanford, W. A., and Ryerson, F. J. (2001) Lead diffusion in apatite and zircon using ion implantation and Rutherford backscattering techniques. *Geochim. Cosmochim. Acta*, **55**, 1663–1673.

Cygan, R. T. and Lasaga, A. C. (1985) Self-diffusion of magnesium in garnet at 750° to 900 °C. *Amer. J. Sci.*, **285**, 328–350.

♠ Dodson, M. H. (1973) Closure temperature in cooling geochronological and petrological systems. *Contrib. Mineral. Petrol.*, **40**, 259–274.

♠ DeVault, D. (1943) The theory of chromatography. *J. Amer. Chem. Soc.*, **65**, 532–540.

McDougall, I. and Harrison, T. M. (1999) *Geochronology and Thermochronology by the ^{40}Ar–^{39}Ar method*. Oxford University Press.

van Orman, J. A., Grove, T. L., Shimizu, N., and Layne, G. D. (2001) Rare-earth element diffusion in a natural pyrope single crystal at 2.8 GPa. *Contrib. Mineral. Petrol.*, **142**, 416–424.

Westrich, J. T. and Berner, R. A. (1984). The role of sedimentary organic matter in bacterial sulfate reduction. *Limnol. Oceanog.*, **29**, 236–249.

Geochemical systems

This chapter looks at the changes that over time affect the geochemical properties of a system or a set of systems, such as the mantle, the crust, or the ocean, when subjected to disturbances whether caused naturally or by human activity. The essential concepts utilized – residence time and forcing – are taken from chemical engineering. Viewing system Earth as a chemical factory composed of reactors, valves, sources, and sinks, has proved to be a simple and robust model. The theory goes by various names, with the "box model" probably the most widely used. We will first set out the principles by describing the behavior of a system with a single reservoir and then go on to generalize the approach.

6.1 Single-reservoir dynamics

Let us begin by considering a lake (Fig. 6.1) containing a mass of water M that we will take to be constant. A river flows through the lake with a rate of flow Q, which we will express in kilograms per year; Q is therefore the same upstream and downstream. We are interested in the balance of a chemical species in the lake. A chemical element i introduced upstream with a concentration C_{in}^i is either lost through the lake outlet or entrained into sediments. The sedimentation rate P is also expressed in kilograms per year. The lake itself is considered homogeneous, being well mixed by turbulent flow and by convection.

The concentration C^i of an element i is therefore the same in the lake and in the downstream outflux. This balance can be written:

$$M\frac{\mathrm{d}C^i}{\mathrm{d}t} = QC_{in}^i - QC^i - PC_{sediment}^i \tag{6.1}$$

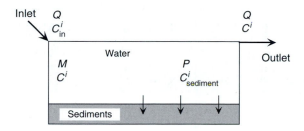

Figure 6.1 Simple box model (a single reservoir) typical of a lake. M: mass of water in the lake, Q: water inflow = outflow, P: sedimentation rate.

and, as usual, we will introduce a partition coefficient D^i to allow for fractionation of the element between solid and water, such that $C^i_{\text{sediment}} = D^i C^i$. Rearranging this gives:

$$\frac{\tau_{\mathrm{H}}}{1 + \beta^i} \frac{\mathrm{d}C^i}{\mathrm{d}t} + C^i = \frac{C^i_{\text{in}}}{1 + \beta^i} \tag{6.2}$$

where $\tau_{\mathrm{H}} = M/Q$ is the time to renew the water in the lake and $\beta^i = PD^i/Q$ is the reactivity of the element relative to sedimentation. This reactivity is zero for an inert tracer, such as a coloring agent or chlorine ion, and increases with the solid/liquid partition coefficient of the element. The term in front of the derivative is the residence time $\tau^i = \tau_{\mathrm{H}}/(1 + \beta^i)$ of element i in the lake, i.e. the mean time that atoms of i spend in the lake before being removed (see Appendix G).

The form of (6.2) is extremely instructive. Time is now an integral part of the balance, contrary to the distillation equations mentioned earlier. If the concentration of i upstream is zero, we obtain by integration the change in a lake initially at concentration $C^i(0)$, possibly by accidental spillage of a pollutant at $t = 0$:

$$C^i(t) = C^i(0)\mathrm{e}^{-t/\tau^i} \tag{6.3}$$

The lake is said to relax from its initial state described by the concentration of the element at $t = 0$. The measurement of relaxation rate is the residence time τ^i, which is shorter for more reactive elements and less than the time it takes to renew the lake water. The residence time τ^i indicates how quickly a disturbance in the balance of element i is damped out by the lake. The corresponding expression for non-zero C^i_{in} is given in Appendix G.

When $t \gg \tau$, relaxation is complete, the derivative in (6.2) vanishes and a constant concentration is reached. This state is known as steady state. The simple relation $C^i(\infty) = C^i_{\text{in}}/(1 + \beta^i)$ shows that the right-hand side is the concentration when the initial condition is fully relaxed (Fig. 6.2). This is called the forcing term. The forcing term, here the input from upstream, shows in which direction the system is heading. Again it can be seen that the final concentration of the lake is lower when the element is more reactive.

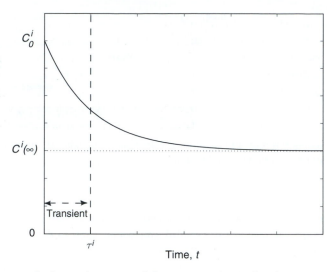

Figure 6.2 Relaxation of a perturbation at time $t = 0$ of the concentration in the element i within a single box system. The residence time τ^i measures the rate at which the perturbation is relaxed. The transient is the regime in which the initial perturbation can still be perceived.

The reader can check from (6.2) that the residence time of element i is simply the ratio of the mass contained in the reservoir to the output of this element at steady state, i.e.:

$$\tau^i = \frac{MC^i(\infty)}{QC^i(\infty) + PC^i_{\text{sediment}}(\infty)} = \frac{\text{mass of element } i \text{ in the reservoir}}{\text{output rate of element } i} \qquad (6.4)$$

Since the system is at steady state, the output rate may conveniently be replaced by the input rate. The residence time of elements in the ocean may thus be evaluated from data of concentration in seawater and rivers in Appendix A and geophysical data in Appendix F. For example, the residence time of strontium is $(1.4 \times 10^{21} \text{ kg} \times 8000 \text{ μg l}^{-1}) / (3.6 \times 10^{16} \text{ kg a}^{-1} \times 70 \text{ μg l}^{-1})$, or 4.4 million years.

A fundamental aspect of reservoir dynamics is that the more reactive pollutants will be eliminated more quickly and their residual concentration will be lower. Reactivity is a function of two factors: the rate of sedimentation (sediment cleans the water) and the fractionation of the element between solid and liquid. Hence the apparent paradox: elements that poison micro-organisms and sediment are eliminated more quickly.

This elementary theory explains some common but essential aspects of the geochemistry of large geological systems. Introducing the definition of τ^i and $C^i(\infty)$ into (6.2) and rearranging, we get:

$$\frac{1}{C^i - C^i(\infty)} \frac{d\left[C^i - C^i(\infty)\right]}{dt} = \frac{d\ln\left[C^i - C^i(\infty)\right]}{dt} = -\frac{1}{\tau^i} \qquad (6.5)$$

which indicates that the relative deviations of element concentrations from their steady-state values fluctuate at a rate scaled by the inverse of the element residence time. Inert elements, i.e. elements whose partition coefficients D^i are very small, have long residence times. Their concentrations therefore remain very stable. These elements are highly concentrated (i.e. they are major elements), their residence time is long, and they display no fast temporal variations. This is the case of strontium in the ocean, for which we have calculated a residence time of 4.4 Ma. By contrast, reactive elements with high partition coefficients D^i and therefore short residence times are very quickly depleted in the reservoir (i.e. they are minor elements). Their short residence time means they readjust quickly after a disturbance, but their concentration level fluctuates wildly about steady state. To use an analogy, a reservoir such as a lake, the ocean, or the mantle, behaves like a low pass filter, eliminating fluctuations in concentration with shorter time characteristics than the residence time of the element. If a time characteristic τ_m of mixing in the reservoir can be defined and the residence time τ^i of an element i is such that $\tau^i > \tau_m$, this element will reside long enough in the reservoir to be well homogenized. If its residence time is shorter than the reservoir mixing time, the element will be drained away in the outflow before it is effectively mixed and will be distributed heterogeneously around the system. For instance, concentrations of Sr (8 ppm) and $^{87}Sr/^{86}Sr$ (0.7093) are homogeneous across the entire ocean.

In the ocean, the export process of most elements is sedimentation and the partition coefficient measures sediment particle/seawater fractionation. In the mantle, the export process is magmatism and the coefficient D^i that measures the magma/residual solid fractionation is the inverse of the partition coefficient usually considered by petrologists. Inert elements or ions are residual, very homogeneous, abundant, and of invariable concentration: this is the case for Mg, Ni, and Cr in the mantle, of Cl, Na, and SO_4 in the ocean, and of N_2, O_2, and Ar in the atmosphere. In contrast, reactive elements are normally at the trace level and exhibit wide variability in space and time: this is the case for the incompatible lithophile elements (Th, Ba, La, Rb, etc.) in the mantle, of Pb, Cd, Zn, and La in the ocean, and of methane and ozone in the atmosphere.

In the absence of any input from upstream, (6.2) can be reformulated as:

$$-\frac{dC^i}{C^i dt} = \frac{1}{\tau_H} + \frac{PD^i}{M} = \frac{1}{\tau_H} + \frac{1}{\tau_S^i} \qquad (6.6)$$

This equation expresses that the total probability of an atom i leaving the system per unit time is the sum of two probabilities: the probability of i exiting downstream, which is equal to the inverse of the renewal time of the water τ_H, plus the probability of i being lost to the sediment, which is the inverse of removal time by the solid $\tau_S^i = M/PD^i$. As a simple but powerful rule, the probabilities of removal through various routes are independent and therefore additive and so are the inverse residence times. We will see below that a third removal mechanism is radioactivity and that the probability of removal by this process is simply the decay constant λ.

A third aspect of residence time is that of the distribution of the "ages" of atoms within the reservoir. To understand the concept of birth and death processes, let us take another analogy, namely the reservoir of a human population. Suppose, for simplicity, that the number of births every year is constant. A healthy population is a poorly "mixed" population since the exit, i.e. death, draws off the elderly preferentially. A trip around a graveyard might show that the average age of death is about 70–80 years, the mean span of our

The distribution of residence times

In a well-mixed system at steady state, the distribution of residence times (the histogram of age groups) is exponential. Let us call $n(t,\theta)$ the number of particles that, at time t, have resided in the system for a time $> \theta$. Using the elementary rules of differentiation, the quantity $dn(t,\theta)$ such as:

$$dn(t,\theta) = n(t,\theta + d\theta) - n(t,\theta) = \frac{dn(t,\theta)}{d\theta}d\theta \qquad (6.7)$$

is the number of particles which have resided in the system for a time comprised between θ and $\theta + d\theta$. This is the number of individuals in the "class of age" θ.

Let us call $p = 1/\tau$ the probability that a particle leaves the system in the unit time and assume that it is independent of the time the particle stayed in the box (particles do not age). The assumption that the system is well mixed therefore amounts to a constant p, independent of θ. We can describe this problem as "Lagrangian" because we actually refer to fluxes along the residence time coordinate that are not very different from the usual fluxes along a directional spatial coordinate. Let us break down on a yearly base, the balance of individual molecules of water in a lake with an inlet and an outlet. At time t, let us say on January 1st, the fraction of water which resided between 10 and 11 years in the lake forms bin 10 ($\theta = 10$). After a year, the bin is flushed and replaced by all the water which had resided between 9 and 10 years in the lake (the bin 9), while the water from bin 10 is either promoted to bin 11 or is lost into the outlet. The number of water molecules $n(t, \theta)$ in bin θ changes because water leaves the system and also because it ages. The rate of change is given by:

$$dn(t,\theta) = -p\, n(t,\theta)\, dt - n(t,\theta + d\theta) + n(t,\theta) \qquad (6.8)$$

where from (6.8):

$$n(t,\theta + d\theta) = n(t,\theta) + \frac{\partial n(t,\theta)}{\partial \theta}d\theta \qquad (6.9)$$

Every year, the residence time of a water volume element that remains in the system increases by one year and, therefore, $d\theta = dt$. We can rearrange the previous equation as:

$$\frac{\partial n(t,\theta)}{\partial t} = -p\, n(t,\theta) - \frac{\partial n(t,\theta)}{\partial \theta} \qquad (6.10)$$

At steady state, the left-hand side of this equation vanishes and the right-hand side may be integrated as:

$$n(t,\theta) = n(t,0)\, e^{-p\theta} = n(t,0)\, e^{-\theta/\tau} \qquad (6.11)$$

which demonstrates our second proposition. By integrating this expression over the entire population, we would find that the mean residence time is simply τ. Using proper statistics, we would also discover that the standard deviation of the residence-time distribution is $\sqrt{\tau}$. The deceptively simple approach of residence-time distributions in a well-mixed reservoir therefore packs a rich amount of information.

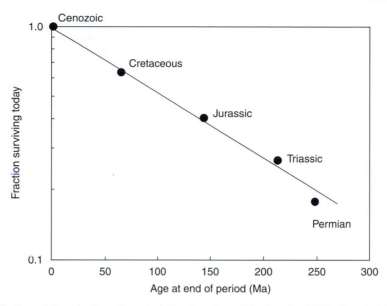

Figure 6.3 Age distribution of fine clastic sediments (after Garrels and Mackenzie, 1971). The straight-line arrangement implies that the sedimentary reservoir is well mixed with a probability of erosion independent of sediment age. The slope of the line indicates that this type of sediment survives on average about 150 million years before being eroded.

time here below. However, the average age of the population is much younger, probably about 30–40 years. Let us imagine now an outbreak of a virus that affects old and young indiscriminately. The population becomes well mixed in the sense that the probability of exiting the system is independent of the individual's age. The gravestones would then show an average age of demise close to the mean age of the population. This latter case is the one most often considered in the dynamics of natural systems where the probability that a particle leaves the system does not increase with the particle's age. It can be shown (see box) that particles in a well-mixed system have an exponential age-group histogram (exit is a Poisson process). Under such conditions, the proportion of individuals that have resided in the system for a time in excess of θ is

$$f(\theta) = e^{-\theta/\tau} \tag{6.12}$$

in which τ is the mean residence time of all the individuals in the system. In a well-mixed reservoir, not all the individuals have the same residence time but they all have the same probability of exit, regardless of their previous history in the system.

Such an example of age distribution is given by the survival of clastic sediments since the end of the Paleozoic (Fig. 6.3). The proportion of sediments surviving beyond a given age decreases exponentially with time. The average age of the sediment is equal to its mean survival time of about 150 million years. This demonstrates the stationary character of the total mass of clays, sandstones, etc. Similar ideas may be used for tracers in the mantle or in the ocean: the "birth" of chemical tracers (e.g. the element Ba or the

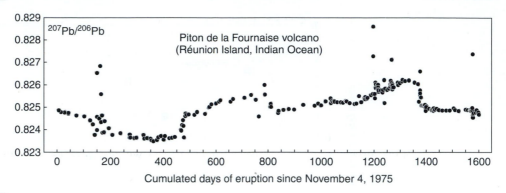

Figure 6.4 ^{207}Pb/^{206}Pb ratios in recent lavas of the Piton de la Fournaise volcano (Réunion Island, Indian Ocean). The rapid fluctuation of this ratio over a time scale of a small number of years shows that the magmatic reservoir holds a volume of lava equivalent to just a few years of eruption. There is therefore no magma chamber larger than a cubic kilometer beneath this volcano (after Vlastelic *et al.*, 2008).

nuclide ^{86}Sr) corresponds to their introduction into the ocean by rivers or into the mantle by subduction zones, while their "death" corresponds to sedimentation (ocean) and extraction into mid-ocean ridges (mantle). Radioactivity corresponds to the death of a radioactive nuclide (e.g. ^{87}Rb) and to the birth of a radiogenic isotope (e.g. ^{87}Sr). These concepts will also be used to explain mixing in the mantle or the ocean at the end of this chapter.

The concept of residence time requires that fluctuations in the input of an element of duration shorter than its residence time do not show up in the output. It has very diverse applications in oceanography and limnology as well as in magmatic petrology or mantle geochemistry. Let us take an example from the magmatic field. An erupting magma reservoir that is continuously replenished can be thought of in the same way as a lake, the upstream being the inflow of fresh magma, the downstream the eruptive activity, and the cumulates the equivalent of sediments. In this context, the compatible elements that are entrained by crystallization are reactive elements and their residence time in the reservoir is shorter than that of the incompatible and therefore inert elements. A reservoir like this will therefore damp out all fluctuations in the concentration of an element whose time characteristic is less than its residence time. The characteristic period of fluctuation of the ^{207}Pb/^{206}Pb ratio in the lava of the active volcano Piton de la Fournaise (on the island of Réunion) is of a few years (Fig. 6.4). As Pb is incompatible ($\beta^i \approx 0$), we thus have an important constraint on the renewal time τ_H of magma in the reservoir supplying the volcano. This time τ_H is of the order of a few years. Given the eruption rate Q, the mass $M = Q \times \tau_H$ of the reservoir can be evaluated. These simple observations limit the volume of the magma reservoir to less than one cubic kilometer. The concept of residence time has greatly altered the way we see volcanic systems: until a few years ago, most volcanoes were once thought to tap immense magma chambers, which, thanks to such observations, now turn out to be much smaller systems of feeder dikes filled with crystal mush.

Let us now consider a radioactive element with a decay constant λ_i. The radioactive decay $\lambda_i M C^i$, can be subtracted from (6.1), modifying (6.2) as:

$$\frac{\tau_H}{1 + \beta^i + \lambda_i \tau_H} \frac{dC^i}{dt} + C^i = \frac{C^i_{in}}{1 + \beta^i + \lambda_i \tau_H} \tag{6.13}$$

This equation applies, in particular, when determining the mixing time of the ocean, which is approximated by the residence time of carbon in the deep ocean. Because the deep ocean is isolated from the atmosphere, the radioactive carbon-14 it acquired during its time at the surface is no longer renewed and so decays within a closed system. Oceanographic measurements indicate that the $^{14}C/^{12}C$ ratio of deep water makes up only 84% of the same ratio in the surface water. Suppose that the ocean composition has reached steady state ($dC^i/dt = 0$). Let us arbitrarily divide the ocean into two reservoirs, one of deep water and one of surface water, which exchange water by vertical mixing. The inflow into the deep reservoir is therefore the downwelling of surface water. Dividing (6.13) for ^{14}C by (6.2) for the stable ^{12}C, we get:

$$\left(\frac{^{14}C}{^{12}C} \right)_{deep} = \frac{1 + \beta^C}{1 + \beta^C + \lambda_{14C} \tau_H} \left(\frac{^{14}C}{^{12}C} \right)_{surface} \tag{6.14}$$

The term for chemical reactivity β^C, which is identical for both carbon isotopes, is small enough compared with unity to be ignored. The reader will check that by introducing the relative values of isotope ratios given above and the value of λ_{14C} from Table 3.1, the renewal time for deep seawater τ_H can be calculated at 1600 years. Because τ_H represents the time between two successive transits of sea water through the deep reservoir, this value is often referred to as the mixing time of the ocean.

6.2 Interaction of multiple reservoirs and geochemical cycles

The dynamics of a multiple-reservoir system differ from those of a single reservoir in different respects. Collective readjustment is invariably faster than it would be if the reservoirs were considered separately. We will see that ignoring this principle would lead to very serious errors.

Let us consider the case of a change in the number of atoms of an element in two reservoirs that exchange matter with one another, for example the set of the "surface water" and "deep water" oceanic reservoirs exchanging material by vertical advection, or of the mantle–continental crust system. Let us call $p^{1 \rightarrow 2}$ the probability for an atom to be transferred from reservoir 1 to reservoir 2 in the unit time, with a similar definition for $p^{2 \rightarrow 1}$. If n_1 and n_2 stand for the number of moles of the element in question in each reservoir, we get the two equations:

$$\frac{dn_1}{dt} = -p^{1 \rightarrow 2} n_1 + p^{2 \rightarrow 1} n_2 \tag{6.15}$$

$$\frac{dn_2}{dt} = p^{1 \rightarrow 2} n_1 - p^{2 \rightarrow 1} n_2 \tag{6.16}$$

Let M_1 and M_2 be the masses of the two reservoirs and Q the mass of material (water, magma) exchanged between them per unit time. As in the previous section we can relate the probability of transfer to the ratio of the flux of element to its amount in the reservoir, for instance:

$$p^{1\to 2} = \frac{1}{\tau_1} = \frac{Q\,(n_1/M_1)}{M_1\,(n_1/M_1)} \tag{6.17}$$

where (n_1/M_1) stands for concentrations and τ_1 for the residence time of the element in reservoir 1. If fractionation takes place during transfer, e.g. upon precipitation of melting, this equation must be replaced by:

$$p^{1\to 2} = \frac{1}{\tau_1} = \frac{Q\,(n_1/M_1)\,D^{1\to 2}}{M_1\,(n_1/M_1)} \tag{6.18}$$

where we recognize at the numerator the product of the material flux Q multiplied by a concentration n_i/M_i and by the partition coefficient $D^{i\to j}$ allowing for possible fractionation of the element as it passes from reservoir i to reservoir j. When material is extracted from the mantle as a basaltic or andesitic magma and incorporated into the continental crust, the magma is chemically fractionated relative to its starting mantle and $D^{\text{mantle}\to\text{crust}}$ increases with the lithophile character of the element. By contrast, when water is exchanged between two parts of the ocean, the coefficients D may be equal to 1.

The balance can therefore be written as two conservation equations:

$$\frac{dn_1}{dt} = -\frac{Q D^{1\to 2}}{M_1} n_1 + \frac{Q D^{2\to 1}}{M_2} n_2 \tag{6.19}$$

$$\frac{dn_2}{dt} = \frac{Q D^{1\to 2}}{M_1} n_1 - \frac{Q D^{2\to 1}}{M_2} n_2 \tag{6.20}$$

In terms of residence times τ_1 and τ_2 of the element in each reservoir, the system becomes simply:

$$\frac{dn_1}{dt} = -\frac{1}{\tau_1} n_1 + \frac{1}{\tau_2} n_2 \tag{6.21}$$

$$\frac{dn_2}{dt} = \frac{1}{\tau_1} n_1 - \frac{1}{\tau_2} n_2 \tag{6.22}$$

It can be seen that these two equations have a zero sum, which reflects the constancy of the total inventory $n_1 + n_2 = N$ of the element in the system. By introducing this condition into (6.21), and rearranging and solving it, we obtain:

$$n_1 = n_{1,0}\, e^{-t/\tau} + N \frac{\tau_1}{\tau_1 + \tau_2} \left(1 - e^{-t/\tau}\right) \tag{6.23}$$

where $n_{1,0}$ is the initial value of n_1. A symmetrical equation would be obtained for n_2. The overall relaxation time τ of the system is defined by:

$$\frac{1}{\tau} = \frac{1}{\tau_1} + \frac{1}{\tau_2} \tag{6.24}$$

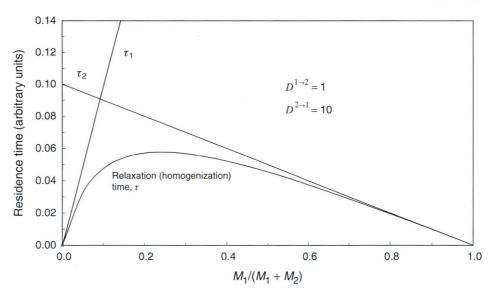

Figure 6.5 Calculation of the homogenization time of an arbitrary element in a two-reservoir system (arbitrary units). Partition coefficients D indicated are those governing the transfer of the element from one reservoir to the other. This homogenization time depends on the relative mass of each reservoir and is less than the time of residence of the element in each reservoir. The cooperative character of exchanges accelerates homogenization compared with what might be expected for readjustment of each reservoir separately.

indicating that τ is shorter than either τ_1 or τ_2 (Fig. 6.5). The steady state is reached and homogenization between the two reservoirs is faster than a simple analysis of the residence time of each of the reservoirs seems to indicate. The falsely intuitive idea that the chemical response time of a system should need longer to adjust than the residence time of its slower reservoir is clearly wrong. The reason for this acceleration can be understood by rewriting (6.21) as:

$$\frac{\tau_1 \tau_2}{\tau_1 + \tau_2} \frac{dn_1}{dt} + n_1 = \frac{\tau_1}{\tau_1 + \tau_2} N \qquad (6.25)$$

and comparing this equation with (6.2). The forcing term arising on the right-hand side from coupling with reservoir 2 interferes with the straightforward relaxation of reservoir 1 and always in the direction of acceleration.

The argument just developed also applies to more complex sets of reservoirs, but we would then need additional tools from linear algebra. We have considered only systems without an external forcing effect; such an effect should, however, be allowed for where river water flows into the ocean or cosmogenic matter into the mantle. In general, this theory underlies the understanding of geochemical cycles, i.e. processes of exchange of each element between the Earth's different reservoirs. The conditions for describing an elementary cycle are: (1) essential reservoirs must be identified and their contents estimated; and (2) fluxes of elements between reservoirs must be evaluated over a sufficient

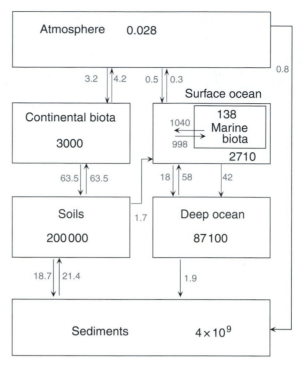

Figure 6.6 The external cycle of phosphorus (Lerman, 1988). Masses (solid type) and fluxes (grey tinted type) are in millions of tons of phosphorus per year.

length of time compared with the relaxation time of the system. The phosphorus cycle is given here (Fig. 6.6) as a good example of a cycle of an element taken from the literature.

One aspect of cycles that is difficult to grasp is that of the time scale at which they apply. It is easy to see that estimating the transportation of an element to the ocean by rivers in flood or from the mantle to the atmosphere by extrapolating from observation of a volcanic eruption gives an inaccurate representation of flows. Catastrophic transfer (or pulse transfer) is particularly difficult to fit into the picture. Another condition may seem implicit to the idea of a cycle, that of the stationary state, or the balanced cycle: an atom is passed many times through all of the reservoirs and the inventory of each reservoir no longer varies. This condition is always an approximation to some extent. Suppose we are interested in the carbon cycle at appropriate time scales for historical climatic variations (100–2000 years). Should we include all carbon transfers between the mantle, crust, ocean, atmosphere, and biomass; and at all time scales, from diurnal (daily) rhythms of the biomass to fluctuations of volcanic emission, at very long geological time scales? At the scale of a century, the effect of diurnal fluctuation is that of the sum of 365×100 cycles and its effect is therefore virtually nil. At the scale of a millennium, the long geological variations will be tenuous. Adding that effect would be the same as arbitrarily introducing a transfer equation equivalent to "non information." Now, it is well known in mathematics that adding an equation of the type $0 = 0$ makes a system of equations insoluble. To ensure the cycle is described properly, we will therefore choose conservatively to draw up an inventory of processes that are important at time scales of 1 to 100 000 years, ignoring faster processes and treating longer-term ones as forcing to be dealt with by separate parameters. The use of a time "window" adequate for all the processes to be considered is an essential condition for the construction of correct geochemical cycles.

It is legitimate to wonder about the relationship between the full representation of chemical transfers by equations describing local fluxes of tracers, in which the concentration of an element is tracked at each point of the system at each moment in time, and the multiple-reservoir model (box or zero-dimension model) with no explicit indications of the position of individual tracers. Both representations use the same physics and represent justifiable and consistent formulations of the conservation principle. The two essential differences are that, in the box model, concentration is averaged over the entire reservoir while the fluxes are integrated over the entire area bounding each reservoir. It could thus be shown that by multiplying the number of reservoirs and making their size tend toward zero, a continuous transition can be made from one representation to the other.

6.3 Mixing and stirring

Let us now try to describe how chemical heterogeneities are destroyed by the convective movements of the mantle or the ocean or by circulation in a water table. Anyone can imagine the fate of a packet of different colored sticks of play dough left in the hands of a small child: from an early stage, where the sticks are distorted but remain identifiable, to a homogeneous bluish-gray dough, passing through a marbled stage with multicolored stripes stretched and folded back on themselves many times (Fig. 6.7). Obviously if a representative sample of the average of the packet is to be taken, it must be sufficiently large compared with the thickness of the stripes. The concept of homogeneity is therefore dependent on the scale of observation or of sampling, and it will always be possible to spot local heterogeneities provided that the medium is observed at sufficiently high resolution.

Mixing and stirring should not be used indifferently. In the case of the play dough, a grown-up person coming early enough to interfere with the entropy-generating toddler can still undo the stretching and folding and with a little bit of work retrieve the stick constituents with their original color intact. The play dough had only been "stirred." If, however, the stretching and folding game goes on for some time and the bluish-gray outcome left to settle for a while, the different ingredients (oil and color) will start to smudge into each other and clean segregation becomes impossible, even as a thought experiment. The irreversibility of the process is a characteristic feature of true mixing.

Where, in an agitated reservoir like the ocean or the mantle, are the effective mixing zones located? Let us consider (Fig. 6.8) a part of the reservoir where displacement is horizontal, velocity $v_x(y)$ is then dependent only on the distance y to the bottom. Two points located initially, one at (x, y) and the other on the same vertical at $(x, y + \Delta y)$, move after a time δt over a slightly different horizontal distance. This difference δl is written:

$$\delta l = v_x(y + \Delta y)\delta t - v_x(y)\delta t = \frac{dv_x}{dy}\Delta y \delta t \qquad (6.26)$$

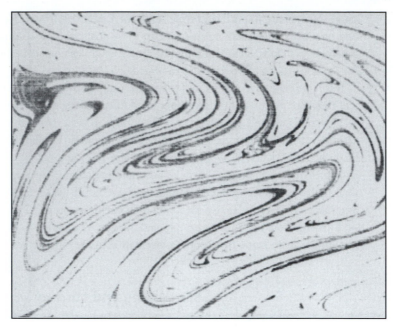

Figure 6.7   Chaotic structures of typical mixtures (after Ottino, 1989).

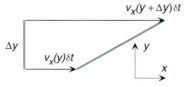

Figure 6.8   Stretching between two points initially Δy apart after time δt; $v_x(y)$ is the horizontal velocity in (x,y).

where the derivative of the velocity $v_x(y)$ is expressed in y. It can be seen that for a given initial vertical separation Δy, the term that governs the horizontal speed of stretching $\delta l/\delta t$ is the vertical variation $\mathrm{d}v_x/\mathrm{d}y$ in horizontal speed, which is known in mathematical jargon as the velocity gradient. The velocity gradient is a measure of the rate of deformation and therefore of mixing. Diverging or converging structures resembling upwelling and subduction, often referred to as poloidal, are two-dimensional and do not do a good job at mixing. Part of the reason is that the flow lines are closed loops, even in three dimensions. Addition of vortex-type zones, such as transform faults, which are often referred to as toroidal because of their doughnut shape, create chaotic flow with open-ended trajectories (tendrils) and provide efficient mixing. The generation of toroidal movements is associated with boundary layers (the coast line and the surface in the ocean, the lithosphere in the mantle) or with lateral changes of viscosity. Convection therefore plays a central role in reducing the size of heterogeneities. Let us think of a heterogeneity as a single stripe, e.g. a lithospheric plate, just introduced into an otherwise homogeneous mantle, or a core of water with distinctive salinity and temperature in the ocean. With time, convection will stretch and fold the original stripe, just like a baker stretches and folds the dough before

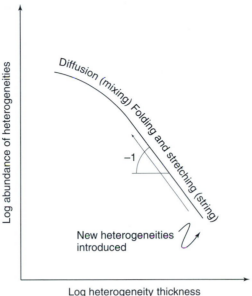

Log abundance of heterogeneities

Diffusion (mixing) Folding and stretching (string)

−1

New heterogeneities
introduced

Log heterogeneity thickness

Figure 6.9 Mixing and stirring. Large-scale heterogeneities (e.g. lithospheric plates) introduced into the mantle are stretched and folded by convection: more abundant and thinner stripes replace the thick original ones. This is the stirring domain and transport is, at least in theory, still reversible. The slope of the log–log plot is −1. When the stripes are reduced so that diffusion becomes a significant transport process, transport becomes irreversible and may be called mixing.

baking bread. Each event doubles the number of layers and reduces its thickness by a factor of two. Let us now think of not one but many lithospheric plates or water cores continuously produced over geological history and injected into the system: each event of stretching and folding removes a layer from its "class of thickness" and adds it to the class of half of the original value. We can again apply our birth-and-death theory to guess the fate of heterogeneities: created by continuous addition of large new objects (young plates), their characteristic sizes trickle down the length scale upon the effect of mantle convection (Fig. 6.9). This is the essence of Batchelor's (1959) theory of mixing. Finally, when the thickness of the stretched and folded layers is reduced to very small length scales (typically 1–10 cm in the mantle), diffusion transport becomes the prevalent smoothing process. Small-scale heterogeneities disappear over time scales of 100–1000 Ma in the mantle and of 10–100 years in the ocean.

Because they are continuously recreated, heterogeneities are unlikely to be evenly distributed, whether stirring by convection is vigorous or not. The most probable outcome in a randomly mixed medium is not a pattern of chemical homogeneity. Let us illustrate this point through a naive comparison. A handful of rice thrown on a checkered table will never produce a regular distribution of grains over the surface. Instead, the grains will appear "randomly" spread (Fig. 6.10). By randomly, we mean that any small square is as likely to receive a rice grain as another. The number of grains one counts on any patch delimited on the table would be found nearly proportional to the surface area of the patch.

Figure 6.10 One thousand grains of rice randomly thrown on a square table. The distribution is not regular, as each small square received a variable number of grains (2–20), but is perceived as random: a grain lands anywhere on the table with equal probability. The number of grains in any patch drawn on the table is proportional to the surface area of the patch (Poisson distribution). Likewise, the mantle is well mixed when the distribution of heterogeneities achieves maximum randomness. A regular distribution of heterogeneities is totally unlikely. Complex patterns of isotopic and chemical heterogeneities are no indication that the mantle is not well stirred.

One more time, we meet our old friend the Poisson distribution. As long as the mutual distance between chemically or isotopically identifiable heterogeneities exceeds the mean diffusion length, vigorous mantle mixing will tend to produce a distribution of chemical and isotopic anomalies that resembles a Poisson distribution. Complex heterogeneity patterns in the mantle do not necessarily indicate imperfect mantle or ocean mixing but are a straightforward consequence of random redistribution of material by mantle convection.

Exercises

1. What is the average time spent (1) by water in the ocean before being renewed by the rivers, and (2) by material in the mantle before being extracted into the oceanic lithosphere? Use the data of Appendix F and neglect density differences.

2. Why can oxygen isotope abundances in carbonate sections and ice cores change abruptly over time scales much sorter than the residence time of water in the ocean (e.g. during the Younger Dryas)? Why do you not expect such abrupt changes to be observed for $^{87}Sr/^{86}Sr$ records?

3. Using the data of Appendix A and F and assuming a steady state, calculate the residence times of the following elements in the ocean: Br, Rb, Mg, Fe, Sr, and Pb. Which elements do you expect to be homogeneously distributed across the ocean? Which elements should show regional or vertical variations? Do you expect the fluctuations of the seawater $^{87}Sr/^{86}Sr$ ratio to be modulated by glacial–interglacial cycles? Why?

4. The composition of the mean mantle can be calculated by removing the amount of elements hosted in continental crust (see exercise in Chapter 1). Using the data of Table 1.4 for the composition of MORB and the appropriate data from Appendix F, calculate the mean residence times of K, Sr, Zr, La, Yb, Th, and U in the mantle before they are extracted into the oceanic crust. Discuss mantle homogeneity vs. heterogeneity for these elements.

5. The diameter of a nearly circular lagoon on a Pacific atoll is 3.5 km and its mean depth 500 m. Water flows through the inlet at a rate $Q = 10^8$ m^3 y^{-1}.
   (i) What is the residence time of water in the lagoon?
   (ii) The lagoon is accidentally contaminated by strontium 90, which has a half-life of 29.1 y. If sedimentation could be ignored, what would be the residence time of this nuclide in the water of the lagoon?
   (iii) Reef growth leaves carbonated sediment on the floor of the lagoon. We assume a sediment density of 2000 kg m^{-3} and a sedimentation rate v of 1 mm y^{-1}. Calculate the sediment flux P in kg y^{-1}. Strontium is scavenged by the sediment with a carbonate–seawater partition coefficient D of 25 m^3 kg^{-1}.
   (iv) What is the residence time of ^{90}Sr in the lagoon in the presence of sedimentation?
   (v) Give an alternative theory in which you replace the residence times by probabilities.

6. Show that, if Sr concentrations do not change very significantly through a sequence of volcanic eruptions, the resorption of a pulse in the $^{87}Sr/^{86}Sr$ ratio can be used to estimate the residence of Sr in the magma chamber (discuss the potential importance of plagioclase on the liquidus). Show that, if the eruption rate is known, the volume of the magma chamber can be calculated. In 1880, an unusual change of $^{87}Sr/^{86}Sr$ of a Hawaiian volcano was resorbed in about 30 years. What was the approximate volume of the magma chamber if the eruption rate was 0.05 km^3 y^{-1}?

7. Strontium has a residence time in the ocean of about 4 My. What is the proportion of Sr atoms that have been in the ocean for more than 20 My? For less than 100 ky?

8. A global hydrogen cycle: we assume that H and D are distributed among two reservoirs, a deep reservoir (the mantle) and a shallow reservoir (the hydrosphere). Use equations (6.19) and (6.20) at steady state and the assumption that there is no residual water left upon melting of the mantle at mid-ocean ridges to demonstrate how the mantle δD ($\approx -85‰$) relates to the D/H fractionation coefficient at subduction zones.

9. If the mean residence time of Nd in the mantle is 6 Ga, what is the proportion of Nd atoms in the mantle that have never been extracted into the ridges?

From reservoir	Masses	To atmosphere	To surface ocean	To deep ocean	To soil	To plants
Table 6.1 A carbon cycle. Masses of C in Gt, fluxes in Gt a^{-1} (Houghton, 2003)						
Atmosphere	780	–	92	0	0	120
Surface ocean	728	90	–	44	0	0
Deep ocean	37 275	0	42	–	0	0
Soil	1500	58	0	0	–	0
Vegetation	550	59	0	0	60	–

10. The two-box ocean model of Broecker and Peng (1982); see Fig. 7.16: the ocean is divided into two boxes, the surface ocean and the deep ocean, separated by the thermocline. Modify equations (6.15) and (6.16) to take into account (1) the input from river flux, and (2) the sedimentation of particles formed in the surface ocean: a fraction of these particles are redissolved below the thermocline, the rest is rapidly exported into the sediments with no re-equilibration with deep water.

11. Another two-box ocean model: again, the ocean is separated into two boxes, but this time horizontally, an Atlantic basin and a Pacific basin. Modify equations (6.15) and (6.16) to take into account (1) the input from river flux into the Atlantic only (a good approximation!), and (2) sedimentation in each basin.

12. Discuss the oceanic cycle of Sr and its isotopic variations. Include a river flux with radiogenic Sr, carbonate precipitation, and exchange of seawater with unradiogenic basalt in ridge-crest hydrothermal systems.

13. A simplified carbon cycle: draw the C cycle with boxes and arrows using the data from Table 6.1. Calculate the carbon residence times in each reservoir. Discuss the pathway of anthropogenic addition of 6.3 Gt a^{-1} of carbon from fossil fuel.

14. Write the equations that were used to draw Fig. 6.5.

15. Let us define a series of one-dimensional "cells" or bins between 0 and 1 (e.g. 0–0.05, 0.05–0.0.10, 0.10–0.15, etc). Use a generator of random deviates from a uniform distribution (e.g. the function RAND in Excel) to produce n pairs (e.g. start with $n = 100$) of values. Each sample in a pair can be labelled "Rb" and "Sr." Sort the pairs, then bin them. The number of values in each bin is the "concentration" of Rb and Sr in each particular cell, and the ratio of these numbers is Rb/Sr. Build the histograms of Rb, Sr, and Rb/Sr. Do the exercise again with a different value of n (e.g. $n = 500$). How do you think the histograms will look when $n \rightarrow \infty$? Discuss different applications to the mantle and the ocean.

16. For readers more motivated by calculations: let us consider a section of the ocean or the mantle as a two-dimensional enclosure $x = [0–1]$, $y = [0–1]$ and the position-dependent velocity field (v_x, v_y) at steady state:

$$v_x(x, y) = \frac{dx}{dt} = \cos(\pi y [t]) \sin(\pi x [t])$$

$$v_y(x, y) = \frac{dy}{dt} = \sin(\pi y [t]) \cos(\pi x [t])$$

so that the x-velocity is zero along the upper and lower edge and the y-velocity is zero along the left and right boundaries. Let us take two points initially positioned at $x_1 = 0.1$, $y_1 = 0.1$ and $x_2 = 0.1$, $y_2 = 0.2$, respectively. How does the distance between these two points change through time ($t = 0.1, 0.2$, etc)? (Hint: remember that x and y are time-independent to infer the position at $t + \Delta t$ from the position at t and proceed by small increments, e.g. $\Delta t = 0.01$; use a spreadsheet or any other software).

17. Discuss why a Poisson distribution of the length-scale of isotopic heterogeneities – and not a homogeneous composition – are indicative of full mixing. Why can we can we assign such a distribution to a "memoryless" process?

References

♠ Batchelor, G. K. (1959) Small-scale variation of convected quantities like temperature in a turbulent fluid, 1, General discussion and the case of small conductivity. *J. Fluid Mech.*, **5**, 113–133.

♠ Bolin, B. and Rodhe, H. (1973) A note on the concepts of age distribution and transit time in natural reservoirs. *Tellus*, **25**, 58–62.

♠ Broecker, W. S. and Peng, T. H. (1982) *Tracers in the Sea*. Palisades: Eldigio.

♠ Eriksson, E. (1971) Compartment models and reservoir theory. *Ann. Rev. Ecol. Syst.*, **2**, 67–84.

♠ Houghton, R. A. (2003) Revised estimates of the annual net flux of carbon to the atmosphere from changes in land use and land management 1850–2000. *Tellus*, **55B**, 378–390.

Lerman, A. (1988) *Geochemical Processes*. Malabar: Kreiger.

Garrels, R. M. and Mackenzie, F. T. (1971) Gregor's denudation of the continents. *Nature*, **231**, 382–383.

Ottino, J. M. (1989) *The Kinematics of Mixing: Stretching, Chaos, and Transport*. Cambridge: Cambridge University Press.

Vlastelic, I. Deniel, C., Bosq., C., *et al.* (2008) Pb isotope geochemistry of Piton de la Fournaise historical volcanism, 1931–1986. *J. Volc. Geotherm. Res.*, **29**, in press.

7 The chemistry of natural waters

The external aspects of geochemical cycles, the phenomena that occur at relatively low temperatures (typically from 0 to $+30\,°C$) in the ocean, the atmosphere, and in rivers, are largely governed by chemical equilibria in solution or at the water–mineral interface. The cycles themselves imply transfer controlled primarily by water–rock interaction (erosion, sedimentation, hydrothermal reactions) and by biological activity. A central role is played by the carbonate system. We will apply these concepts to the geochemistry of erosion and of the ocean, with a discussion of the impact of these cycles on climates in particular.

7.1 Basic concepts

A few important concepts that are part of college chemistry are required.

1. Acidity is the concentration $[H^+]$ (mol kg^{-1}) of protons in a solution. The exact form, H^+ or H_3O^+, in which these protons occur is of little significance. A scale of acidity is defined by the potential pH of the protons in the solution, such that pH $= -\log [H^+]$. At $25\,°C$, pure water has a pH of 7. A lower pH indicates an acidic solution and a higher pH a basic solution.

2. Ion behavior is dictated by the dissociation of acids and bases. In an acid–base reaction, the acid is the proton donor and the base is the acceptor. A strong acid such as HCl or a strong base such as NaOH become completely dissociated to produce Cl$^-$ and Na$^+$ ions, which behave essentially like inert species and are of relevance only in terms of charge. Weaker acids become partly dissociated by releasing one, two, or possibly more

protons. In this way, carbonic acid H_2CO_3 dissociates into a bicarbonate ion HCO_3^- and then into a carbonate ion CO_3^{2-} by losing first one and then two protons. The extent of dissociation is given by the mass action law:

$$\frac{[HCO_3^-][H^+]}{[H_2CO_3]} = K_1 \qquad (7.1)$$

$$\frac{[CO_3^{2-}][H^+]}{[HCO_3^-]} = K_2 \qquad (7.2)$$

Both constants of the first (K_1) and second (K_2) dissociation vary with temperature and, to a lesser extent, with salinity. The notation $pK = -\log K$ is commonly used. A special case of dissociation is that of water by the reaction:

$$H_2O \Leftrightarrow H^+ + OH^- \qquad (7.3)$$

for which the dissociation constant is:

$$\frac{[H^+][OH^-]}{[H_2O]} = K_{H_2O} \qquad (7.4)$$

In the case where the solution is dilute, the $[H_2O]$ concentration of water in the solution is essentially constant and this relation can be re-written in its more familiar form:

$$[H^+][OH^-] = K_w = 10^{-14} \qquad (7.5)$$

3. Ion complexation is the association of ions carrying charges of opposite sign. In the ocean, for instance, the copper ion Cu^{2+} may be surrounded by different anions OH^-, Cl^-, HCO_3^-, which form different species of copper, but also by humic acids from the soil. In this sense, H_2CO_3 and HCO_3^- may be viewed as carbonate complexes of the proton. Complexation also obeys the mass action law with its successive constants.

4. Redox reactions relate to electron exchange: a reductant gives up electrons to an oxidant. Oxidation of Fe^{2+} into Fe^{3+} is a common result of electron acceptance by oxygen atoms:

$$Fe^{3+} + e^- \Leftrightarrow Fe^{2+} \qquad (7.6)$$

Here Fe^{3+} is the electron acceptor and Fe^{2+} the donor. Although neither free electrons nor free protons exist in solutions, we can still assign to them a share of the energy system. The ΔG of redox reactions is not really in use and it is customary to refer to an electron activity, actually to the negative decimal logarithm of this activity as pe. Alternatively, the electron activity is measured by an electrode potential E_H in volts, measured relative to a reference electrode with pe and E_H related by the relationship $E_H = 0.059$ pe at 25 °C. Oxidation of organic carbon (coal, petroleum, bitumen) by atmospheric oxygen is among the most commonly employed of artificial energy-producing processes, but it also occurs naturally when sediments with a high detrital carbon content are eroded:

$$\begin{array}{ccccc} C & + & O_2 & \Leftrightarrow & CO_2 \\ \text{sediment} & & \text{air} & & \text{atmosphere} \end{array} \tag{7.7}$$

5. Atmospheric gases are scarcely soluble in water, except for CO_2. The solubility of gases is described by Henry's law, which establishes a simple proportionality between the partial pressure P_i of the gas i above the solution and its concentration in the solution. For CO_2, we write:

$$[H_2CO_3] = k_{CO_2} P_{CO_2} \tag{7.8}$$

where k_{CO_2} is the coefficient of solubility. The concentration of the H_2CO_3 species in natural waters is very low and the species actually present is dissolved CO_2, but for all practical purposes, the present notation is sufficient.

6. The solubility of solids precipitating from solutions is expressed using another coefficient of the mass action law, the product of solubility K_s. For carbonate precipitation $Ca^{2+} + CO_3^{2-} \Leftrightarrow CaCO_3$, the saturation condition is written:

$$\left[Ca^{2+}\right]\left[CO_3^{2-}\right] = K_s \tag{7.9}$$

7. The condition of electrical neutrality: this condition is normally written by calculating the charge balance of fully dissociated species Alk, which is known as alkalinity (not to be confused with basicity, which characterizes a solution with pH > 7):

$$Alk = \left[Na^+\right] + \left[K^+\right] + 2\left[Ca^{2+}\right] + 2\left[Mg^{2+}\right] + \cdots$$
$$- \left[Cl^-\right] - \left[NO_3^-\right] - 2\left[SO_4^{2-}\right] - \cdots \tag{7.10}$$

But since the neutrality of the solution must be maintained, we can also write:

$$Alk \approx \left[OH^-\right] - \left[H^+\right] + \left[HCO_3^-\right] + 2\left[CO_3^{2-}\right] \tag{7.11}$$

in which the minor borates and phosphates have been omitted. Alkalinity being a concentration of electrical charges, it is expressed in equivalent per kilogram (eq kg^{-1}). This is a measure of the neutralizing power of a solution: by adding HCl to a highly alkaline solution, the solution bubbles intensely as the carbonate ions are driven off and replaced by Cl^- ions. Notice that while the pH of a solution can vary with temperature or CO_2 pressure, alkalinity does not change and for this reason it is said to be conservative. Alkalinity is conservative like $[Na^+]$ or $[Cl^-]$ but unlike $[HCO_3^-]$ or $[OH^-]$. The sum $[OH^-] - [H^+]$, known as caustic alkalinity, is normally negligible compared with carbonate alkalinity $[HCO_3^-] + 2[CO_3^{2-}]$.

8. All the thermodynamic constants depend on temperature and, to a lesser extent, on pressure. Those analogous to an equilibrium coefficient may be expressed by a similar law to (2.11).

7.2 Dominance diagrams

The purpose of this section is to introduce some graphic methods used to assess which species dominate solutions in conditions of given acidity (pH) and oxidation state (pe). Let us consider the following acid–base equilibrium:

$$HSO_4^- \Leftrightarrow SO_4^{2-} + H^+ \quad (\log K = -2.0) \tag{7.12}$$

We follow the conventional presentation of the reaction as a dissociation. We can write the constant of the equilibrium as:

$$\frac{\left[SO_4^{2-}\right]\left[H^+\right]}{\left[HSO_4^-\right]} = K \tag{7.13}$$

Taking the logarithm, we obtain:

$$pH = -\log K + \log \frac{\left[SO_4^{2-}\right]}{\left[HSO_4^-\right]} \tag{7.14}$$

Clearly, for $pH > pK = -\log K$, we have $\left[SO_4^{2-}\right] > \left[HSO_4^-\right]$, while the opposite is true for $pH < pK$.

The same principle holds for redox reactions, for example:

$$Fe^{3+} + e^- \Leftrightarrow Fe^{2+} \quad (\log K = +13.0 \quad \text{or} \quad E_H = 0.77 \text{ V}) \tag{7.15}$$

(note that this time we consider that the electron combines with the oxidant). Writing the equilibrium constant and taking the logarithm, we obtain:

$$pe = \log K + \log \frac{Fe^{3+}}{Fe^{2+}} \tag{7.16}$$

(note the ratio in the order of oxidant/reductant on the right-hand side). We note $pe_0 = \log K = 13$. When $pe > pe_0$ (or $E_H > 0.77$ V), we conclude that $Fe^{3+} > Fe^{2+}$, and $Fe^{3+} < Fe^{2+}$ otherwise.

Figure 7.1 shows these two examples of dominance diagrams. When the two types of dominance diagrams are combined, the useful pe–pH diagram (or, equivalently, the E_H–pH diagram) is obtained. This is normally complicated by the fact that oxo-anions, such as SO_4^{2-} or NO_3^-, require a proton to achieve reduction, which makes redox reactions and the redox dominance diagrams pH-dependent as in:

$$SO_4^{2-} + 9H^+ + 8e^- \Leftrightarrow HS^- + 4H_2O \quad (\log K = 34) \tag{7.17}$$

The geochemically important example of the sulfur system, with H_2S, HS^-, S_2 (solid sulfur), SO_4^{2-}, and HSO_4^- as the dominant species, is shown in Fig. 7.2.

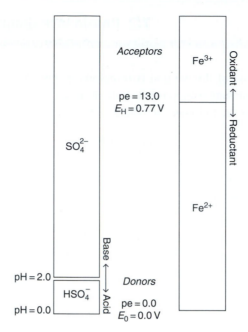

Figure 7.1 Dominance diagram for an acid–base reaction (left) and a redox reaction (right). The acid is the proton donor and the reductant is the electron donor, while the base and the oxidant are their respective acceptors. Note the equivalent potential scale used for redox reactions, which is no longer used for acid–base reactions. The boundaries between the dominance domains are the pK, which for the redox reaction is referred to as pe$_0$. The coupling of electron and proton exchange shifts both boundaries in a linear way.

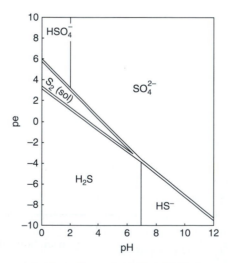

Figure 7.2 A pe–pH dominance diagram for the sulfur system. The double lines represent a change in the oxidation state of the elements. Note that some boundaries are parallel to the y-axis, such as that separating SO_4^{2-} and HSO_4^-, because they only involve proton exchange. The boundary between SO_4^{2-} and HS^- is a coupled electron and proton exchange and therefore shows as segment of straight–line with a slope equal to the proton/electron ratio (see Exercise 1). To obtain the usual E_H–pH diagrams, use $E_H = 0.059$ pe at 25 °C.

7.3 Speciation in solutions

An important problem in the geochemistry of solutions is the distribution of the chemical components among the various species. A simple example is that of the distribution of carbonates as H_2CO_3, HCO_3^-, and CO_3^{2-}. Calculating the speciation of a solution of known composition involves evaluating the abundance of different chemical species under prescribed conditions of temperature, pressure, pH, and other factors. Let us make a very simple simulation of the deep ocean isolated from the atmosphere after the formation of bottom waters and which receives carbonate tests of sedimented foraminifera. We introduce m_{CO_2} moles of CO_2 per liter into pure water, and then a very small known quantity of calcite, such that the molarity of calcite in water is m_{CaCO_3}, and leave the calcite to dissolve. Let us now inquire into the abundance of the species present in the solution. The CO_2 acidifies the water allowing the calcite to be dissolved.

We choose two components, Ca^{2+} and CO_2. The six possible species are OH^-, H^+, Ca^{2+}, and the carbonates H_2CO_3, HCO_3^-, and CO_3^{2-}. The concentration of H_2O is high enough compared with the rest of the dissolved species to be invariant in the process, and so we can eliminate it both as a species and a component. If we wish to determine the abundances of the six species, we have two mass-balance equations, one for the sum of CO_2:

$$\left[H_2CO_3\right] + \left[HCO_3^-\right] + \left[CO_3^{2-}\right] = m_{CO_2} + m_{CaCO_3} \tag{7.18}$$

and the other for the sum of calcium ions:

$$\left[Ca^{2+}\right] = m_{CaCO_3} \tag{7.19}$$

and one equation for electrical neutrality:

$$\left[HCO_3^-\right] + 2\left[CO_3^{2-}\right] + \left[OH^-\right] = \left[H^+\right] + 2\left[Ca^{2+}\right] \tag{7.20}$$

We therefore need another three equations (given in the form of mass action laws) to determine the six unknowns. We choose, of course, (7.1) and (7.2) for carbonic acid dissociation, plus (7.5) for water dissociation.

This seemingly natural choice is in fact arbitrary. There is nothing to prevent us from replacing one of these equations by a combination of them. Thus, either of the two equations for carbonic acid dissociation could be replaced by the product of the two:

$$\frac{\left[CO_3^{2-}\right]\left[H^+\right]^2}{\left[H_2CO_3\right]} = K_1 K_2 \tag{7.21}$$

It can be seen that, in the general case, speciation calculations involve resolving systems that include non-linear equations and therefore call on advanced numerical techniques. There are many software packages that perform these calculations with a high degree of sophistication allowing for many complexes and the presence of solid or gaseous phases. Such software is economically significant for the exploration and management of water, oil, and geothermal resources. It can also be used for understanding weathering and the genesis of mineral deposits.

7.4 Water–solid reactions

Water has the effect on the minerals of igneous or metamorphic rocks of putting the more soluble cations into solution and producing residual clay minerals. A notable exception is quartz. A first type of reaction controlling interaction between the lithosphere and hydrosphere essentially involves reactions of proton and cation exchange such as:

$$2NaAlSi_3O_8 + \quad 2H^+ \quad + H_2O \Leftrightarrow$$
$$\text{(albite)} \qquad \text{(solution)}$$

$$Al_2Si_2O_5(OH)_4 + 4SiO_2 + \quad 2Na^+$$
$$\text{(kaolinite)} \qquad \text{(silica)} \quad \text{(solution)} \tag{7.22}$$

Since albite is a very common constituent of crustal rocks, this reaction can also be seen as limiting the solubility of kaolinite, one of the most important clay minerals in soils. In the absence of feldspar, kaolinite solubility is controlled by a different proton-exchange reaction:

$$\tfrac{1}{2}Al_2Si_2O_5(OH)_4 + \quad \tfrac{7}{2}H_2O \quad \Leftrightarrow \quad H_4SiO_4 \quad + Al(OH)_4^- + \quad H^+$$
$$\text{(kaolinite)} \qquad \text{(solution)} \quad \text{(solution)} \quad \text{(solution)} \quad \text{(solution)} \tag{7.23}$$

The complexation of ions in solution makes the calculation of mineral solubility more complicated. Ferric iron hydroxide $Fe(OH)_3$, one of the important minerals of soils, is a good example. Ferric iron speciation in near-neutral fresh water (chlorine would produce strong complexes) is controlled by the species Fe^{3+}, $Fe(OH)_2^+$, and $Fe(OH)_4^-$. The concentration of the dissolved species $Fe(OH)_3$ is essentially zero. We can write the following dissolution reactions for the $Fe(OH)_3$ mineral:

$$Fe(OH)_3(s) + 3H^+ \Leftrightarrow Fe^{3+} + 3H_2O \tag{7.24}$$

$$Fe(OH)_3(s) + H^+ \Leftrightarrow Fe(OH)_2^+ + H_2O \tag{7.25}$$

$$Fe(OH)_3(s) + H_2O \Leftrightarrow Fe(OH)_4^- + H^+ \tag{7.26}$$

where $Fe(OH)_3(s)$ stands for solid hydroxide and the other species are in solution. Using the equilibrium constants suitable for these two equations and considering that the concentration of a pure solid phase is unity, we obtain:

$$\frac{\left[Fe^{3+}\right]}{\left[H^+\right]^3} = 10^{3.2} \tag{7.27}$$

$$\frac{\left[Fe(OH)_2^+\right]}{\left[H^+\right]} = 10^{-2.5} \tag{7.28}$$

$$\left[Fe(OH)_4^-\right]\left[H^+\right] = 10^{-18.4} \tag{7.29}$$

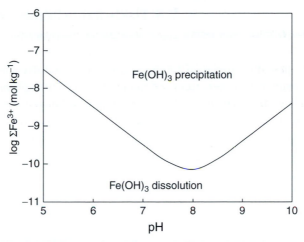

Figure 7.3 Ferric iron hydroxide saturation in pure water as a function of pH. The minimum in the solubility curve results from the dominance of different hydroxide complexes at different pH values.

which enables us to calculate the total ferric content ΣFe^{3+} as:

$$\Sigma Fe^{3+} = \left[Fe^{3+} \right] + \left[Fe\,(OH)_2^+ \right] + \left[Fe\,(OH)_4^- \right]$$

$$= 1.6 \times 10^3 \left[H^+ \right]^3 + 3.2 \times 10^{-3} \left[H^+ \right] + \frac{4.0 \times 10^{-19}}{\left[H^+ \right]}$$

The results are plotted in Fig. 7.3. Ferric iron solubility is extremely small and, because of variable complexation by the different hydroxides, ΣFe^{3+} goes through a minimum for pH $\approx$ 8.

In other reactions, CO_2 is consumed by silicate dissolution, such as:

$$
\begin{array}{cccc}
Mg_2Si_2O_6 & + & 4CO_2 & + & 2H_2O \\
\text{(pyroxene)} & & \text{(atmosphere)} & &
\end{array}
$$

$$
\begin{array}{cccccc}
\Leftrightarrow & 2SiO_2 & + & 4HCO_3^- & + & 2Mg^{2+} \\
& \text{(silica)} & & \text{(solution)} & & \text{(solution)}
\end{array}
\tag{7.30}
$$

The CO_2-consumption reactions are not fundamentally different from the proton-exchange reactions, but show the overlapping relationship between erosion, the carbon cycle – particularly marine carbonates and the CO_2 pressure of the atmosphere – and therefore, by way of the greenhouse effect, of climate. Proton-exchange reactions are very sensitive to the pH of solutions, since, allowing for a unit concentration of solids, the mass action law, for example for (7.22), requires a constant $[Na^+]/[H^+]$ ratio in the solution. Both acid rain, charged with sulfuric acid by oxidation of sulfur compounds at altitude, and groundwater, charged with humic acids, will attack rock since their protons will be able to displace the Na^+, Ca^{2+}, or other ions of the minerals. Complexing anions (OH^-, Cl^-, CO_3^{2-}, SO_4^{2-}, PO_4^{3-}, humic, and fulvic acids) also play a critical role in mineral solubility and transport. These reactions are greatly dependent on temperature. In contrast to limestone dissolution that recycles the fossil alkalinity produced since the origin of the Earth, the chemical erosion of silicates (Eqs. (7.22) and (7.30)) creates fresh alkalinity. Carbon dioxide outgassed from the mantle will eventually be stored in newly formed sedimentary carbonates.

7.5 Electrolyte chemistry

Most inorganic species in solutions are ionized and develop a local electrostatic potential, which creates new forms of energy storage. These phenomena add a certain level of complexity to the theoretical understanding of solution chemistry and force us to use new parameters known as "activities" instead of concentrations. It also accounts for the remarkable behavior of particles in natural waters such as gels and flocculation. Let us start by recalling some elementary concepts of electricity. Imagine that we wire each of two metal plates to the positive or negative outlets of a battery. This action drives electric charges of opposite signs at the surface of each plate. We hold the two plates apart by two springs attached to the wall and measure the forces which tend to bring the two plates together. The field between the two plates is described by the standard laws of electrostatics (Coulomb attraction). Let us now introduce a sheet of mica in between the two plates. The effect is to reduce the force that pulls the plates together. Mica is a dielectric medium: the electrostatic field created by the plates induces ion pairs in the mica (dipoles) to orient themselves so as to create a secondary field opposing the primary field of the plates. The ratio of the force in vacuum to the force with the mica sheet present is known as the dielectric constant ε.

Pure water is a strong dielectric. The H_2O molecules are polar: because of the angle maintained between the two OH bonds, the hydrogen atoms pull the covalent electron they each share with the oxygen. The water molecules therefore show a positively charged apex and orient themselves against the ambient field, e.g. the field created by surfaces. All of these dipoles reduce the distance at which charges are "felt" in the solution: pure water therefore screens the mutual attraction or repulsion of neighboring ions. Because of thermal agitation, the dielectric constant of water decreases when temperature increases. Ions in solution also interfere with the dielectric properties of the solvent: several H_2O molecules attach themselves to each ion to form what is known as a solvation shell. These water molecules are no longer available to counteract the ambient electrostatic field. Figure 7.4 shows the example of the hydrated Cr^{6+} ion with its shell of six water molecules. Adding salt to a solution therefore decreases its dielectric constant ε. The mean distance over which

$Cr(H_2O)_6^{6+}$

Figure 7.4   The hydrated octahedron $Cr(H_2O)_6^{6+}$. The strong field created by the Cr^{6+} cation attracts water molecules which are no longer available to contribute to the dielectric effect of water.

the electrostatic field is transmitted in a medium is known as the Debye length λ_D and, in an electrolyte solution, can be shown to be:

$$\lambda_D = C\sqrt{\frac{\varepsilon RT}{e^2 I}} \tag{7.31}$$

with e being the charge of the electron and C a constant. In this equation, I is the ionic strength $\frac{1}{2}\sum c_i z_i^2$ of the solution, where c_i and z_i stand for the concentration and charge, respectively, of ion i. The potential at the distance r of an ion considered as a point charge will be:

$$V(r) = V(0)\,e^{-r/\lambda_D} \tag{7.32}$$

The energy stored in the shell around the ion is an integral part of the total energy budget and therefore affects the thermodynamic properties of ions in solutions with respect to uncharged solutes. This effect is described in chemistry textbooks as the Debye–Hückel theory.

Adding salt to a solution also has another effect, which we have already met: it reduces the Debye length and therefore the thickness of the charged fringe around ions and molecules and lets small particles approach each other at short distance and coalesce. This is know as flocculation and is responsible for the removal of iron and other elements from estuaries.

7.6 Biological activity

Biological activity has conditioned the chemistry of the ocean and the atmosphere since earliest times. It is characterized by exploitation of the environment to manufacture soft and hard body parts, and to maintain the metabolism of organisms. The primary plant producers extract carbon from the surrounding medium, atmospheric CO_2 in the case of land plants, and carbonates dissolved in seawater in the case of algae, and reduce it with solar energy (photosynthesis). The higher elements of the prey–predator chain, the food chain, recycle the primary reduced carbon. Organisms run on real reduced-carbon batteries. Respiration provides the oxygen for combustion and eliminates the CO_2 produced. While the reduced carbon, in the form of carbohydrates and lipids, acts as a repository of energy, it is also used to build soft parts, notably proteins. In addition to carbon, the production of organic matter requires nitrogen, phosphorus, and trace elements (particularly Fe, Mg, Zn), which are essential for manufacturing many vital enzymes.

For locomotion or defense, organisms form hard parts, based essentially on silica SiO_2 for diatoms, calcite $CaCO_3$ for foraminifera and invertebrates, and on phosphate for vertebrates. Carbonate precipitation consumes alkalinity. In the seas, calcite and silica are the main biogenic components of sediments and stand in contrast to the detrital material carried by rivers and the wind, and which is composed largely of clays, hydroxides, and quartz.

7.7 The carbonate system

By using methods from solution chemistry, it is possible to calculate the speciation of carbonates in water, for example the $\alpha_{HCO_3^-}$ fraction of carbonate dissolved as bicarbonate ions. This parameter is defined as:

$$\alpha_{HCO_3^-} = \frac{\left[HCO_3^-\right]}{\left[CO_3^{2-}\right] + \left[HCO_3^-\right] + [H_2CO_3]} \tag{7.33}$$

Dividing this equation by $[H_2CO_3]$, utilizing the two dissociation constants of (7.1 and 7.2), and multiplying by $[H^+]^2$ gives:

$$\alpha_{HCO_3^-} = \frac{K_1\left[H^+\right]}{K_1 K_2 + K_1\left[H^+\right] + \left[H^+\right]^2} \tag{7.34}$$

with similar expressions for $\alpha_{CO_3^{2-}}$ and $\alpha_{H_2CO_3}$. Carbonate speciation therefore only depends on $[H^+]$. If these expressions are plotted against pH, three dominance zones can be identified, separated by the pK_1 (6.4) and pK_2 (10.3) of carbonic acid (Fig. 7.5). At low pH values, the surplus of protons leaves all the carbonates as carbonic acid H_2CO_3, while at high pH the deficit of protons favors CO_3^{2-}; HCO_3^- is the dominant ion in the intermediate pH zone. Natural water is weakly acidic or weakly basic. Seawater, with a pH of 7.6–8.0, is largely dominated by HCO_3^- ions.

The carbonate system in a solution in contact with the atmosphere is controlled by the chemical variables P_{CO_2}, pH, alkalinity $Alk \approx \left[HCO_3^-\right] + 2\left[CO_3^{2-}\right]$, the sum of available carbonates $\Sigma CO_2 \approx \left[HCO_3^-\right] + \left[CO_3^{2-}\right]$ (omitting H_2CO_3 and atmospheric CO_2), and

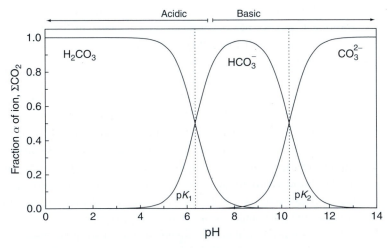

Figure 7.5 Speciation of dissolved carbonates as a function of the pH of the solution (see (7.34)). Notice that the predominance of each carbonate species changes when pH values equal to pK are crossed. Much natural water has a pH close to neutral and is therefore dominated by the bicarbonate ion HCO_3^-.

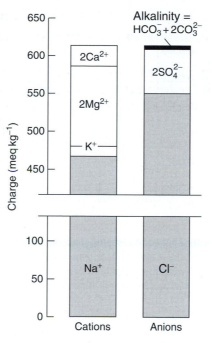

Figure 7.6   The composition of seawater in number of charges (equivalent) per unit weight. Alkalinity (2 meq kg^{-1} in seawater) is the deficit of charge carried by totally dissociated (therefore chemically inert) anions compared with the charge carried by totally dissociated cations. In a solution, this deficit, which is essentially balanced by carbonates, is constant. It is a measure of its capacity to neutralize strong acids (after Broecker, 1994).

the physical variables of temperature and, to a lesser extent, pressure. These variables are not independent and we seek to work with conservative variables, which exclude pH and carbonate concentrations. Combining the expressions for Alk and ΣCO_2 gives:

$$\left[CO_3^{2-}\right] = Alk - \Sigma CO_2$$

$$\left[HCO_3^-\right] = 2\Sigma CO_2 - Alk \qquad (7.35)$$

In seawater (Fig. 7.6), alkalinity varies from 2.0 to 2.4 meq kg^{-1} (the equivalent is a number of charges expressed in molar units). Some 90% of dissolved carbonate is in the form $[HCO_3^-]$. Alkalinity is much lower, or even negative, in rain water and river water. By combining these expressions with those describing the solubility of CO_2 and the dissociation of carbonic acid, we obtain the very important equation:

$$P_{CO_2} = \frac{K_2}{k_{CO_2} K_1} \frac{(2\Sigma CO_2 - Alk)^2}{(Alk - \Sigma CO_2)} \qquad (7.36)$$

This expression is plotted for the ocean–atmosphere system in Fig. 7.7, where CO_2 pressures are expressed in ppmv (parts per million volume) and it is taken that at 25 °C, k_{CO_2} is approximately equal to 0.029 moles per kg and per atm. The pH of seawater can

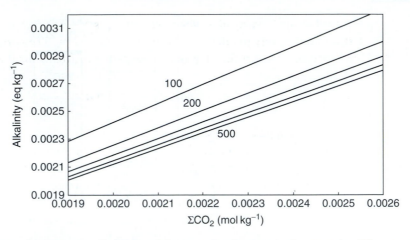

Figure 7.7 The relationship between alkalinity and the sum of carbonates in the ocean for different values of CO_2 content in the atmosphere, expressed in parts per million volume on the straight lines (Eq. (7.36)). The assumed equilibrium temperature is 15 °C. Today, the atmosphere contains about 350 ppmv CO_2. Ocean and atmosphere are close to equilibrium.

be calculated either in surface water at a known pressure P_{CO_2}, or in deep water for a fixed value of ΣCO_2, by introducing the values of $\left[HCO_3^-\right]$ and $\left[CO_3^{2-}\right]$ obtained through (7.35) into the dissociation equation (7.1 and 7.2).

Let us consider a different example relevant to our oceanic system in which the calcium concentration of the ocean is fixed by carbonate saturation and seawater maintained at a constant atmospheric pressure of CO_2. Equations (7.18) and (7.19) have to be replaced by two equations expressing saturation in calcite:

$$\left[Ca^{2+}\right]\left[CO_3^{2-}\right] = K_s \tag{7.37}$$

and CO_2 solubility:

$$[H_2CO_3] = k_{CO_2} P_{CO_2} \tag{7.38}$$

The total concentration ΣCO_2 of carbonate ions becomes:

$$\Sigma CO_2 = \frac{K_s}{[Ca^{2+}]}\left(1 + \frac{\left[H^+\right]}{K_2} + \frac{\left[H^+\right]^2}{K_1 K_2}\right) \tag{7.39}$$

while P_{CO_2} is given by:

$$P_{CO_2} = \frac{K_s}{K_1 K_2 k_{CO_2}} \frac{\left[H^+\right]^2}{[Ca^{2+}]} \tag{7.40}$$

The carbonate system therefore has two degrees of freedom which represent different geological conditions. Controls by suitable pairs of parameters, such as constant alkalinity and ΣCO_2 (deep ocean) or calcite saturation and P_{CO_2} (surface ocean) represent typical cases. Two of the parameters are imposed and all of the other variables follow from there.

Reconstructing the composition of the ancient oceans therefore comes down to finding geochemical tracers (proxies) for evaluating one or another of these parameters. Much hope is pinned on the capacity of isotopic concentrations of boron measured from the calcareous tests of foraminifera to reflect the pH of the early oceans. The boric acid of seawater is a weak acid and boron is found in two different states governed by the reaction:

$$H_3BO_3 + H_2O \Leftrightarrow B(OH)_4^- + H^+ \tag{7.41}$$

We note the apparent dissociation constant of H_3BO_3 as K_B ($pK_B = 8.75$) and the product of the equilibrium constant of this reaction by the $[H_2O]$ molarity of water, which is very close to 55.6 in all natural waters. The proportion of the two species clearly depends on the pH of the solution and we can formulate the fraction of boron stored for each of them, for example $\varphi_{H_3BO_3}$, in the usual way:

$$\varphi_{H_3BO_3} = \frac{[H_3BO_3]}{[H_3BO_3] + \left[B(OH)_4^-\right]} = \frac{\left[H^+\right]}{\left[H^+\right] + K_B} \tag{7.42}$$

Clearly, the sum of $\varphi_{H_3BO_3}$ and $\varphi_{B(OH)_4^-}$ is unity. Moreover, an isotopic equilibrium equation between the two boron carrier species can be written:

$$\frac{\left(^{11}B/^{10}B\right)_{H_3BO_3}}{\left(^{11}B/^{10}B\right)_{B(OH)_4^-}} = \alpha^B(T) \tag{7.43}$$

where $\alpha^B(T)$ is the isotopic fractionation coefficient between the two species of boron present in the solution. The residence time of boron in seawater being several million years, it can be considered that neither its content nor its overall isotopic composition change between glacial and interglacial periods, which alternate over a time scale of 100 000 years. The mass balance condition for the $^{11}B/^{10}B$ ratio can therefore be written as:

$$\left(\frac{^{11}B}{^{10}B}\right)_{ocean} = \varphi_{H_3BO_3} \left(\frac{^{11}B}{^{10}B}\right)_{H_3BO_3} + \varphi_{B(OH)_4^-} \left(\frac{^{11}B}{^{10}B}\right)_{B(OH)_4^-} \tag{7.44}$$

Only the charged borate $B(OH)_4^-$ seems to be incorporated in the calcite of foraminifera. The previous two equations can therefore be combined in the form:

$$\left(\frac{^{11}B}{^{10}B}\right)_{foram} = \left(\frac{^{11}B}{^{10}B}\right)_{B(OH)_4^-} = \frac{\left(^{11}B/^{10}B\right)_{ocean}}{\varphi_{H_3BO_3}\alpha^B(T) + \left(1 - \varphi_{H_3BO_3}\right)} \tag{7.45}$$

If we know the $^{11}B/^{10}B$ ratio of the ocean from measurement of present-day seawater, the growth temperature of foraminifera (given by oxygen isotopes), and isotopic fractionation $\alpha^B(T)$ (0.978 at 20 °C), we can deduce $\varphi_{H_3BO_3}$ and therefore, in principle, the pH of the oceans in which the foraminifera formed (Fig. 7.8). This method assumes that we have identified species for which isotopic fractionation of boron occurs reproducibly, which is a significant limitation for the time being.

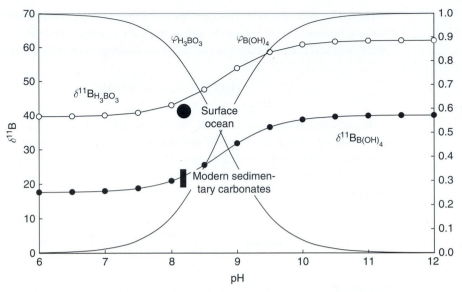

Figure 7.8 Isotopic composition of dissolved boron and boron in sedimentary carbonates as a function of the pH of seawater. The $\delta^{11}B$ values are defined with reference to a standard boric acid (SRM 951). The fractions φ and $\delta^{11}B$ refer to each species H_3BO_3 and $B(OH)_4^-$. There is an isotopic fractionation of ^{11}B of 22 per mil between the two species. The $\delta^{11}B$ of $B(OH)_4^-$ incorporated in the sedimentary carbonate is a measure of the pH of the ocean from which limestone forms.

7.8 Precipitation, rivers, weathering, and erosion

The hydrological cycle is well understood (Fig. 7.9). Evaporation caused by intense solar heating of the oceans between the tropics produces water vapor. When the hot, moist air rises, the water vapor condenses into fine droplets forming clouds, but only falls to the ground as rain when the droplets coalesce to attain sufficient size. The masses of hot air migrate toward the poles and as they cool precipitation progressively drains the moisture from the atmosphere. This can be seen from the measurement of isotopic concentrations of hydrogen and oxygen in precipitation, i.e. rain (meteoric) water and snow (Fig. 7.9). Because heavy isotopes concentrate preferentially in liquid water and ice, progressive condensation of atmospheric vapor produces a very clear poleward depletion of precipitation in the heavier isotopes of hydrogen and oxygen. A correlation ensues between δD (a measure of the $^2H/^1H$ ratio) and $\delta^{18}O$ of rain or snow, corresponding to this progressive depletion of deuterium (D or ^2H) and ^{18}O in accordance with the linear approximation of fractional condensation (see Chapter 3):

$$\frac{d\delta D}{d\delta^{18}O} = \frac{\alpha^D(T) - 1}{\alpha^{18O}(T) - 1} \tag{7.46}$$

where coefficients α measure the given isotope fractionation between liquid and vapor at temperature T of the clouds. Strictly, this expression should be corrected to allow for

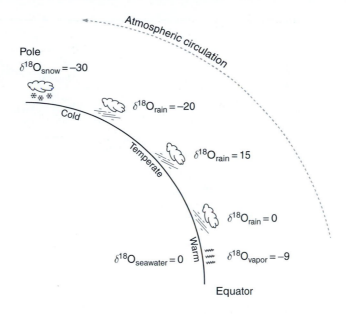

Figure 7.9 The atmospheric water cycle and the $\delta^{18}O$ (in ‰) of various types of precipitation. Vapor formed by evaporation of the warm surface ocean between the tropics migrates with the atmospheric circulation towards the poles. Incremental precipitation in rain at low latitude and in snow at high latitude progressively depletes atmospheric water vapor, rain, and snow in ^{18}O. A similar poleward depletion of the atmosphere takes place for deuterium with respect to hydrogen, thereby producing the relationship of Fig. 7.10.

kinetic effects during phase changes. This expression yields the equation of the Global Meteoric Water Line as $\delta D = 8\,\delta^{18}O + 10$ (Fig. 7.10). Rain waters from temperate regions and polar ice are depleted with respect to seawater in the light isotope by up to several percent for oxygen and up to tens of percent for hydrogen. We can infer from the Rayleigh distillation equation that the amount of precipitation decreases exponentially down the GMWL: the amount of water reaching the poles is a tiny fraction of the water lost in transit. This relationship does not pass through the composition of seawater because of kinetic fractionation during evaporation of warm surface water, while condensation in clouds takes place at much lower temperatures under near-equilibrium conditions. A somewhat similar distribution is found with altitude: as air rises in mountain areas, the various types of precipitation become depleted in D and ^{18}O.

This relationship implies that almost all underground water is recycled rain water. Geothermal springs lie on a horizontal line through the isotopic composition of the rainfall of the location. This indicates that the groundwater is rain water in which oxygen has exchanged its isotopes with the surrounding rock; as the rock is virtually devoid of hydrogen, δD is left unchanged. Therefore, there is no such a thing as "juvenile" spring water, i.e. water that comes from the depths of the Earth without ever having seen the surface. The same is true of hydrothermal springs.

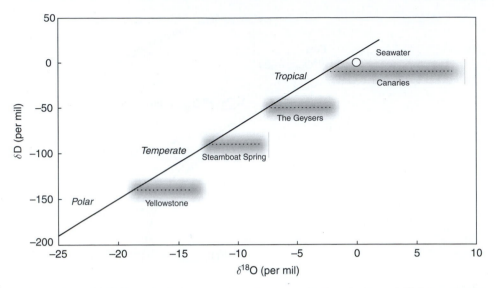

Relationship between the isotopic concentration of hydrogen (δD) and oxygen ($\delta^{18}O$) in meteoric waters (solid line). Liquid precipitation depletes the atmospheric water vapor in heavy isotopes as the clouds move poleward. Most groundwaters plot along the meteoric line ($\delta D = 8\delta^{18}O + 10$) demonstrating their meteoric origin and negligible interaction with minerals in the first kilometer of the crust. Many geothermal springs (dotted lines) have the same δD as local rain water. As there is practically no hydrogen in the crust, this indicates their meteoric origin, with the $\delta^{18}O$ variations reflecting exchange with crustal mineral oxygen at temperatures of about 75–350 °C.

The mineral contribution of sea spray to rain water smoothly decreases with distance from the coast. This gradient involves Na^+ and Cl^- ions above all, the readily identifiable so-called cyclical ions. The remainder of the water infiltrates or runs off, contributing to erosion.

Chemical erosion (weathering) can be formulated as a set of dissolution reactions. Silicates formed at high temperature in igneous or metamorphic environments (olivine, pyroxenes, feldspars, micas) are metastable in the low-temperature conditions of the surface. They react with precipitation to form clay minerals, that are the stable form of silicates, hydroxides of the most insoluble ions in their oxidized form (Fe^{3+}, Al^{3+}), while the most soluble ions (Na^+, K^+, Ca^{2+}, Mg^{2+}) are released in the run-off and river water. Silica is scarcely soluble and quartz from any origin is largely left untouched by erosion, but the level of $[SiO_2]$ in rivers is comparable to the concentration of other major elements.

It is kinetic barriers that preserve high-temperature mineral assemblages such as those of granite and gneiss over very long geological times at the Earth's surface, whereas equilibrium thermodynamics implies that feldspars and many other silicates should have turned into clay. Erosion studies must therefore pay due attention to dissolution rates. Most erosion processes are dissolution/reprecipitation reactions resembling the albite destruction described by (7.22). Normally, dissolution is the rate-limiting step, and the nature of dissolved cations is not critical in controlling the reaction rate, with the notable exception

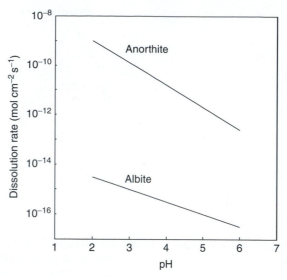

Figure 7.11 Dependence of albite and anorthite weathering rates on pH. Acid waters dissolve feldspars faster than neutral waters. These rates typically increase by a factor of ten when temperature increases by 25 °C.

of H^+. Quite commonly, rain-water pH values fall in a range of 4–6. Acid waters have the greatest erosion potential because abundant H^+ displaces the major soluble ions (Na^+, K^+, Ca^{2+}, Mg^{2+}) constituting the rock-forming minerals. Once normalized to surface area and after establishment of steady state, most mineral dissolution rates R may adequately be described by the equation:

$$\ln R = \ln R_0 + n \, \text{pH} \tag{7.47}$$

where R_0 and n are constant for a specified mineral. Temperature dependence of the dissolution rate is very critical: warm climates promote chemical erosion. This dependence takes the usual form:

$$\ln R = A - \frac{E}{RT} \tag{7.48}$$

where A is a constant and E the activation energy of this particular reaction. Values of n between 0 and -1 are common, while E usually falls between the activation energy of ionic diffusion in solution and the energy required to break the silicate bond. Weathering rates typically increase by a factor of ten when temperature increases by 25 °C. The pH-dependence property of the weathering rates is illustrated for feldspars in Fig. 7.11.

The major features of river chemistry may be summarized with a small number of parameters. After subtraction of cyclic ions, the remainder of the dissolved load of fresh water can be divided into carbonate alkalinity (from dissolution of silicates by the reactions described above or from redissolution of limestone), sulfate and chloride from dissolution of evaporites, and silica from dissolution of silicates (Fig. 7.12). These quantities vary greatly with the site where water erodes the soil. In addition to the dissolved load, rivers transport abundant mineral (clay and iron hydroxides) and organic colloids.

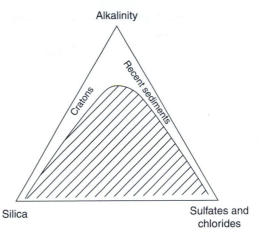

Figure 7.12 Compositional distribution of the dissolved charge of rivers among its three main components: silica, alkalinity (mainly carbonates), and chlorides and sulfates derived mostly from evaporites. These compositions are closely controlled by the nature of the substrate (granitic cratons, sedimentary platform). The hatched zone contains no data (after Stallard and Edmond, 1983).

7.9 Elements of marine chemistry

Rivers carry their dissolved load to the ocean. It is practical to group all of this added mineralization under the heading of erosion-related alkalinity flux. Many elements do not pass the filter of estuaries; as discussed above, at this point, fresh water and salt water mix, causing a substantial decrease in the dielectric constant of the water. Massive precipitation of colloids takes place, with the absorbtion of many elements such as the transition elements, rare-earths, etc. This phenomenon is amplified by the richness of estuaries in organic matter.

The ocean surface is characterized by abundant life, sustained throughout the depth of water where light can penetrate (known as the photic zone, some 50 m deep) by primary photosynthetic production by algae. This primary productivity allows other planktonic forms or larger organisms to develop by grazing and predation. As indicated above, organic matter concentrates light isotopes, particularly ^{12}C relative to ^{13}C and ^{14}N relative to ^{15}N, and the intensity of this productivity can be measured with the increased $\delta^{13}C$ and $\delta^{15}N$ of surface water compared with deep water (Figs. 7.13 and 7.14).

Biological activity consumes the surplus alkalinity supplied by rivers, notably in the form of $CaCO_3$, which is one of the major constituents of the hard parts of the micro-fauna (foraminifera). Biological activity is limited only by the availability of other elements required for tissue construction (P, N), for metabolism (Fe), and for locomotion and defense (SiO_2). It is a remarkable fact that oceanic organisms, including phyto- and zooplankton, consume these constituents in fairly constant proportions, particularly the $C:N:P$ proportions which are universally $115:15:1$ in organic matter (Redfield ratio). These proportions are believed to correspond approximately to a fairly general reaction, the

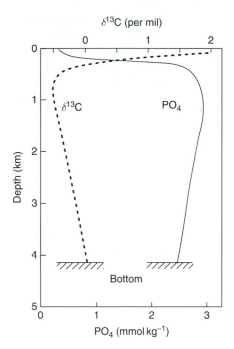

Figure 7.13 Characteristic distribution of a nutrient (phosphate) and of the $\delta^{13}C$ isotopic composition of carbon dissolved in seawater by depth. Surface biological activity depletes the water in nutrients and in ^{12}C. Dissolution of dead debris and biological excreta below the thermocline produces the opposite effect. When primary organic productivity at the surface increases, these contrasts between surface water and deep water also intensify (after Broecker, 1994).

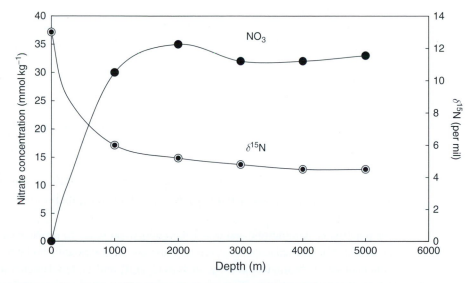

Figure 7.14 Variation in nitrate and $\delta^{15}N$ content in the waters of the Southern Ocean: intense surface biological activity depletes the dissolved nitrate and, as with carbon (Fig. 7.13), preferentially removes the light isotope, here ^{14}N (after Sigman *et al.*, 1999).

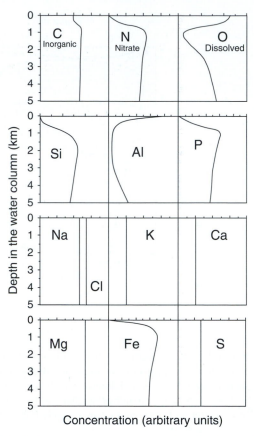

Figure 7.15   Elements used by biological activity, such as C, N, P, Si (nutrients), are depleted at the surface. Residual major elements, such as Na, Cl, K, Mg, are not affected by biological activity. Aluminum is a detrital element introduced into the thermocline by airborne particles. Dissolved oxygen is equilibrated with atmospheric gases near the surface but is significantly depleted below the thermocline as a result of the oxidation of falling organic debris.

Redfield–Ketchum–Richards (RKR) reaction, which leads to the production of biological material:

$$106CO_2 + 16HNO_3 + H_3PO_4 + 122H_2O$$
$$\Leftrightarrow (CH_2O)_{106} (NH_3)_{16} H_3PO_4 + 138O_2$$

This reaction is endothermic and uses solar energy (photons) to produce carbon and nitrogen in a highly reduced state that later may liberate large amounts of chemical energy. The elements used by biological activity, such as C, N, P, and Si, are depleted in surface water and are known as nutrients. They contrast with the residual major elements, such as Na, Cl, K, and Mg, which are not affected by biological activity (Fig. 7.15). In the section on

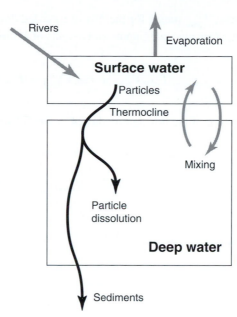

Figure 7.16 The two-reservoir model of the ocean of Broecker and Peng (1982).

chromatography, we saw that the reactivity of some ions (Cd, Ba, Fe, rare-earths, Th) relative to biological particulate matter rich in nitrate, phosphate, and silica accelerates their transfer from the surface to the ocean floor, even if they are not directly involved in the biological cycle. Such ions therefore display a nutrient-like character with depletion in the surface water and regeneration in deep water.

The waste matter and carrion produced by surface organisms sink through the ocean where they are recycled by other life forms. In the layer of water down to 100–1000 m, known as the thermocline because temperature varies very quickly, nitrogen, phosphorus, and other nutrients are redissolved. This process is accompanied, of course, by the consumption of oxygen (respiration), which is therefore depleted at such depths relative to the remainder of the ocean (oxygen minimum), and by the release of CO_2.

Some 95% of organic matter is recycled in this way before it sinks to great depths. Nutrients are recycled through the thermocline many times before being incorporated into the sediment. The importance of recycling has been demonstrated by Broecker and Peng (1982) in a simple but illustrative way (Fig. 7.16). Let us consider an ideal ocean made of two layers, one on top of the other; surface water and deep water, separated by the thermocline. Elements are carried from the continents to the sea by rivers and are removed from the deep ocean by sedimentation. Let us call F_{river} the input of river water (kg m^{-3} s^{-1}) into the ocean and C_{river}^i the concentration (mol kg^{-1}) of river water in element i. F_{river} is compensated by evaporation so that the amount of surface water remains unchanged. We call F_{mix} the amount of water exchanged per unit time between the surface and deep-water layers, and C_{deep}^i and $C_{surface}^i$ the concentration of element i in deep and surface water, respectively; P^i_{part} stands for the flux of i carried downward by sinking particles. We first write that, at steady state, inputs and outputs of i must be equal so that:

$$F_{river}C^i_{river} + F_{mix}C^i_{deep} = F_{mix}C^i_{surface} + P^i_{part} \qquad (7.49)$$

The fraction g of the downgoing flux of i carried by particles is therefore:

$$g = \frac{P^i_{part}}{F_{mix}C^i_{surface} + P^i_{part}} \qquad (7.50)$$

or:

$$g = 1 - \frac{\dfrac{F_{mix}}{F_{river}}\dfrac{C^i_{surface}}{C^i_{river}}}{1 + \dfrac{F_{mix}}{F_{river}}\dfrac{C^i_{deep}}{C^i_{river}}} \qquad (7.51)$$

Because, at steady state, dissolved elements that enter the ocean in rivers must leave it as sediments, the fraction f of settling particles that eventually exit the deep-water reservoir as sediments is such that:

$$f P^i_{part} = F_{river}C^i_{river} \qquad (7.52)$$

and therefore:

$$f = \frac{1}{1 + \dfrac{F_{mix}}{F_{river}}\left(\dfrac{C^i_{deep}}{C^i_{river}} - \dfrac{C^i_{surface}}{C^i_{river}}\right)} \qquad (7.53)$$

The quantity $1 - fg$ indicates the proportion of the riverine input of i to the ocean that is recycled through the thermocline instead of being sedimented. Using the carbon residence time in the deep ocean calculated from (6.14) as an estimate of F_{mix}, we obtain $F_{mix}/F_{river} \approx 30$. For phosphates, we use the values $C^i_{deep}/C^i_{river} = 3$ and $C^i_{surface}/C^i_{river} = 0.15$, which emphasizes the depletion of surface waters in nutrients. We therefore obtain $g = 0.95$ (95% of the riverine input of phosphate is exported to the deep ocean as particles), $f = 0.01$, and $fg = 0.01$, which indicates that $\approx 1\%$ of the riverine phosphate is actually exported to sediments while 99% is recycled through the thermocline by upwelling. For silica we would obtain 95%. An alternative statement is that a phosphorus atom is recycled 100 times and a silicon atom 20 times through the thermocline before they find their way into the sediments.

The downward transfer of nutrients is therefore achieved by settling of the particles and not by the sinking of surface waters depleted by biological activity. In contrast, upward transfer of nutrients is most efficiently effected by deep-water upwelling. This mechanism explains why deep-water upwellings off the coasts of Mauritania, Peru, Namibia, etc., are associated with intense primary productivity.

The recycling of soft parts is accompanied by the redissolution of calcite from carbonate tests. Alkalinity and ΣCO_2 increase while pH decreases with depth, typically from values of 8.2 in surface water to 7.8 in deep water. As discussed earlier, the "surface-water" and "deep-water" boxes may, by way of fractionation throughout the thermocline, maintain a sharp contrast in spite of some nutrients having very long overall residence times.

The deep ocean is cold and isolated from the atmosphere, contrary to surface water. It is undersaturated in $CaCO_3$ whereas surface water is saturated, explaining why living organisms have found an energy advantage in using this compound preferentially for the fabrication of hard parts. The saturation limit, or lysocline, and the "carbonate compensation depth" (CCD), which is the depth below which carbonates are not observed in sediments, currently lie at a depth of about 4500 m. Above this level, carbonate sediment is abundant, as on either side of the mid-ocean ridges, which may rise to just 2500 m below the surface; at greater depths, all calcite is dissolved and only residue is deposited, very slowly, as red clay, the composition of which is much affected by atmospheric dust.

In addition to vertical chemical variations, horizontal variations are important too. This is partly because the largest rivers (e.g. the Amazon, Mississippi, and Congo) flow into the Atlantic, which therefore receives a greater supply of nutrients than the Pacific, hence its rich biota. The largest mass of deep water forms in the North Atlantic by downwelling of the very salty water of the Gulf Stream, made even denser at high latitudes by the formation of the ice cap (Fig. 7.17). This is the North Atlantic Deep Water (NADW). Other masses form near the Antarctic, such as the Antarctic Bottom Water (AABW) in the Weddel Sea. This very cold water, which is less salty than the NADW as it is less subjected to evaporation, lies at the bottom of the main oceans. The NADW begins its long journey along the coast of the Americas to the Indian Ocean and the Pacific Ocean by way of the circum-Antarctic current (Fig. 7.18). There it rises to the surface and returns to the Atlantic via Indonesia and Madagascar. The whole oceanic circulation system is driven by temperature and salinity contrasts (thermohaline circulation) and is often referred to, for obvious reasons, as the oceanic "conveyor belt." (Fig. 7.18) In Chapter 6 we calculated the age for renewal of the deep water at 1600 years, which is approximately the duration of a cycle. During this deep journey, the water that forms at the surface, and hence is poor in nutrients but rich in oxygen, receives surface organic input. As the debris raining from the surface is dissolved at depth, deep water progressively gains nutrients and alkalinity, and loses oxygen. It also gains CO_2 through the effects of respiration, which is reflected by the smaller carbonate sediment production in the Pacific than in the Atlantic. Generally, these characteristics clearly distinguish the old water of the deep Pacific from the young Atlantic waters.

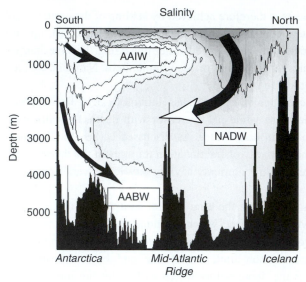

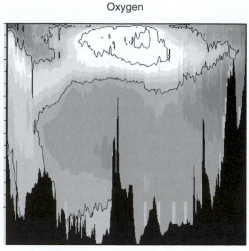

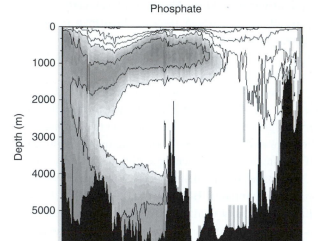

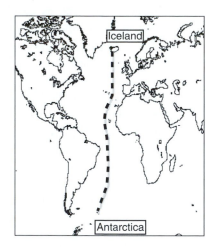

Figure 7.17 Formation and circulation of Atlantic water. The cross-sections follow the dashed line shown in the bottom right panel. The Gulf Stream, rich in salt because of surface evaporation in the intertropical zone, cools in winter at high latitudes and sinks to form the North Atlantic Deep Water (NADW). This new deep water moves southward at a depth of about 3500 m along the coast of America. It is salty and oxygenated but depleted in nutrients, such as phosphate, by the intense biological activity of the North Atlantic. The Antarctic Bottom Water (AABW) forms in the Weddel Sea. This very cold water is less subjected to evaporation and receives melt water from the Antarctic ice sheet: it is therefore less saline than NADW. It is also richer in nutrients, especially phosphate, as it recycles old water returned from the Pacific by the circum-Antarctic current. It is the most dense of all water masses and composes the bottom water of the main oceans. Antarctic Intermediate Water (AAIW), diluted by rain in the southern mid latitudes, is lighter. Figures drawn using OceanAtlas.

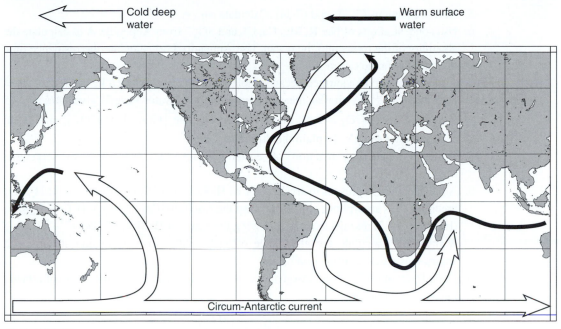

Cold deep
water

Warm surface
water

Circum-Antarctic current

Figure 7.18 The great conveyor belt of ocean circulation. Cold deep water forms in the North Atlantic by downwelling of the very salty water of the Gulf Stream and travels along the coast of the Americas. It then merges with the circum-Antarctic current, which feeds the Indian Ocean and the Pacific Ocean. The return flow of warm surface water from the Pacific passes through the Indonesia straits into the Indian Ocean, and circulates around the tip of South Africa into the Atlantic where it joins the Gulf Stream.

Exercises

1. Show that, for a Red–Ox reaction $Ox + ne^- + mH^+ \Leftrightarrow Red$, we can write $n\left(pe - pe_0\right) + mpH = \log [Ox]/[Red]$.

2. Discuss how the variables ΣCO_2, pH, *Alk*, and P_{CO_2} are controlled in the following cases: (1) a parcel of surface water in equilibrium with the atmosphere at constant P_{CO_2}; (2) a parcel of surface water saturated in calcium carbonate; (3) a parcel of water in the deep ocean. Assume that Ca^{2+} is the only metallic cation present in solution and find, as in Section 6.2, the breakdown of the system into its components, species, and their relating equations.

3. Discuss what happens to a glass of sparkling mineral water when HCl (or a lemon twist) is added to it. Show that a plot of $[H^+] = 10^{-pH}$ vs. the concentration of the HCl added is a way of titrating the alkalinity (this is the principle of Gran titration extensively used in hydrochemistry).

4. What is wrong with the following statement: in an atmosphere of increasing P_{CO_2}, more carbonates are added to seawater, so more calcium carbonate is precipitated? Why are some springs with CO_2-rich waters turning into petrifying fountains?

5. Derive equations (7.33) and (7.34). Calculate $\alpha_{H_2CO_3}$.

6. Use concentrations of Na, K, Mg, Ca, Cl, and SO_4^{2-} from Appendix A to calculate the alkalinity of the rivers and the ocean. Use Appendix F to calculate the residence time of alkalinity in the ocean.

7. A measurement on a surface ocean sample gives $Alk = 2.35$ meq kg^{-1} and $\Sigma CO_2 = 2.15$ mmol kg^{-1}. A similar measurement on a deep-water sample gives $Alk = 2.45$ meq kg^{-1} and $\Sigma CO_2 = 2.40$ mmol kg^{-1}. Calculate $\left[CO_3^{2-}\right]$, $\left[HCO_3^{-}\right]$, and pH in each seawater sample. Explain the differences.

8. Write a weathering reaction of enstatite (Mg_2SiO_6) by water to serpentine $Mg_3Si_2O_5(OH)_4$.

9. Aqueous silica is present in fresh water as H_2SiO_3 and $HSiO_3^-$. Silicic acid H_2SiO_3 precipitates as amorphous silica following the reaction SiO_2 (solid) $+ H_2O = H_2SiO_3$ (log $K = -2.7$) and dissociates following the reaction $H_2SiO_3 = HSiO_3^- + H^+$ (log $K = -9.6$). Calculate the abundance of each soluble species in equilibrium with amorphous silica as a function of pH.

10. Complexation of Zn^{2+} in seawater. We define a complexation constant β_n of a metal ion M in solution as:

$$\beta_n = \frac{[ML_n]}{[M][L]^n}$$

For Zn^{2+}, the decimal logarithm of complexation constants are (1) for OH^-: 5.0 ($n = 1$), 11.1 ($n = 2$), 13.6 ($n = 3$), 14.8 ($n = 4$); and (2) for CO_3^{2-}: 10.0 ($n = 1$). What are the relative proportions of the main species of Zn in seawater for a water sample in which $Alk = 2.35$ meq kg^{-1} and $\Sigma CO_2 = 2.15$ mmol kg^{-1} (assume that the second dissociation constant of carbonic acid is 9.0). Hint: write the sum of all the Zn species and factor [Zn^{2+}], calculate [Zn^{2+}]/ΣZn, then the other species.

11. What are the sources of alkalinity in river water? Which one can be considered as recycled and which is juvenile?

12. Show how Eq. 7.46 giving the slope of the meteoritic δD–$\delta^{18}O$ correlation in meteoritic waters can be derived from Eq. 2.31.

13. From the definition of the isotopic fractionation factor $\alpha^{18}O$ between liquid water and vapor, show that $D^{16}O \approx 1$, and $D^{18}O \approx \alpha^{18}O$, where D stands for partition coefficient. For fractionation of oxygen isotopes during water vapor condensation, use Eq. 2.28 to show that $\left(^{18}O/^{16}O\right)_{res} \approx \left(^{18}O/^{16}O\right)_0 f^{\alpha^{18}O-1}$, where f is the fraction of original ^{16}O (and therefore of original water vapor) left in the atmosphere. What is the corresponding relationship in δ units? Establish similar relationships for the D/H ratio. Remember that liquid water is enriched in the heavier isotope.

14. The west coast of the Americas is fringed by elevated coastal mountain ranges. Explain how, in a regime of west winds, the isotopic composition of oxygen and hydrogen in precipitations changes with elevation.

15. A core of salty seawater depleted in nutrients and relatively oxygenated is observed in the South Atlantic at a depth of 2200 m about 200 km east off the Brasilian continental slopc. What is the origin of the core water?

Depth (m)	Temp. (°C)	Salt (psu)	O_2 (ml l^{-1})	PO_4 (μmol l^{-1})	SiO_3 (μmol l^{-1})	NO_3 (μmol l^{-1})
0	25.62	35.05	4.84	0.16	4.8	0.1
10	25.49	35.07	4.81	0.11	4.4	0.2
20	25.30	35.07	4.82	0.11	4.3	0.2
30	24.92	35.08	4.90	0.12	4.1	0.2
50	23.70	35.10	5.01	0.10	4.3	0.2
75	22.03	35.07	5.06	0.11	4.8	0.4
100	20.55	35.01	5.00	0.15	5.4	0.8
125	19.31	34.94	4.86	0.16	4.7	1.3
151	18.35	34.89	4.79	0.24	5.4	2.1
201	17.04	34.79	4.73	0.36	7.8	4.0
252	16.15	34.71	4.74	0.42	8.2	6.7
302	15.38	34.64	4.66	0.47	10.3	7.0
403	13.15	34.46	4.46	0.78	16.1	13.1
504	10.59	34.25	4.26	1.29	27.6	18.8
605	7.97	34.12	3.31	1.89	44.1	26.1
706	6.07	34.13	2.35	2.38	65.3	33.8
807	4.95	34.18	1.55	2.81	87.8	38.8
908	4.26	34.25	1.28	2.91	107.6	41.5
1009	3.86	34.34	1.03	3.06	118.6	42.4
1111	3.44	34.40	1.05	3.06	126.1	42.9
1212	3.16	34.46	1.09	3.06	131.7	42.5
1313	2.96	34.49	1.23	3.05	139.1	42.3
1415	2.76	34.51	1.43	2.97	141.7	42.5
1516	2.60	34.53	1.57	3.03	146.4	41.8
1770	2.23	34.58	1.90	2.87	151.4	41.5
2024	2.00	34.62	2.21	2.89	157.1	38.9
2533	1.72	34.65	2.65	2.74	157.7	37.9
3043	1.58	34.67	2.90	2.67	155.4	37.3
3554	1.51	34.68	3.31	2.61	155.6	36.4
4067	1.47	34.69	3.50	2.57	152.5	35.7
4580	1.47	34.69	3.78	2.56	148.2	35.0
5095	1.46	34.70	3.83	2.51	142.9	35.3
5610	1.52	34.69	3.99	2.48	138.6	35.4

Table 7.1 Water-column properties in the Pacific at 26°N and 160°E

16. Discuss the properties of the water column in the Pacific at 26° N and 160° E (Table 7.1) by plotting temperature, salt, O_2, and the nutrients vs. depth. Identify the thermocline, the oxygen minimum. What is the halocline? Compare the profile of the soft (N, P) and hard (Si) nutrients.

17. Silicon and erbium (Er) concentrations have been analyzed in samples taken at different depths of the water column at a South Atlantic station (Table 7.2). Plot Si vs.

Table 7.2 Silica and erbium concentrations at different depths of the water column at a South Atlantic station (Bertram and Elderfield, 1993)

Depth (m)	SiO_3 (mmol kg^{-1})	Er (pmol kg^{-1})	depth (m)	SiO_3 (mmol kg^{-1})	Er (pmol kg^{-1})
241	3.6	3.49	2332	57.8	6.09
331	6.3	3.66	2581	57.4	5.6
418	10.1	3.74	2832	58.8	5.18
495	13.8	3.87	3082	61.2	5.75
565	16.7	3.97	3330	65.6	5.89
741	25.6	4.14	3532	74.1	6.81
839	31.3	4.31	3737	84.7	6.71
1082	47	4.67	3945	95.9	6.53
1273	59.6	4.92	4202	104.7	7.44
1466	61.5	5.64	4458	108.4	8.1
1657	66.2	5.35	4700	110.5	7.3
1841	63.2	5.24	4995	111.3	7.99
2088	60.3	5.58			

depth and explain the observations. Plot Er vs. Si and explain why an element of no biological interest can be correlated to Si.

18. What thickness of $CaCO_3$ sediment spread over the entire surface of the ocean would it take to remove 100 ppmv of CO_2 from the atmosphere? Use data from Appendix F and a specific gravity of 2700 kg m^3 for $CaCO_3$.

19. Explain why upwelling of deep water, such as under equatorial latitudes, reintroduces CO_2 into the atmosphere.

20. What is the potential paleoceanographic use of the following records: (1) $\delta^{18}O$ in pelagic foraminifers? (2) $\delta^{18}O$ in benthic foraminifers? (3) the $\delta^{13}C$ difference between benthic and pelagic foraminifers of the same age?

21. Discuss the potential climatic changes and the depth of CCD upon (1) a nearly instantaneous surge in atmospheric CO_2 triggered by a massive sub-aerial volcanic eruption; (2) the ensuing surge in riverine alkalinity flux resulting from the weathering of the lava flows; (3) an increase in primary productivity; and (4) a strong decrease in oceanic thermohaline convection.

22. The solubility of calcite varies in the ocean with depth. Since [Ca^{2+}] is essentially constant over the residence time of this element (≈ 1 My), this variation is expressed as $\left[CO_3^{2-}\right] = 90\,e^{0.16(z-4)}$, where z is the depth in kilometers and solubility is expressed in μmol kg^{-1}. Using the assumption that a change in the depth of the CCD is nearly equivalent to a change in the depth of the lysocline (saturation level) and that the relative change of $\left[HCO_3^-\right]$ can be neglected with respect to the relative change of $\left[CO_3^{2-}\right]$, estimate the relative change of P_{CO_2} associated with a shallowing of the CCD by 1 km. (Hint: consider using equations (7.2) and (7.40)).

References

Bertram, C. J. and Elderfield, H. (1993) The geochemical balance of the rare earth elements and neodymium isotopes in the oceans. *Geochim. Cosmochim. Acta*, **57**, 1957–1986.

Broecker, W. S. (1994) *Greenhouse Puzzles*. Palisades: Eldigio.

Broecker, W. S. and Peng, T.-H. (1982) *Tracers in the Sea*. Palisades, Eldigio.

Sigman, D. M., Altabet, M. A., McCorkle, D. C., Francois, R., and Fisher, G. (1999) The $\delta^{15}N$ of nitrate in the Southern Ocean: consumption of nitrate in surface waters. *Global Geochem. Cycles*, **13**, 1149–1166.

Stallard, R. F. and Edmond, J. M. (1983) Geochemistry of the Amazon. 2. The influence of geology and weathering environment on the dissolved load. *J. Geophys. Res.*, **88**, pp. 9671–9688.

The purpose of this Chapter is a short review of how biological processes affect the geochemical pathways that would prevail in the absence of life and how they contribute to the production of specific components that can occasionally form the bulk of some biological material such as oil. Expertise in biogeochemistry requires a strong background in biology and biochemistry and also some understanding of how biomass interacts as a whole with the mineral world. Background is well beyond the scope of the present book and we will hence try to restrict ourselves to the simplest of concepts.

8.1 The geological record

Oxidized rocks, limestones, cherts, and phosphates contain the biological materials with the most spectacular contribution to the geological record. Modern limestones are largely formed by the accumulation of carbonate tests of foraminifera and unicellular algae such as coccolithophores. Diatom frustules contribute massive amounts of silica to sediments at the bottom of the Southern oceans. The gigantic phosphorites of Africa represent fossil hard parts (teeth and bones) or their remobilization by diagenetic fluids: they are mined to produce fertilizers for agriculture. On the sea floor, these three types of rocks are often associated with each other in areas rich in nutrients, continental platforms, wind-driven upwellings of deep seawater such as next to the coasts of Morocco and Peru and the older seawater from the Southern oceans.

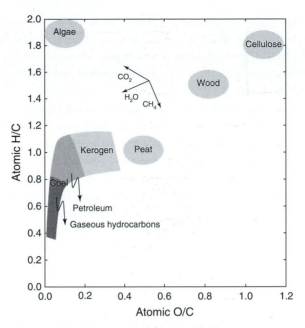

Figure 8.1 A van Krevelen (1950) plot representing the relationship between the major types of organic substances present in the geological record. The arrows indicate how loss of CH_4 ($x = 0$; $y = 4$), CO_2 ($x = 2$; $y = 0$), and H_2O ($y/x = 2$) affects compositions in this diagram. Loss of H_2O and CO_2 from the original organic products generates kerogen. Upon further heating, loss of liquid hydrocarbons, and subsequently of gaseous hydrocarbons inclusive of methane, leaves coal as the main residue (modified from Killops and Killops, 2005).

Reduced carbon is another important remnant of biological activity. Different types of organic material are found in rocks:

- *Humic substances* are common components of soils and waters. The precursors of these very complex compounds are diverse organic compounds resistant to biodegradation such as lignin and tannins.
- *Kerogen* consists of a variety of polymers and macromolecules formed during diagenesis by microbial degradation and condensation of humic substances, which make the products progressively more insoluble in diagenetic fluids. Different types of kerogens are distinguished, which may eventually lead to the formation of liquid hydrocarbons and coal. Bitumen is the fraction of carbon compounds that can be extracted from a rock by liquid solvents.
- *Liquid hydrocarbons* (petroleum) evolve by loss of water, carbon dioxide, and methane from some types of kerogen upon heating at temperatures of 60–150 °C (the oil "window"). The nature of the end-product depends on the rate of transformation.
- *Gaseous hydrocarbons* such as methane evolve from kerogen and coal at temperatures of 150–230 °C.
- *Coal* is the residual carbon-rich material left behind after hydrocarbons have left the rock. The most abundant humic coals derive from vascular plants, whereas sapropelic coal forms from fine-grained organic particles.

The maturation of carbon-rich material can be best presented in a van Krevelen (1950) plot representing the H/C ratio as a function of the O/C ratio (Fig. 8.1), in which loss of water, carbon dioxide, and methane are easily represented. The remains of vascular plants have high H/C and O/C ratios; those of algae, high H/C and low O/C ratios. Removal of H_2O and CO_2 by heating drives organic material into the field of kerogen and then coal. Subsequent removal of liquid hydrocarbons and methane drives the residue into the field of low-H coal.

It is a classically held view that major coal deposits formed in periods following lowering of sea levels, which were the most prone to deposition in landlocked basins. In contrast, deposition of the largest oil source rocks coincided with transgressive periods and increased biological productivity. The largest fraction of mature organic matter in the crust is actually stored, not in coal, the second largest reservoir, and not in oil either, but in methane hydrates: at pressures exceeding 100 atmospheres and temperatures lower than $18\,^{\circ}C$, methane forms liquid hydrates or clathrates which are found in very large amounts beneath the permafrost of continents at high latitudes and in the sediments of the marine continental slopes.

8.2 Some specifics of biological activity

In geological cycles, biological compounds are rarely considered as a source or a sink of elements simply because biomass is only a small mass fraction of geological reservoirs, such as the ocean or even soils. The relationship of life to geochemical pathways is a bit like that of the beaver to the river: it is there to mine available energy that would otherwise have been dissipated with no profit to any living thing. There is clearly more to life than building a lodge to protect offspring against predators, but the principle is still valid: life is opportunist and gets a free ride on untapped energy flows. And just as the beaver has a strong impact on the river and associated habitats, life affects the ocean, the atmosphere, and leaves fairly strong footprints in the chemistry of sediments. A plankton-free ocean would certainly have a chemistry very different from our familiar seawater. We will see that life may even be held responsible for some remarkable geodynamic features such as no less than the stabilization of granitic continents.

Two characteristics of organic material are fairly striking. First, it rises incredibly high on the redox scale. Trading electrons between very different redox potentials is the hallmark of biological activity. Life is about pushing C and N towards the most reduced state possible and then using a multistage transfer of high-energy electrons between a reduced primary electron donor (PED) and a terminal electron acceptor (TEA) to produce mechanical work and build structural and functional molecules. The cell builds up the only energy form it can use, adenosine-triphosphate or ATP, using external inputs, so that new proteins can be synthesized for maintenance and mechanical work performed. For autotrophs (plants), the primary source of energy is solar electromagnetic radiation in the visible part of the spectrum. Pigments, most notably chlorophyll, capture this energy in a process called photosynthesis, which may be either oxygenic (cyanobacteria, algae, plants release oxygen) or anoxygenic. In most biological systems, energy is stored as

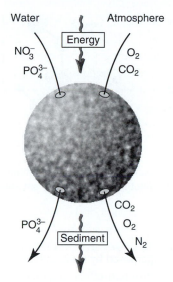

Organic matter, carbonate, phosphate,
silica, sulfide

Figure 8.2 Autotrophic life as a simple chemical plant. Nutrients are absorbed from the aqueous environment (seawater or groundwater) while CO_2 needed for photosynthesis and O_2 needed for respiration are ultimately derived from the atmosphere. Gaseous waste is also made of CO_2 and O_2.

carbohydrates (sugar). The oxygenic synthesis of glucose may be represented by the formal reaction:

$$6H_2O + 6CO_2 \Leftrightarrow C_6H_{12}O_6 + 6O_2 \qquad (8.1)$$

Heterotrophs burn reduced carbon prepared by other organisms: they eat food and breathe.

Breakdown of glucose to pyruvate $CH_3-CO-COO^-$ leads to ATP production in the absence (fermentation) or in the presence (respiration) of a terminal electron acceptor, such as sulfate or nitrate. Respiration is far more efficient than fermentation to produce ATP, which explains why aerobic conditions and dissimilative reduction of nitrate, sulfate, and iron are life's 'first choice'. Alternative sources, notably the energy of chemical reactions (chemotrophs), are occasionally used.

Cell metabolism requires some major constituents and trace metals from its environment. Obtaining C from CO_2 and H from H_2O is rarely an issue as long as a source of energy is present, but nitrogen must be tapped from either the atmosphere or from dissolved nitrates. For phosphate groups, the only source is the PO_4^{3-} ion released into natural waters by weathering of continents. Oxygen is the waste of photosynthesis, while CO_2 is the waste of respiration. Nitrate processing eventually releases nitrogen and some phosphate also is eliminated in the ambient liquid medium (Fig. 8.2).

A second property of biomolecules is their widespread chirality (handedness): a molecule can exist under two configurations which are the mirror image of each other

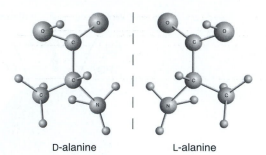

D-alanine   L-alanine

Figure 8.3 Example of chirality: alanine, which is an amino acid. The two enantiomers are mirror images of each other and cannot be superposed by any rotation. Unlabelled atoms are hydrogen. Biological alanine is L.

but cannot be superposed by rotation (Fig. 8.3). These two configurations, known as enantiomers, have a different effect on polarized light and are prefixed by D, for dextrogyre, for rotating light to the right, or L (levogyre) otherwise. Biomolecules are either non-chiral or 100 percent L or D (homochirality), most sugars being D and 19 out of the 20 amino acids being L (glycine is not chiral). The mechanism that produced homochirality in the first place is relevant to the origin of life but is not understood.

Some additional defining effects of biogeochemistry may be summarized as follows:

1. The most abundant elements in the biomass are reduced carbon, hydrogen, nitrogen, and phosphate. Fossil biomolecules are of particular economic importance since, as every living creature, humans like to turn reduced carbon into energy.
2. Biological activity affects sediment composition through the deposition of biominerals (carbonate, phosphate) or by interfering with diagenesis.
3. Over the Earth's history, biological activity has modified the composition of the atmosphere (notably by adding oxygen and removing carbon dioxide) and of the ocean. It is still today an essential parameter in climatic regulation.
4. Biological activity maintains chemical gradients such as those of nutrients in the water column.
5. Some of the original biomolecules, mostly insoluble lipids, are left untouched by diagenesis. They are true geochemical fossils and as such are called biomarkers.

We will now address some of these properties.

8.3 The chemistry of life

Living creatures can be divided into eukaryotes, which shelter their genes in a nucleus (e.g. plants and animals), and prokaryotes with no nuclear membrane. The second group is clearly divided by genetic analysis into archæa, which often live in unusual or extreme

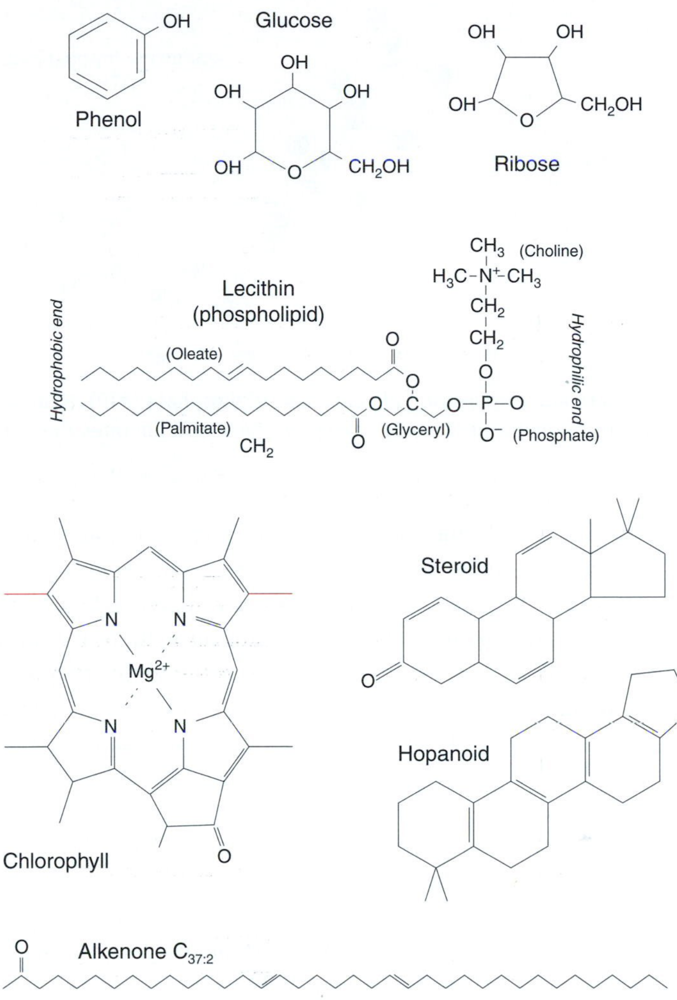

Figure 8.4   The structure of some important components of biological material and biomarkers.

environments, and (eu)bacteria. We will start this chemical tour by reviewing the main constituents of organic matter (Fig. 8.4):

1. Carbohydrates ('sugars') derive their name from their generic formula $C_n(H_2O)_n$. The common and simple sugar glucose $C_6H_{12}O_6$ is a monosaccharide whose ring structural form leads to polysaccharides via condensation reactions. The commonest of these latter is cellulose, a ubiquitous component of cell walls, notably in plants but also in chitin used by arthropods and insects as shell material. Carbohydrates are the primary reserve of energy and are used to produce adenosine triphosphate (ATP).

2. Amino acids are organic (carboxylic) acids in which one carbon is attached to an amino NH_2 group. They contain most of the nitrogen present in organic matter. Polymerization of amino acids via their NH_2 group produces proteins. Enzymes are proteins with a catalytic function, i.e. molecules which reversibly provide the little increment of energy that helps a particular reaction go over the activation energy barrier.

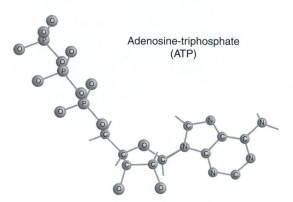

Adenosine-triphosphate
(ATP)

Structure and composition of adenosine-triphosphate (ATP). Hydrolysis of this compound to form adenosine-diphosphate (ADP) and phosphoric acid liberates the energy needed by the cell.

3. Lipids include the fats and the glycerides, or esters of glycerol esters with one or more fatty acids. When PO_4^{3-} groups attach themselves to one chain or more of carboxylic acids, they are known as phosphoglycerides. Tetracyclic compounds made of four pyrrole units (a five-membered ring with a nitrogen atom opposing two double carbon bonds) are biologically important as they form pigments known as porphyrines: chlorophylls (Mg) and hemoglobin (Fe) are closely related compounds. Terpenes are a broad class of lipids based on the isoprene chain (C_5H_8), which has one unsaturated bond at each end. Many form important compounds with multiple cyclic bonds (sterols, hopanes).

4. Aromatic alcohols (phenols) condense to form lignin, the second most abundant component in plants.

5. Nucleotides contain nitrogen-bearing compounds, a pentose sugar (ribose or deoxyribose), and a phosphate. Nucleic acids are polymers of nucleotides.

The universal energy currency of cellular metabolism is ATP, a compound which combines an amino acid known as adenine, a sugar called ribose, and phosphate groups (Fig. 8.5). Hydrolysis of ATP gives adenosine-diphosphate (ADP) and phosphoric acid by releasing a very large amount of energy.

A distinction should be made between living organisms and fossil organic matter. For all practical purposes, the mass inventory of living cells is dominated by an outer plasmic membrane and an inner medium called cytoplasm, which can be either a gel or a sol depending on environmental conditions (Fig. 8.6). Vegetal cells are also wrapped in an additional film of cellulose and some plants even develop a rigid framework of lignin. Lignin and cellulose are the main initial constituents of coal. Genetic material dominated by polynucleotides (RNA = ribonucleic acid and DNA = deoxyribonucleic acid) is either loose in the cytoplasm of prokaryotic cells (e.g. bacteria) or hosted in the nucleus, in mitochondria (animals), or in chloroplasts (plants) of eukaryotic cells. In general, membranes are composed of two layers of phospholipids made at one end of hydrophilic phosphoglycerides and at the other end of hydrophobic fatty acids, which mostly consist of long chains of hydrocarbons (Fig. 8.7). The cytoplasm is a complex solution of

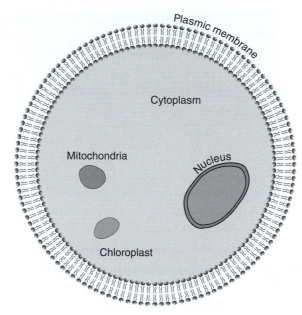

Figure 8.6 The major elements of a cell are the cytoplasmic liquid and the lipidic membrane. Eukaryotes store their genetic material in the nucleus, which is also isolated from the cytoplasm by a membrane. Energy conversion is ensured by mitochondria, probably a former symbiotic bacteria. Chloroplasts are only present in plant cells and are used for photosynthesis.

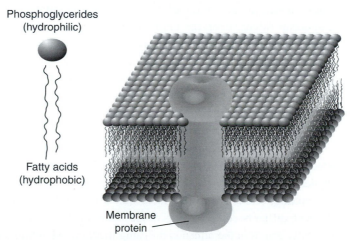

Figure 8.7 A simple membrane is made from a double layer of phospholipids, long molecules made from a hydrophilic head of phosphoglyceride and a long tail of hydrophobic fatty acid. Large proteins are used to ferry various components, such as H^+ or K^+, through the membrane.

proteins, glucides, and about one percent of inorganic ions, notably K^+, Na^+, Mg^{2+}, Ca^{2+}, Fe^{2+}, Cl^-, PO_4^{3-}, and SO_4^{2-}. Again, the polynucleotides are a repository of PO_4 and amino NH_2 groups, and nitrogen-containing cycles. We have already met the Redfield ratio which states that the C:N:P proportions in plankton are in the proportions of 115:15:1. Table 8.1 shows the inorganic composition of a sample of dry plankton material.

Table 8.1 Elemental composition of the marine plankton (Martin, 1973). Concentrations in $\mu g\,g^{-1}$ except for Na, K, Mg, Ca which are in $mg\,g^{-1}$

	Na	K	Mg	Ca	Sr	Ba
Phytoplankton	14	1.3	1.6	0.7	147	33
Zooplankton	11	1	1	0.2	132	16

	Al	Fe	Mn	Cu	Zn
Phytoplankton	110	224	6		
Zooplankton	94	199	11	6	100

8.4 Biominerals

What comes around goes around. In a world with no biological activity, sediments would probably be deposited on the ocean floor on average at pretty much the same rate as they are in our biological world. Alkalinity, phosphate, and dissolved silica liberated from the continents by erosion would be transported from the continents to the oceans and create huge carbonate, phosphate, and silica deposits. Iron and manganese would come out of black smokers at the bottom of the ocean and precipitate as metalliferous sediments near mid-ocean ridges. Even the most soluble compound, NaCl, has a residence time in the ocean of 300 My: elements can only accumulate in the ocean for as long as solubility is not exceeded or as long as they are not scavenged by detrital particles. In the end, what precipitates on the ocean floor is what came through estuaries and other inputs. Beyond the accumulation of organic matter discussed in the previous section, the role of biology with respect to inorganic elements is somewhere else: it changes the site and sometimes the mineral form under which minerals leave the ocean.

How life separates elements and recombine them into insoluble compounds to fit its own needs may vary. One of the main functions of membranes is to use energy to temporarily defeat the second principle of thermodynamics and the destructive power of diffusion and mixing: building a hydrophobic wall between the inside and the outside of a cell, or between different compartments of the same cell brings diffusion to a halt simply because it locally removes the aqueous diffusing medium. Membrane proteins nailed through the membrane take over and act as guarded gates to transfer elements individually (Fig. 8.7). They are fueled by ATP, the standard energy source, and therefore do not have to comply with the normal rules of diffusion. How proteins can be so selective with respect to particular elements is not a trivial question. Coordination is of course an important parameter, but the simple ligands, such as citrate, lactate, and oxalate, have the same order of preference, known as the Irving–Williams stability order, which for divalent ions is $(Mg^{2+}, Ca^{2+}) < Mn^{2+} < Fe^{2+} < Co^{2+} < Ni^{2+} < Cu^{2+} > Zn^{2+}$. This means for instance that, under similar conditions, the Cu^{2+} lactate and citrate will be more stable than the Fe^{2+} lactate and citrate, respectively. Other parameters are used by specific proteins to capture a particular ion, such as the charge, the ionic radius, the pH, the concentration of the ligand, the redox potential, multiple complexation, and even the spin of the element. The cell utilizes this complex arsenal to manipulate the crystallization pathway of some particular minerals.

A distinction can be made between minerals formed as a by-product of cell metabolism or activity (biologically induced mineralization) and minerals that are an integral part of the organism physiology (biologically controlled mineralization):

1. *Biologically induced mineralizations.* Cyanobacteria and their stromatolite constructs are one of the earliest examples of an organism precipitating calcite. Precipitation of iron and manganese oxyhydroxides is frequent as a consequence of bacteria creating a reducing environment which concerts Fe^{2+} and Mn^{2+} into low-solubility Fe^{3+} and Mn^{4+}. It is also known that bacteria play a strong role in the precipitation of silica from hydrothermal springs. Apatite forms upon decay of biological matter. A form of pyrite, called framboidal for its raspberry-like appearance, is an example of biologically mediated sulfide precipitation.

2. *Biologically controlled mineralizations.* The most important biominerals of this type are carbonates (foraminifera) and silica (diatoms and radiolarians), which cover broad expanses on the sea floor. These minerals are mostly used to confer protection and mobility to the organisms. A spectacular case is intracellular precipitation of nicely aligned magnetite crystals in so-called magnetotactic bacteria.

8.5 Biological controls on the ocean–atmosphere system

Consequences of this biological activity on the chemistry of the ocean–atmosphere system are very significant. Today, the alkalinity flux through the ocean is regulated by the precipitation of organic calcite. Phytoplankton acts on the atmosphere by the photosynthetic production of oxygen, which is balanced by the respiration of both the phyto- and the zooplankton. What limits the process is not very well understood: it is believed, as suggested by John Martin, that the availability of micronutrients, notably Fe and Zn, may keep productivity in check. The story goes that during a seminar at Woods Hole, John Martin rose and said "Give me a half tanker of iron, and I will give you an ice age." The project IRONEX, designed to test this assumption, supports this view: south of the Galapagos Islands, in the Eastern Pacific, in waters naturally rich in major nutrients because of the very low latitude, addition of 445 kg of iron triggered a strong algal bloom with a clearly visible change in chlorophyll levels and measurable effects on dissolved CO_2.

Export of organic matter towards the sediment acts as a CO_2 pump and a source of O_2 for the ocean–atmosphere system. The fate of buried carbon controls the long-term evolution of atmospheric oxygen. If sedimentary C is oxidized, either because it is oxidized during diagenesis or because the sediment is shoved up against continental margins, reaction with atmospheric oxygen converts it back to carbon dioxide. This phenomenon can be seen at work today in the natural oil seeps of the Persian Gulf. If sediments are subducted with oceanic lithospheric plates, carbon is injected into the mantle and the atmosphere inherits the leftover oxygen irreversibly.

8.6 Diagenetic transformation of organic material

Mud deposited on the ocean floor normally contains enough dead organic matter to provide food for a burrowing fauna of fish, worms, and mollusks. Even so, almost all these creatures depend on oxygen being available in the interstitial water. Only the top few centimeters of the mud, known as the bioturbated layer, will provide such an oxygenated environment. This layer is recognizable from its burrows and its chaotic character. The top part of the oxygenated sediment is thus mixed to the extent that the finer sedimentary structures disappear. Imagine for a moment that we are observers on the sea floor, resisting burial, and living long enough to observe the sediment sinking beneath our feet. We would see the bioturbated layer acting like a moving average on the sedimentary record (electronics aficionados might liken it to a low-pass filter). If, however, the bottom water is particularly poor in oxygen, usually because of a great abundance of organic matter, burrowing animals cannot breathe, bioturbation is impossible, and the sedimentary record conserves even the finest details down to the infra-millimetric scale (the rocks are then often referred to as laminites).

Below the bioturbated layer, the dissolved oxygen of the interstitial water is fully used up. Other organisms, microbes, take over. To maintain their metabolism they are able to oxidize organic matter by reducing other dissolved components, such as nitrates to nitrogen (denitritrification), solid MnO_2 and $Fe(OH)_3$ to soluble Mn^{2+} and Fe^{2+}, and sulfates to hydrogen sulfide. These reactions are also known as dissimilatory reactions because they do not involve assimilation of their products by cellular material. These biochemical reactions occur in the space of a few tens of centimeters below the water–sediment interface. The sequence of early diagenetic reactions in the uppermost layers below the sediment–water interface is controlled by the redox potential of the possible diagenetic reactions. We are going to assume that dead organic matter is represented for simplicity by CH_2O (formaldehyde), a very simple compound which could be a precursor of sugars and which is oxidized into HCO_3^-, the stable form of carbon in solution. Another form of "food" is the acetate ion released by acetogenic bacteria, probably the most abundant organic substrate in oxygen-poor sediments. The dominance plot of the different redox species is shown in Fig. 8.8 for a pH $\approx$ 7: it indicates the sequence of reduction reactions, which is simply that of decreasing pe. Right below the interface, the oxic layer is well ventilated by the bioturbation and organisms use normal respiratory oxidative mechanisms to turn organic matter into carbon dioxide plus water, but also ammonium into nitrites followed by nitrates, a process known as nitrification. When all the dissolved oxygen is used up and cannot any longer be replenished by burrowing animals, bacteria turn to increasingly stable electron acceptors, nitrate, Mn^{3+}, Fe^{3+}, and SO_4^{2-} to oxidize organic matter. The sequence of the most important dissimilatory reactions is shown in Fig. 8.9 and is determined by the energy they liberate. It can be seen from this sequence how interstitial solutions become increasingly reducing with depth. Under these conditions, nitrate is reduced to nitrogen (denitrification), which is lost to the ocean through pore waters, while ferric iron and manganese accumulated on the sea floor as hydroxides are progressively reduced and dissolved. Most of the reduced Mn^{2+} is also returned to bottom seawater

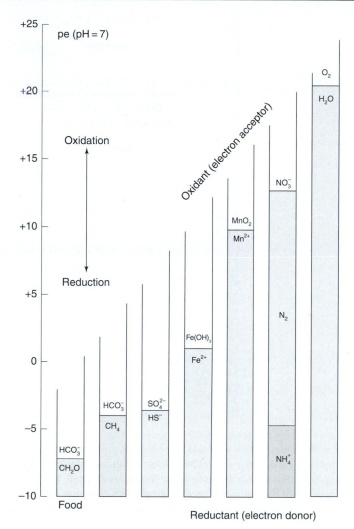

Figure 8.8 Dominance plot of the major species present in sediments and pore waters and which can act as either electron donor (reductant) or acceptor (oxidant). Low-pe organic matter CH_2O (food) is supposed to be present in excess. The energy driving its oxidation (upward) into HCO_3^-, is obtained from the dissimilatory reduction (downward) of the oxidants shown at the high-pe end of each bar.

with pore water where, upon re-oxidation, it contributes to the small diagenetic component of polymetallic nodules. In contrast, ferrous iron reacts with the sulfur liberated by the dissimilatory reduction of sulfate and precipitates as pyrite FeS_2, a particularly common diagenetic mineral in sedimentary rocks with a high organic matter content. At even greater depth and if temperature is adequate, leftover organic material may be reduced to methane. All these reactions are biologically mediated. Overall, pore waters seeping out from sediments rich in organic matter and nitrogen are not only highly reduced but also rich in metals. Because porous flow is largely controlled by the rate of burial, this is certainly an important control factor of these processes.

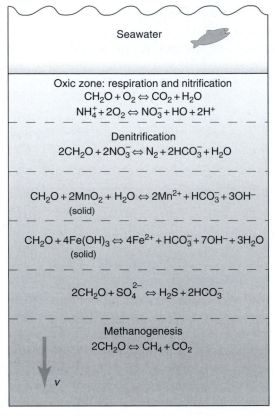

The section shows:

Seawater

Oxic zone: respiration and nitrification
$$CH_2O + O_2 \Leftrightarrow CO_2 + H_2O$$
$$NH_4^+ + 2O_2 \Leftrightarrow NO_3^- + HO + 2H^+$$

Denitrification
$$2CH_2O + 2NO_3^- \Leftrightarrow N_2 + 2HCO_3^- + H_2O$$

$$CH_2O + 2MnO_2 + H_2O \Leftrightarrow 2Mn^{2+} + HCO_3^- + 3OH^-$$
(solid)

$$CH_2O + 4Fe(OH)_3 \Leftrightarrow 4Fe^{2+} + HCO_3^- + 7OH^- + 3H_2O$$
(solid)

$$2CH_2O + SO_4^{2-} \Leftrightarrow H_2S + 2HCO_3^-$$

Methanogenesis
$$2CH_2O \Leftrightarrow CH_4 + CO_2$$

v

Figure 8.9 The successive reactions of early diagenesis below the sediment–water interface. Interstitial water from the upper layer is ventilated by burrowing animals. Bacteria use the dissimilatory reduction of nitrate, solid MnO_2, and $Fe(OH)_3$, then sulfate still present in interstitial water to oxidize dead organic material, here represented by the molecule CH_2O. The inception of methanogenesis requires that the sediments were originally rich in organic material. The section represents only a few tens of cm but the whole pile is moving downwards at the rate of sedimentation.

Denitrification is associated with progressive loss of N_2, sulfate reduction with progressive sulfide precipitation, and methanogenesis with methane seeping: Rayleigh distillation amplifies isotopic fractionation between the reduced and the oxidized reservoirs of N, S, and C with respect to equilibrium values and leads to major isotopic shifts in the residue. Most of the methane produced by methanogene, however, is converted back into organic matter by methanotrophic archæa and never reaches the sediment–water interface. Methanotrophs are often associated with acetogenic bacteria which release acetate as an end-product of fermentation. Such associations in which a species utilizes the poisonous by-product of another are known as consortia.

Dead organisms and fecal pellets are highly unstable and prone to utilization as biological fuel by other organisms. Hydrolysis of the phosphate groups and oxidation of the amino group take place very rapidly during decay, while most carbohydrates such as cellulose disappear during fermentation. At low temperature, the most resistent part of the original organic components are (a) the long aliphatic chains of lipids, saturated (no double

C–C bond) or unsaturated, derived from membrane material and (b) the aromatic compounds derived from the phenol rings of lignin. On land, this forms the base of humic substances, notably humic and fulvic acids, which are heterogeneous mixtures of amino acids, sugars, proteins, and aliphatic compounds. Fulvic acids are soluble in water under all conditions, humic acids are insoluble at pH < 2.

During diagenesis, humic compounds evolve into kerogen which, upon heating between 50 and 120 °C evolves into the hydrocarbons of crude oil. Oil is normally preserved when it seeps out from its primary site of formation towards a porous reservoir. Further heating induces the cracking of hydrocarbon chains with loss of methane (natural gas) and eventually, at temperatures in excess of 300 °C, leaves pure carbon as graphite (coal or anthracite) in the rock. Since methane (CH_4) has the highest possible hydrogen/carbon ratio, two convenient maturation indices of organic products are the O/C and H/C ratios (Fig. 8.1). Normally the terms used are: gas, for C_2 to C_4 hydrocarbons; light oil, for C_4 to C_{14}; and oil, beyond that point.

8.7 Biomarkers

Easily degradable organic matter, such as proteins and carbohydrates, is unlikely to fare well with recycling and diagenesis. The fate of insoluble molecules is certainly more promising. Among those, some fatty acids, long-chain alkanes and alkenes, and terpenes often pass diagenesis largely unharmed even if they lose some of their functional groups. These geochemical fossils can be matched with existing organisms or their ancestors, and used to interpret ancient microbial ecosystems. They can also have been left by unknown organisms and reveal long-gone creatures and metabolic pathways. Biomarkers can be used to reconstruct the evolutionary tree of a particular group of microbes. Occasionally, the discovery of new biomarkers has led to the discovery of unsuspected organisms and metabolic processes, such as the "anammox" bacteria capable of combining ammonia with nitrites into N_2. There is, however, much ambiguity in interpreting these complex molecules since more than one group of bacteria or algae may produce similar substances.

Among the most successful findings, let us quote archeol as a compound produced by methanogenic archæa and crocetane produced by methanotroph archæa, which were essential in understanding the consortia built around methanogens and methanotrophs. Aromatic carotenoids are pigments which betray the presence of bacteria selectively catching light at particular wavelengths to achieve the phototrophic oxidation of sulfides. Some forms of terpenes, known as hopanes, have only recently been related to cyanobacteria. Other well-identified but orphan compounds remain without a known progenitor. Alkenones are produced by coccolithophores to regulate cytoplasm viscosity and the relative abundance of related varieties is commonly used to infer ancient sea-surface temperatures and pressures of atmospheric carbon dioxide.

Because of their remarkable preservation, biomarkers are also used for isotopic analysis: the $\delta^{13}C$ of alkenones is believed to be a very sensitive, and unfortunately unique, indicator of the carbonate speciation in the surface ocean and therefore of the CO_2 pressure in the atmosphere.

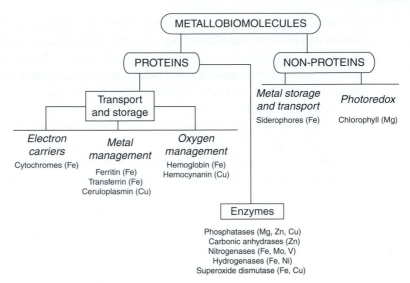

Figure 8.10 Some biological molecules containing metals and their biological functions (after Fenton, 1995).

8.8 Metals in organic matter

Metallic deposits of probable biological origin are known, such as the Precambrian banded iron formations (BIF). It is suspected that this is not an isolated case and the emerging field of the stable isotope geochemistry of metals may soon be able to come up with biomarkers of its own. Among the trace metals that contribute to different functions in the cell, some may contribute to the sedimentary record. For example:

1. Iron enters multiple molecules such as porphyrin, a ring of nitrogen-bearing chelates. Porphyrin groups are present for electron transfer in cytochrome and for dioxygen transport in hemoglobin. Proteins containing both S and N (ferredoxin) perform more specialized functions such as photophosphorylation during photosynthesis.
2. Because of its property of super-acidity, Zn is one of the most prevalent trace elements of life and has been found in over 600 proteins, notably associated with the immune system. It is found in carbonic anhydrase, which dramatically increases the rate of conversion of CO_2 into HCO_3^-.
3. Copper is present in cytochrome oxidase for electron transfer in the energy cycle and replaces iron in hemocyanin for dioxygen transport in some species.
4. Magnesium concentration in the cytoplasm is in the percent range. It is the core cation of chlorophylls.
5. Potassium is used to maintain gradients and controls transfers across the cell membranes (biological pumps).

Some metallobiomolecules are listed in Fig. 8.10.

References

Fenton, D.E. (1995) *Biocoordination Chemistry*. Oxford: Oxford, University Press.

Killops, S. and Killops, V. (2005) *Introduction to Organic Geochemistry*, 2nd edn. Oxford: Blackwell.

Martin, J. H. and Knauer, G. A. (1973) Elemental composition of plankton. *Geochim. Cosmochim. Acta*, **37**, 1639–1653.

♠ van Krevelen, D.W. (1950) Graphical statistical method for the study of structure and reaction processes of coal. *Fuel*, **29**, 269–284.

9 Environments

A major achievement of low-temperature geochemistry is its ability to provide estimates of variables such as ocean temperature, atmospheric composition and pressure, erosion intensity, and biological productivity. These estimates come through geochemical observables, known as proxies, which can be related with some confidence to a variety of parameters of our environment. The understanding of ancient climates, oceans, atmospheres, and biological activity would be very poor in the absence of these proxies and would remain qualitative and highly speculative. The derived environmental information, however uncertain it may be, can always be tested against predictions and with the help of improved observations can be continuously improved.

Let us first briefly review some of the most important environmental proxies for modern environments (< 65 Ma).

1. As shown by Dansgaard in 1964, mean annual air temperature can be determined (or estimated) from the mean δD or $\delta^{18}O$ value of the local precipitation (rain or snow).
2. The amount of ice locked up in polar regions is derived from the average $\delta^{18}O$ value of seawater.
3. The temperature of deep oceanic water can be obtained from the $\delta^{18}O$ values of benthic foraminifera.
4. The surface ocean $\delta^{18}O$ is perturbed by evaporation, precipitation, and continental run-off. The sea-surface temperature (SST) can instead be obtained from the Mg/Ca and Sr/Ca ratios in the carbonates produced by organisms living in the photic zone, typically corals, pelagic foraminifera, and coccolithophores. The $\delta^{18}O$ values of fish

tooth enamel (phosphate), which is more resistant to diagenetic modification than carbonates, are a useful temperature proxy. Sea-surface temperature is also obtained from the relative abundances of alkenones extracted from sediments.

5. As shown in Chapter 7, the pH of the ocean may be constrained by boron isotopes in carbonates. Again, the pH of surface waters also varies with evaporation, precipitation, and continental run-off. Deriving oceanic pH values from the rare deep-sea corals holds much promise.

6. The pressure of carbon dioxide is fairly difficult to constrain. It has been suggested recently that the $\delta^{13}C$ of alkenones is a reliable proxy of P_{CO_2}.

7. Recent sea-level fluctuations can be tracked by dating corals that live right beneath the surface of the ocean. Dating is most precisely obtained from ^{230}Th disequilibrium.

8. Local biological productivity can be constrained by the difference in $\delta^{13}C$ between pelagic and benthic organisms. If diagenesis is not too severe, $\delta^{15}N$ can also be useful. Productivity has also be estimated using the concentrations of biolimited elements, notably Cd.

9. The $^{87}Sr/^{86}Sr$ ratio of marine carbonates is an excellent proxy of erosion rates and alkalinity fluxes: today, abundant radiogenic Sr from the old basement of the Himalayas is carried to the ocean by the powerful rivers from south and southeast Asia.

10. The isotope compositions of neodymium contrast between the Atlantic and the Pacific oceans because the most important rivers flow into the Atlantic. Neodymium (Nd) isotopes can be used to trace mixing between Atlantic and Pacific waters, whereas exchange between seawater and mid-ocean ridges and alteration of young volcanic rocks are particularly active in the Pacific.

11. Sulfur isotopes are fractionated by about 20 per mil between sulfate and sulfide. The values of $\delta^{34}S$ therefore are a useful indicator of the relative amount of evaporites and iron sulfide locked up at a given time in the sedimentary mass.

12. We will see in Chapter 12 why the Earth's mantle and crust are seriously depleted in platinum-group elements, and in particular in iridium (Ir). Nearly all meteorites are very rich in these elements. Large meteorite impacts are best fingerprinted by the presence in sediments of a thin layer of material rich in iridium corresponding to the time of impact.

This list is certainly not exhaustive but already gives a good feeling for what geochemistry can do about past environments and how much work is involved in understanding the modern climatic fluctuations and the major events that shaped the composition of the atmosphere and the oceans over geological time.

9.1 Phanerozoic climates

Atmospheric carbon dioxide is a greenhouse gas. When the Sun's rays strike the ground, they are reflected but, because the Earth is much cooler than the Sun, at longer wavelengths, in the infrared. We can feel this radiation when we walk along a paved road at night after a sunny summer's day. Homonuclear diatomic molecules such as N_2 and O_2 do not absorb infrared, visible, or ultraviolet solar radiation because they remain symmetrical with a zero electric dipole, no matter how hard they are hit by photons. Such collisions are described as elastic. Nitrogen and oxygen therefore do not block the sunlight, they are optically transparent over most of the solar electromagnetic spectrum, and do not contribute to warming. In contrast, the symmetry of H_2O, CO_2, CH_4, and O_3 vibrations changes when photons collide with these molecules: new vibrational and rotational patterns reflect absorption of electromagnetic radiation and the whole gas phase warms up, at least as long as atmospheric pressure is strong enough for frequent molecular collisions. The reflected radiation is therefore absorbed by H_2O, CO_2, CH_4, and ozone and warms up the atmosphere. This is the greenhouse effect. The history of the Earth's climate is therefore for a large part that of its atmospheric CO_2.

9.1.1 Quaternary climates

This effect can be seen clearly in the composition of Quaternary sediments where the extent of carbonate preservation varies with the depth of deposition in keeping with the pattern of glaciations (Fig. 9.1) that recur every $\approx 100\,000$ years. Carbonate preservation clearly decreases and CCD rises during the high-CO_2 peaks of interglacial periods. This pattern is similar to the pattern of fluctuations in marine oxygen isotopes as recorded by the $\delta^{18}O$ of the calcite of deep-water foraminifera. During glacial times, a greater quantity of ice with very negative $\delta^{18}O$ values is stored in the polar ice caps and seawater oxygen is enriched in ^{18}O (Fig. 9.2). Because the enrichment of ^{18}O in rain and snow is stronger during cold periods, the glacial $\delta^{18}O$ maxima in seawater and the carbonates in isotopic equilibrium with it (maximum ice volume) are the mirror image of $\delta^{18}O$ minima in ice cores.

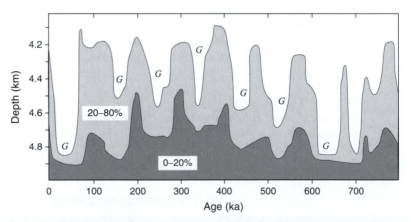

Figure 9.1 Preservation of carbonates in Quaternary sediments at various depths. The different shades of grey refer to the percentages of carbonate preservation shown. The zone between the two curves mainly depicts the transition of the carbonate compensation depth (CCD). During glacial periods (G), the CO_2 content of the atmosphere was lower, the ocean less acidic, the CCD sank, and the calcareous fraction of sediments increased (after Farrell and Prell, 1989).

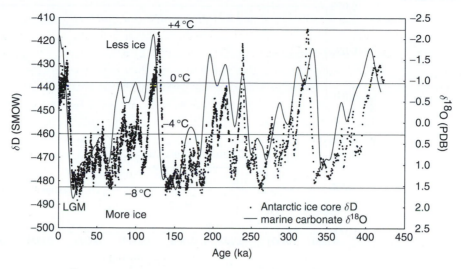

Figure 9.2 Fluctuation of $\delta^{18}O$ (relative to PDB standard) in the carbonate of benthic foraminifera (SPECMAP database). Increased storage of water in polar ice during glacial periods increases the ^{18}O isotope concentration of the ocean. The $\delta^{18}O$ curve follows (although with a small time lag) the temperature curve deduced from the δD values measured in the Vostok ice core in Antarctica (Petit *et al.*, 1999). The horizontal lines show the temperature shifts in the polar region deduced from the δD values in ice. LGM = last glacial maximum. The periodic aspect of fluctuations is caused by astronomical forcing (Milankovic cycles).

Lower average temperatures during the ice ages are explained by the astronomical modulation of insolation and its amplification by a lower level of CO_2 in the atmosphere, and therefore a weaker greenhouse effect. Plotting P_{CO_2} vs. ΣCO_2 for different pH values for seawater saturated in calcite as given by (7.39) and (7.40), we can see the effect of changing P_{CO_2} on the precipitation or dissolution of calcium carbonate (Fig. 9.3). A reduced P_{CO_2} decreases ΣCO_2, makes the ocean less acidic and therefore enhances carbonate preservation, which is reflected by a deepening of the CCD. The intervals during which limestone is abundant at great depths are cold periods.

The short-term variability in the isotopic composition of oxygen recorded by foraminifera is essentially indicative of the abundance of polar ice: temperatures at high latitudes vary periodically with orbital forcing, especially the precession of the equinoxes (19 000 and 23 000 years), the variation in the inclination (obliquity) of the Earth's axis of rotation on the ecliptic (41 000 years), and the eccentricity of the Earth's orbit around the Sun (100 000 years). These are the famous Milankovic cycles regulating the alternating pattern of Quaternary glacials and interglacials (Fig. 9.2).

9.1.2 Mesozoic and Cenozoic climatic trends

In contrast, the long-term variability of oxygen isotopes in benthic (deep-sea) foraminifera reveals the secular cooling of the deep ocean waters. The temperatures recorded by the $\delta^{18}O$ values fell by some 10–12 °C since the Paleocene optimum (55 Ma) (Fig. 9.4).

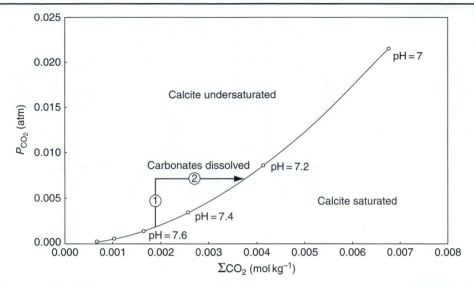

Figure 9.3   Effect of a change in the P_{CO_2} on the dissolution and precipitation of carbonates. The curve represents the pressure of carbon dioxide in the atmosphere above a solution saturated in calcite at different values of ΣCO_2. ① Sudden P_{CO_2} increase. ② Reaction of the excess dissolved CO_2 with carbonate sediments. The opposite effect (precipitation of carbonates) would be obtained for a drop in P_{CO_2}.

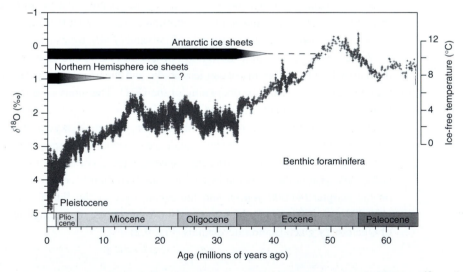

Figure 9.4   The cooling of deep water since the Paleocene ($\approx$ 55 Ma) optimum is recorded by the $\delta^{18}O$ value (here reported relative to the PDB standard) of benthic foraminifera. An uncertainty of several degrees results from our ignorance of the volume of polar ice in the Tertiary (3% of the total hydrosphere today): the temperature scales on the right assume a world without ice. The broad fluctuation range over the last million years or so is due to the glacial–interglacial oscillations (figure redrawn from Zachos *et al.*, 2008).

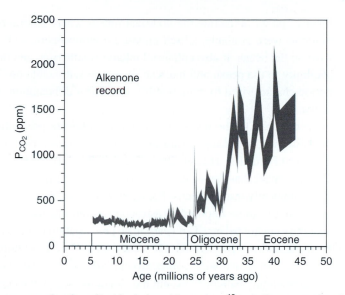

Figure 9.5 Atmospheric pressure of carbon dioxide deduced from the $\delta^{13}C$ of alkenones. The global trend of oceanic cooling since the end of the Cretaceous is particulary visible for the deep ocean (see previous figure). It reflects a drop in atmospheric CO_2 and therefore a decline of the greenhouse effect (Pagani *et al.*, 2005).

Inasmuch as $\delta^{13}C$ of alkenones is a reliable indicator of atmospheric pressure of carbon dioxide, this cooling is probably associated with a drop in P_{CO_2} by a factor of four (Fig. 9.5). The strong decline of the greenhouse effect over the last ≈ 55 Ma therefore explains the global cooling of the planet and the advent of the "icehouse" thermal regime.

Carbon dioxide is injected into the atmosphere by volcanoes. Erosion then transfers it to the hydrosphere by reactions similar to those described by (7.30) above. It is then removed from the ocean by precipitation of carbonate and by sedimentation in the form of reduced carbon (kerogen, petroleum, or coal). Some of the reduced organic carbon in the sediment is re-oxidized as CO_2 by oxygen in the air when rocks are exposed by erosion; the naturally occurring oil seeps of hydrocarbon-rich areas of California and the Persian–Arabian Gulf are evidence of this. Human activity is accelerating this process through the burning of fossil fuels in power plants and transport. Some of the sedimentary carbonates are removed from the surface system via subduction zones, but we do not know exactly how much of this CO_2 is returned to us by decarbonation at depth and orogenic volcanism in those parts of the world. It appears, however, that the quantity of carbonates at the Earth's surface has increased significantly over geological time.

How atmospheric P_{CO_2} can have changed so much over 70 Ma is still somewhat under discussion. One hypothesis is that fluctuations in volcanic emissions of CO_2 forced the system. The Cretaceous was indeed a period of intense volcanic activity. The rapidly widening rifts of the dislocating Pangea supercontinent were preceded by strong bursts of volcanic activity such as the Ontong–Java oceanic plateau in the Western Pacific, the Caribbean–Columbia plateau, the Rajmahal and Deccan traps in India, and the Paraña–Etendeka traps of South America and South Africa. The greenhouse effect was very intense and we all picture the luxuriant forests where dinosaurs roamed. Provided enough nutrients were available, a high pressure of atmospheric CO_2 increased primary productivity in the ocean. It also enhanced intense weathering, and thus induced strong fluxes of alkalinity to the ocean and massive limestone sedimentation. The Cretaceous is indeed a period characterized by remarkable carbonate sedimentation, in particular on continental shelves.

So how can we be invoking a reduced carbonate precipitation during the high P_{CO_2} bouts of the Quaternary and apparently the opposite for the Cretaceous? This is actually a matter of kinetics. The residence time of alkalinity in the ocean is $\approx 100\,000$ years. We also remember that the mean time of oceanic overturn is 1600 years by today's standards, i.e. geologically instantaneous. For temperature and P_{CO_2} fluctuations on this time scale or faster, such as those imposed by astronomical forcing, the incoming and outgoing fluxes of alkalinity are not balanced and carbonate precipitation reflects short-term external changes. For time scales >100 ka, like the steady high P_{CO_2} conditions of the Cretaceous, steady state is achieved and what comes around goes around: the abundant carbonate series precipitated on the seafloor represent the alkalinity lost by the ocean and therefore must have been compensated for by a corresponding alkalinity flux to the ocean. The conclusion remains that the Cretaceous was a time of high chemical erosion rate and therefore of high P_{CO_2}. Observations on different time scales therefore provide opposite conclusions on P_{CO_2} changes.

An alternative explanation of climate control, suggested by the isotope geochemistry of strontium in seawater, is that of tectonic forcing. In marine carbonates, the $^{87}Rb/^{86}Sr$ ratio is so low that it has barely varied with time since deposition (see (3.6)). Consequently, $^{87}Sr/^{86}Sr$ remains frozen near its original value and therefore at the $^{87}Sr/^{86}Sr$ value of seawater (from which the limestone is derived) at the time of sedimentation. The reason for this is that the concentration in the alkali element Rb (whose behavior is very similar to that of K) is very low compared with that of the alkaline-earth element Sr (whose behavior is very similar to that of Ca). It has been observed that the $^{87}Sr/^{86}Sr$ ratio of seawater as recorded in marine carbonates has increased since the Cretaceous (Fig. 9.6). This increase is evidence of the increased input of continental strontium with its far higher $^{87}Sr/^{86}Sr$ ratio (0.711 on average) than that of strontium derived from the weathering of basalts (0.702). We will see later that this difference reflects the high Rb/Sr ratio of the continental crust. This value points to increased erosion, while the evidence of carbon isotopes and the general cooling of climates and of the ocean since the Cretaceous argue against increased emissions of volcanic CO_2. This suggests, then, more efficient erosional processes, and it has been proposed that the collision between India and Asia, and the resulting uplift of the Himalayas in the Tertiary, were responsible for the decline in atmospheric CO_2. The high elevation of the Himalayas and the monsoon regime that dominates this area favor particularly intense mechanical erosion. South and southeast Asia therefore account for a large fraction of the sediment produced in the world. In contrast to the dissolved loads of rivers flowing into the Atlantic, suspended solids and sediments are carried to the Indian Ocean (e.g. along the Ganges and Indus) or to the Pacific (e.g. along the Mekong and Yangtze). This second scenario emphasizes the consumption of CO_2 and explains climatic cooling by the more intense erosion resulting from the uplift of the Himalayas.

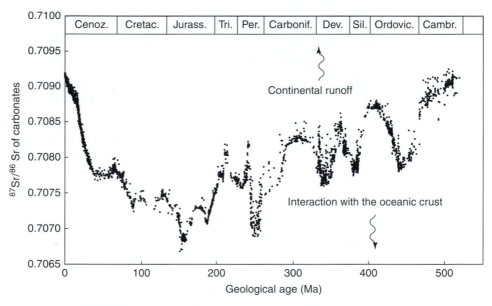

Figure 9.6 Evolution of the $^{87}Sr/^{86}Sr$ ratio in marine carbonates over geological time. The very low Rb content of these rocks ensures that the record of the ratio has not been significantly altered by radioactive ingrowth since deposition. The rapid increase in this ratio since the end of the Cretaceous is attributed to uplift of the high-$^{87}Sr/^{86}Sr$ Himalayas, which accelerated the input to the ocean of radiogenic strontium from the ancient basement of the mountain range. This indicates tectonic forcing of the rate of erosion with enormous climatic consequences.

9.1.3 Biogeochemical catastrophes in the Phanerozoic

The biogeochemical evolution of our planet is perturbed by occasional events perturbing the long-term trends described in the previous section. At about 65 Ma, i.e. at the boundary between the Cretaceous and the Paleocene, the Earth was impacted by a large asteroid (recently identified as the impactor of the Chixculub crater in the Yucatan) which triggered such huge atmospheric and oceanic effects that most living groups, and in particular dinosaurs and ammonites, disappeared in the aftermath of the collision. The correlation between the so-called K/T boundary mass extinction and a large meteoritic impact was first identified by the presence all over the world of a thin layer of sediments rich in iridium, a metal from the platinum family, that is orders of magnitude more abundant in meteorites than crustal rocks. Other mass extinctions, such as that taking place at the boundary between the Permian and the Triassic, have been ascribed to similar planetary-wide catastrophes.

   The geochemical consequences of the K/T boundary event are quite spectacular: the $\delta^{13}C$ of the bulk carbonate deposited either at the bottom of the ocean, as depicted in Fig. 9.7 for the South Atlantic (Shackleton and Hall, 1984), or on land, decreases instantaneously by some 2–3 per mil. It has also been found that the $\delta^{13}C$ difference between benthic and pelagic organisms which reflects the $\delta^{13}C$ gradient in the water column (Fig. 7.13) disappears, which indicates a nearly instantaneous collapse of global biological productivity.

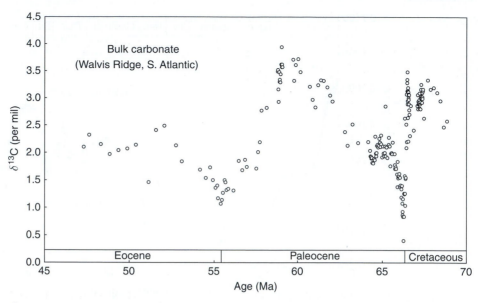

Figure 9.7 Carbon isotope compositions of the bulk carbonate in sediments from the Deep Sea Drilling Project hole 237 cored on the Walvis Ridge, South Atlantic (Shackleton and Hall, 1984). The sharp drop of the δ^{13}C values at the Cretaceous-Paleocene boundary and the lack of a C isotope difference between benthic and pelagic species (not shown) indicate that burial of organic material with negative δ^{13}C suddenly came to an end, thus signaling the failure of organic productivity. In contrast, the sharp decline of the δ^{13}C values at the end of the Paleocene is visible in both benthic and pelagic species: it is believed that this event represents the sudden release of methane from isotopically light methane hydrate trapped in sediments into the atmosphere.

Other events have a different geochemical signature and the associated mass extinction hence presumably must be due to different effects. The Paleocene–Eocene boundary at 55 Ma (Fig. 9.7) is also characterized by a sharp δ^{13}C decline, but this time both the benthic and pelagic organisms are shifted by approximately the same amount. It is believed that large amounts of isotopically light carbon were suddenly released into the ocean and the atmosphere. A common interpretation is the destabilization, for reasons that still remain to be established, of very large masses of methane hydrates present in sediments and permafrost and the liberation of large amounts of lethal methane gas into the atmosphere.

A third type of biogeochemical catastrophes which, however, have rarely led to extinctions of the same amplitude as meteoritic impacts, are the oceanic anoxic events (OAEs). These are episodes of deposition of sulfide-rich anoxic sediments rich in heavy metals known as black shales and resembling the organic-matter-rich, oxygen-starved laminites deposited at the bottom of the Black Sea today. The best known of these short events took place during the Toarcian (183 Ma), the early Aptian (120 Ma), and at the Cenomanian–Turonian boundary (93 Ma). It is not well understood whether they represent oxygen deficiency in the water column such as at the bottom of the Black Sea, or rather increased burial of organic material.

The unstable character or the redox conditions of the oceanic system is best demonstrated by the secular evolution of sulfur isotopes since the Cretaceous (Fig. 9.8).

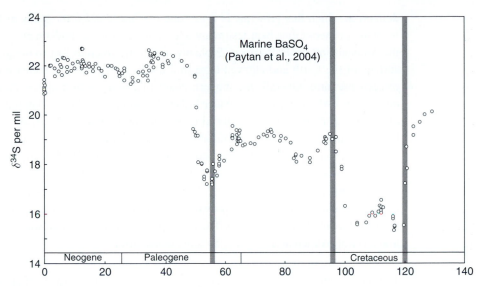

Figure 9.8   Sulfur isotope compositions of marine barites, $BaSO_4$, throughout the Cretaceous and the Tertiary (Paytan *et al.*, 2004). The shaded bars represent OAEs (Oceanic Anoxic Events). The sharp changes in $\delta^{34}S$ reflect strong variations in the burial rates of sedimentary pyrite and global changes in the redox system of the ocean.

Paytan *et al.* (2004) analyzed the sulfur isotope compositions of marine barites ($BaSO_4$), a mineral oversaturated in the surface ocean but undersaturated at depth. As for most sulfates, sulfur isotope fractionation upon barite precipitation is very small. The obvious advantages of this mineral over more common evaporites are its clearly marine origin and its shown immunity to diagenetic reworking. By about 125 Ma, the $\delta^{34}S$ decreased by 4 per mil, oscillated up and down for most of the Cretaceous and the Paleocene, and sharply increased to modern marine-sulfate values at about 55 Ma (22‰); it has stayed nearly constant ever since. There is probably no correlation with OAEs. Although such large variations must reflect variable burial rates of sedimentary pyrite, the fast rate of change, in particular at 55 Ma, suggests either very large sulfur fluxes or fairly low sulfate concentrations in the ocean.

9.2 The rise of atmospheric oxygen

9.2.1 The 2.1 Ga crisis

Charles Leyll's uniformitarianism states that "the present is key to the past," which asserts that fundamentally the same geological processes operate today as in the distant past. In modern language, we could simply state that the Earth has long since reached some form of steady state. Although this view proved helpful to elucidate some geological observations, the memory of the first geological times did not just disappear overnight. If there is a domain in which this principle is worth questioning, it is for the study of the distant past for which observations bear testimony that many geological phenomena were different in the Archean and the Proterozoic. Probably one of the changes that so uniquely affected the surface of the Earth and allowed terrestrial life to thrive was the increase of atmospheric oxygen pressure. Overall, the modern chain of events is as follows: (i) photosynthesis dissociates carbon dioxide to produce reduced biological carbon, (ii) burial and subduction of biological matter remove reduced carbon to the mantle, and (iii) the residual oxygen builds up in the atmosphere. To this should be added the dissociation of atmospheric water by solar ultraviolet radiation in the upper atmosphere: loss of hydrogen, too light to be held back by terrestrial gravitation, contributes to the increase of atmospheric oxygen. We will return to this process in Chapter 12. The modern biological processes by which oxygen is released into the atmosphere probably appeared with cyanobacteria (now fossilized as stromatolites) before 3.0 Ga ago. The atmosphere before that time was characterized by a low level of oxygen as attested to by three observations:

1. Banded iron formations (BIF) are known from Archean terranes of all continents. The archetypal 2.5 Ga old Hamersley formation in Australia is one of the largest iron deposits in the world. These are sediments made of mm-thick laminae of quartz and magnetite Fe_3O_4, which can be traced over very long distances and record some distinctive astronomical cycles. Most of these rocks are devoid of detrital minerals, which also hints at a pelagic depositional environment. As $Fe(OH)_3$ is essentially insoluble in oxic waters, it has been suggested that the deep ocean was rich in soluble Fe^{2+} and silica, most likely introduced by submarine geothermal activity, and that locally oxidizing conditions, possibly biologically mediated, would lead to massive precipitation of iron hydroxides and silica (Holland, 1973). Banded iron formations are rare after 1.8 Ga but make a remarkable comeback in the late Proterozoic ($\approx$ 600–800 Ma).
2. Paleosols, 2.2 to 2.4 Ga old, which are rich in chlorite, sericite (a mica similar to muscovite produced by weathering), and quartz are known in Canada and South Africa. They show a characteristic loss of iron unknown in modern soils, which demonstrates that groundwater was oxygen depleted and that Fe was in its soluble ferrous form.
3. Minerals prone to atmospheric oxidation and subsequent dissolution, most conspicuously uraninite UO_2, are essentially absent from detrital rocks. Large uraninite deposits, however, are common in Archean rocks such as the 3.0-Ga-old Witwatersrand conglomerates (South Africa). They demonstrate that the atmospheric oxygen level must have been low.

For reasons which still remain to be explained, these features vanished by 2.1–1.9 Ga; this is broadly interpreted as indicating an abrupt rise in atmospheric oxygen pressure. This idea recently received strong support from sulfur isotope analyses in Archean sediments. Mass fractionation of the 32, 33, 34, and 36 isotopes of sulfur in recent rocks back to 2.1 Ga is perfectly mass-dependent, which means that, for example, variations of the $^{34}S/^{32}S$ ratio are twice the variations of $^{33}S/^{32}S$. In other words, the quantity $\Delta^{33}S$ calculated as

$$\Delta^{33}S = \delta^{33}S - 0.5\,\delta^{34}S \tag{9.1}$$

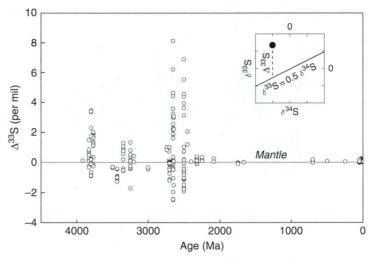

Figure 9.9 Evidence for mass-independent isotope fractionation of sulfur isotopes in sediments (Farquhar *et al.*, 2000). The value of $\Delta^{33}S = \delta^{33}S - 0.5\ \delta^{34}S$ is a measure of the deviation from the normal mass-dependant isotope fractionation behavior. Mass-independent isotope fractionations prior to ≈ 2.45 Ga indicate photolysis of atmospheric SO_2 by ultraviolet radiation and disproportionation of sulfur compounds into reduced and oxidized species. After that time, the rise of atmospheric P_{O_2} led to the formation of the ozone layer at the base of the stratosphere and protected SO_2 from UV photolysis.

is very near zero. It was shown by Farquhar *et al.* (2000) (Fig. 9.9) that prior to 2.45 Ga, the S isotopic abundances deviated very substantially from the mass-dependent conditions or, in other words, that $\Delta^{33}S$ was very different from zero. The interpretation calls for different sulfur and oxygen cycles over geological times. In the modern atmosphere, ultraviolet radiation does not reach the lower atmosphere: it is absorbed by the ozone layer located in the stratosphere at ≈ 25 km above the ground before SO_2 pressure becomes significant. Modern atmospheric SO_2 is therefore oxidized to sulfate and is washed down with precipitation, all the reactions being characterized by regular mass-dependent isotope fractionations. In contrast, prior to 2.45 Ga, the oxygen atmospheric pressure was too low to produce a significant ozone layer. As shown by experiments, mass-independent isotope fractionation results from the breakdown of SO_2 by ultraviolet radiation (UV photolysis) and its subsequent recombination as reduced and oxidized compounds. The UV photolysis of Archean atmospheric SO_2 therefore produced a variety of compounds that do not follow a mass-dependent fractionation behavior, hence making non-zero $\Delta^{33}S$ values a strong marker of the ozone layer and therefore of the total oxygen pressure. It was recently found that $\Delta^{36}S$ correlates with $\Delta^{33}S$, which makes the case of mass-independent fractionation even stronger.

The isotopic record of sulfur and nitrogen provides additional clues about the evolution of the pre-Phanerozoic atmosphere:

1. Average values of sulfur isotopes of sedimentary pyrite show mantle-like $\delta^{34}S$ values (~ 0) from the oldest samples well into the late Proterozoic (≈ 600 Ma). This reflects the

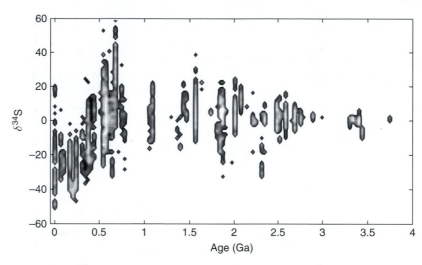

Figure 9.10 Evolution of the $\delta^{34}S$ values in pyrites from sediments (compilation courtesy of Don Canfield) over the geological ages. The values remain close to zero until ≈ 600 Ma, which suggests that until then the bulk of oceanic sulfur was dominantly in the form of sulfide. The rise of P_{O_2} in the late Proterozoic shifted sulfur speciation in seawater from sulfide to sulfate, yet the incoming fluxes maintained the $\delta^{34}S$ of this reservoir at mantle values. Sulfur isotopes in < 600-Ma-old pyrite are substantially fractionated with respect to marine sulfate (Canfield, 2004).

fact that oceanic sulfur is dominantly of mantle origin (volcanic eruptions, hydrothermal activity). Because sulfur isotopes fractionate between sulfates and sulfides by about 20 per mil (Chapter 3), this record indicates the lack of competing reservoirs of sulfate and sulfur (Fig. 9.10) prior to the Phanerozoic. This suggests that, prior to the Phanerozoic, most of the oceanic sulfur was not present as sulfate but was dominantly in the form of sulfide delivered by magmas and hydrothermal activity and went to the sediments with no isotopic fractionation.

2. The nitrogen isotopes of sedimentary kerogen show a strong increase in $\delta^{15}N$ values from slightly negative mantle-like values prior to 2.7 Ga to positive values (up to + 10) over the late Archean, Precambrian, and Proterozoic. These data can be interpreted as indicating that nitrification and denitrification did not operate in the early Archean the way they operate today.

3. In contrast, the $\delta^{13}C$ values of ancient kerogen are in the range of their modern counterparts.

9.2.2 The Snowball Earth and the emergence of Metazoans

A second major oxygen crisis leading to another major increase in atmospheric P_{O_2} took place during the Late Proterozoic. The BIFs that had disappeared from the surface of the Earth after 1.85 Ga returned on different continents between 800 and 600 Ma, at localities such as Rapitan, Canada, and Urucum, Brazil, and give the first indication that the atmospheric system was severely tipped out of balance with the return of oxygen deficiency. These layers are overlain by rather poorly dated Neoproterozoic glacial deposits. These cold spells are known as the Sturtian (≈ 730 Ma), Marinoan (≈ 640 Ma), and Gaskiers (≈ 590 Ma) glaciations and are of worldwide extension. The deposits are characterized by laminated layers of fine clays and siltstones, known as diamictites, in which are embedded large clastic fragments; some of them are angular, others rounded and decorated with striations. These rocks are identified beyond doubt as glacial deposits. Above the diamictites occur discordant massive layers of dolomite-rich carbonates, known as the cap-carbonates. Associated with these dolostones, or on top of them, phosphorites are commonly found, which contain wonderfully preserved multicellular organisms, e.g. in Doushantuo in China. These are the first incontrovertible Metazoans, also endowed with body symmetry and motility. With these layers began the Ediacaran radiation and the advent of multicellular life. Although its members were unequally successful with respect to their descendance, the Ediacaran fauna was incredibly rich and seemed to bloom very rapidly in the aftermath of the Gaskiers event.

Although many of these geological observations were made up to 40 years ago, it is only during the last decade that we have been able to start making sense of them. In the late 1980s, paleomagnetic data on some of the iron-rich glacial layers showed that glaciers reached sea level even under tropical and even equatorial latitudes. Such a surprising observation is unlikely to reflect a tilted terrestrial spin axis, because the equatorial bulge of our planet, like the bulge of a top, is stabilized near its current position by the attraction of the Moon. The term of Snowball Earth was coined in 1992 by Joe Kirschvink to reflect the extensive glaciation of the planet to very low latitudes, possibly sealing off most of the ocean surface from sunlight. With extensive ice blanketing, increased reflection of the solar radiation cooled the Earth even more (in jargon, the albedo became very high), and if it were not for the greenhouse effect, the Earth could have remained frozen for eons. These glacial bouts probably lasted millions of years, long enough for most previous life forms to become extinct. It is believed that volcanic emissions of CO_2 saved our planet from the ice doom: by progressively building up in the atmosphere, CO_2 captured more and more of the solar radiation, surface temperature increased and the ice quickly melted. Weathering under high atmospheric P_{CO_2} acting on a barren crust triggered an alkalinity rush to the ocean and led to the massive sedimentation of carbonates.

Carbon isotopic compositions of the cap-carbonates were found to be quite informative. The $\delta^{13}C$ values nose-dive by more than 15 per mil over a very narrow sedimentary interval (Fig. 9.11). One of the early interpretations appealed to the nearly complete cessation of photosynthetic activity and to the demise of the biomass as a result of the ice blocking sunlight. Carbon would have the isotopic composition of mantle inputs, i.e. ≈ -7 per mil. The discovery of $\delta^{13}C$ lower than mantle values rather suggests that an oceanic reservoir of isotopic light carbon was being oxidized. Destabilization of gas hydrates has been suggested to provide enough carbon with very negative $\delta^{13}C$ values, but such speculation still remains to be tested. Regardless of which process caused the carbon isotope shift, it left the terrestrial ocean and the atmosphere much more oxidized than in the pre-glacial times. It is believed that access to higher P_{O_2} allowed multicellular organisms to increase their metabolism and their motility so as to lead the way to the thriving Phanerozoic life.

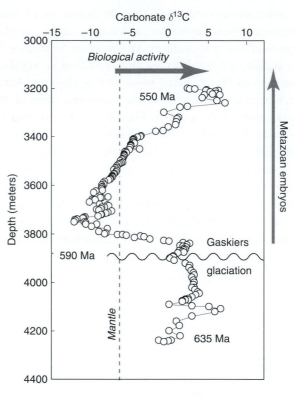

Negative excursion of δ^{13}C values after the Gieskes glaciation in Oman (data from Fike *et al.*, 2006). The sediments are discordant over the Marinoan glacial deposits ($\approx$ 635 Ma). The Gieskes glaciation ($\approx$ 590 Ma) locally misses diamictites but can be recognized with the cap-carbonates in the middle of the section. The δ^{13}C values lower than those of the mantle indicate that a carbon reservoir rich in ^{13}C, possibly gas hydrates, was being oxidized. Oxidation coincides with the radiation of multicellular organisms (Metazoan).

As previously alluded to, oxidation of the ocean also resulted in the oxidation of oceanic sulfur into sulfate, which clearly shows in the burial of pyrite with negative δ^{34}S values (Fig. 9.10). The reason why such a massive rise of oxygen in the atmosphere and the ocean took place remains so far poorly understood.

9.3 The geochemical environment of the origin of life

In a famous experiment, Miller and Urey (1959) triggered electrical discharges in a balloon containing ammonia, methane, and liquid water. After a few hours, the solution became very rich in amino acids, which are the building blocks of proteins. After first being hailed as a breakthrough illuminating our understanding of the origin of life, this experiment was heavily criticized for misrepresenting the early Earth atmosphere, which, critics said, was largely dominated by CO_2 and N_2, and therefore too oxidizing to allow for methane and ammonia to be stable. Miller and Urey's work fell into near oblivion for decades. This criticism incorrectly assumes that all atmospheric gases would be well mixed and everywhere in equilibrium. The actual situation is very different since, although the atmosphere was dominated by carbon dioxide and nitrogen, along mid-ocean ridges, hydrogen, methane, and ammonia were produced in abundance by serpentinization reactions. Production of serpentine results from the combination of two reactions between water and olivine, a major mineral of basaltic rocks and a solid solution of Fe and Mg components. First, iron from the Fe-olivine component (fayalite) is oxidized by water to produce silica, magnetite, and hydrogen:

$$3Fe_2SiO_4 + 2H_2O \Leftrightarrow 2Fe_3O_4 + 3SiO_2 + 2H_2$$
$$\text{(olivine) (solution) (magnetite) (quartz) (gas)} \tag{9.2}$$

Second, silica liberated in this reaction reacts with water and Mg-olivine to form serpentine:

$$3Mg_2SiO_4 + SiO_2 + 4H_2O \Leftrightarrow 2Mg_3Si_2O_5(OH)_4$$
$$\text{(olivine) (silica) (solution) (serpentine)} \tag{9.3}$$

Hydrogen liberated by serpentinization reacts with CO_2 and N_2 to give methane and ammonia:

$$4H_2 + CO_2 \Leftrightarrow CH_4 + 2H_2O$$
$$3H_2 + N_2 \Leftrightarrow 2NH_3 \tag{9.4}$$

The Mid-Atlantic Ridge today shows how this may have happened: reaction of water with hot igneous material at temperatures $< 600\,°C$ produces hydrogen, which in turn reacts with CO_2 and N_2 to produce methane and ammonia, respectively. Whether hydrogen was lost from the atmosphere and methane and ammonia oxidized by atmospheric gases is not relevant to such a dynamic system: the source of reducing gas was enormous and steady, and favorable geological sites must have been ubiquitous at the bottom of the ocean. The floor of mid-ocean ridges, and also the ocean floor on top of the magma ocean in the early stages of the Earth's evolution, must have represented major production sites for methane and ammonia. The lack of an ozone layer probably allowed solar ultraviolet radiation to reach the surface of the Earth. Irradiation of the surface of the ocean by solar UV wherever deep oceanic upwellings would degas dissolved methane and ammonia into the atmosphere provided natural conditions for the formation of proteins and the emergence of life.

How early life could resist oxidizing conditions and find nutrients to keep going is another issue. Today, the main terrestrial productivity is concentrated at the surface of the oceans. The most important nutrients, nitrates and phosphates, are, however, brought to the sea by the rivers. Mid-ocean ridges produce no nitrate, but we do not know which nitrogen source was used by early organisms, presumably dissolved nitrogen or ammonia. In contrast, modern ridges are a net sink for phosphorus, and if rivers were to dry out overnight, this element so critical for life would rapidly disappear from the ocean and, with it, oceanic productivity. It is likely that, at some point, the emergence of continents, another hallmark of our planet with respect to its neighbors, started carrying renewable resources to the ocean for life to grow stronger and stronger. A water world may not have bred life with the same efficiency as our dual world of oceans and continents which is so unlike other planets. The early emergence of plate tectonics may have been instrumental in defining the conditions for long-lasting biological activity.

Continents may have been paid back for the help they provided life. We have seen and will discuss again in the next chapter that biological activity strongly interferes with diagenesis. It actually increases the efficiency of erosion, helps process more detrital sediments, and in the long run increases the proportion of hydrous material, such as clay minerals, which are much more prone to melting than their anhydrous equivalents. Biological activity may therefore have strengthened the production of orogenic magmas and eventually the amount of continental crust produced. The feedback loop between plate tectonics and the emergence of life is likely to have endowed our planet with the main features of its biogeochemical dynamics.

Exercises

1. In the range 0–20 °C, the vapor pressure of water at saturation $P_{H_2O}^{sat}$ changes with temperature T as $\ln P_{H_2O}^{sat} = -5365.37\, T^{-1} + 26.06$, where pressure is in Pa and temperature in K. Assume that atmospheric water vapor forms above the ocean at low latitudes at 15 °C and calculate a relationship between the residual fraction f of water vapor, as given by the last exercise of Chapter 3, and temperature. Assuming that $1000 \ln \alpha^{18}O = 1.0779\ 10^6\ T^{-2} - 2.796$, infer a relationship between the $\delta^{18}O$ values of rain water and their precipitation temperature. Find appropriate linear approximations to all these equations.

2. Use the Mg/Ca ratios from Table 9.1 to calculate the sea-surface temperature of the Eastern Tropical Pacific over the last glacial cycle. Use $T(°C) = -0.667x^2 + 7.76x + 8.73$ with $x = $ Mg/Ca. Which cycles can you identify?

Table 9.1 Mg/Ca in *Globigerinoides ruber* from the Eastern Tropical Pacific for different ^{14}C calendar ages (Lea, 2004)

$t(y)$	Mg/Ca	$t(y)$	Mg/Ca	$t(y)$	Mg/Ca	$t(y)$	Mg/Ca
1800	2.99	27800	2.32	73500	2.61	116400	2.86
4400	3.07	32200	2.59	78600	2.79	121500	3.20
6900	3.16	39800	2.56	83600	2.78	125400	3.53
9500	3.06	44100	2.55	88700	2.55	131000	3.19
12000	2.84	50900	2.53	93700	2.63	134100	2.79
14300	2.72	55000	2.63	98800	2.87	138000	2.54
16500	2.52	58600	2.53	101300	2.95	141400	2.51
18800	2.38	65700	2.40	106300	2.93	144900	2.52
23300	2.44	69200	2.45	111400	2.78	148400	2.63

3. Use data from Appendices A and F to compute a modern value of the sulfate residence time in the ocean. Draw a sulfur cycle involving erosion, rivers, the ocean, ridges, and the burial of sedimentary pyrite. Examine Fig 9.8 and propose possible interpretations to the fast $\delta^{34}S$ oscillations in the Cretaceous and the Paleocene.

References

Canfield, D. E. (2004) The evolution of the Earth surface sulfur reservoir. *Am. J. Sci.*, **304**, 839–861.

♠ Dansgaard, W. (1964) Stable isotopes in precipitation. *Tellus*, **16**, 436–468.

Farquhar, J., Bao, H. M., and Thiemens, M. (2000) Atmospheric influence of Earth's earliest sulfur cycle. *Science*, **289**, 756–758.

Farrell, J. W. and Prell, W. L. (1989) Climatic change and $CaCO_3$ preservation: an 800 000 year bathymetric reconstruction from the Central Equatorial Pacific Ocean. *Paleoceanogr.*, **4**, 447–466.

Fike, D. A., Grotzinger, J. P., Pratt, L. M., and Summons, R. E. (2006) Oxidation of the Ediacaran Ocean. *Nature*, **444**, 744–747.

Holland, H. D. (1973) Oceans – possible source of iron in iron-formations. *Econ. Geol.*, **68**, 1169–1172.

Kirschvink, J. L. (1992) Late Proterozoic low-latitude global glaciation: the Snowball Earth. In *The Proterozoic Biosphere* (ed. J. W. Schopf and C. Klein), pp. 51–52. Cambridge: Cambridge University Press.

Lea, D. W. (2004) The 100, 000 year cycle in tropical SST, greenhouse forcing, and climate sensitivity. *J. Clim.*, **17**, 2170–2179.

♠ Miller, S. L. and Urey, H. C. (1959) Organic compound synthesis on the primitive Earth, *Science*, **130**, 245–251.

Pagani, M., Zachos, J. C., Freeman, K. H., Tipple, B., and Bohaty, S. (2005) Marked decline in atmospheric carbon dioxide concentrations during the Paleogene. *Science*, **309**, 600–603.

Paytan, A., Kastner, M., Campbell, D., and Thiemens, M. H. (2004) Seawater sulfur isotope fluctuations in the Cretaceous. *Science*, **304**, 1663–1665.

Petit, J. R., Jouzel, J., Raynaud, D., *et al.* (1999) Climate and atmospheric history of the past 420 000 years from the Vostok ice core. *Nature*, **399**, 429–439.

♠ Shackleton, N. J. and Hall, M. A. (1984) Carbon isotope data from Leg 74 sediments. *Init. Repts. DSDP*, **74**, 613–619.

Zachos, J. C., Dickens, G. R., and Zeebe, R. E. (2008) An early Cenozoic perspective on greenhouse warming and carbon-cycle dynamics. *Nature*, **451**, 279–283.

Mineral reactions

In this chapter we take a look at the chemical and mineralogical changes accompanying the formation of sedimentary and metamorphic rocks. Marine mud is consolidated into sediment through early diagenetic reactions. These sediments and the other rocks that form the bedrock may be affected by percolating hot water, mineralized to varying degrees, and acids. This is hydrothermal metamorphism. When rock is dragged deep down by subduction and thereby heated and dehydrated, it is transformed, producing a great variety of different mineral assemblages. This process is termed "metamorphism." Some of these thermal processes concern the transformation of organic matter, whose ultimate products are the fossil fuels such as natural gas, petroleum, and coal.

The principal geochemical issues raised by these processes are to identify the nature of the rock before its transformations, the physical conditions (temperature and pressure) of the transformations, and the nature and intensity of exchanges between the transformed rocks and the interstitial solutions. Once again the essential analytical tool is thermodynamics and readers may wish to refresh their knowledge of this by referring to Appendix C.

Transformations are often controlled by water pressure and temperature. Let us take the example of the important reaction whereby muscovite (white mica) disappears from gneiss and schist, and which characterizes the entry of metamorphic rocks into granulite facies:

$$KAl_3Si_3O_{10}(OH)_2 + SiO_2 \Leftrightarrow KAlSi_3O_8 + Al_2SiO_5 + H_2O$$
$$\text{(muscovite)} \quad \text{(quartz)} \quad \text{(K-feldspar)} \quad \text{(sillimanite)}$$

$$(10.1)$$

Equilibrium between pure solid and gaseous phases (Appendix C) allows us to write:

$$\ln P_{H_2O} = \frac{\Delta H}{RT} + \text{constant} \qquad (10.2)$$

where ΔH is the heat (enthalpy) of the reaction. Other reactions involve other gases, such as carbon dioxide, but they are treated in the same way.

The behavior of oxygen is particularly important. The redox reaction between two types of common oxides in igneous and metamorphic rocks:

$$6Fe_2O_3 \Leftrightarrow 4Fe_3O_4 + O_2$$
$$\text{(hematite)} \quad \text{(magnetite)} \qquad\qquad (10.3)$$

is equivalent to the sum of the two half-reactions:

$$6Fe_2O_3 + 2e^- \Leftrightarrow 4Fe_3O_4 + 2O^- \qquad (10.4)$$
$$2O^- \Leftrightarrow O_2 + 2e^- \qquad (10.5)$$

This shows that redox reactions are merely a trade in electrons between acceptors (oxidants such as oxygen) and donors (reducing agents such as Fe^{2+}). There is nothing wrong, for example, in considering ferrous iron Fe^{2+} as a complex of ferric ion Fe^{3+} with an electron. As the vast majority of rocks are good electrical insulators, electrons must be conserved throughout mineralogical reactions and phase changes. Appealing to an externally imposed oxygen "fugacity" would be misleading: each and every reduction reaction, i.e. any gain in electrons by one species, must be offset by loss of electrons by another species. No local surplus or deficit of electrons is permitted and species having more than one state of oxidation (above all Fe, C, and S) must engage in balanced electron exchanges. As with water, oxygen pressure can be formulated by the mass action law. For example, for the relationship between hematite and magnetite, we write the equation:

$$\ln \frac{[Fe_3O_4]^4_{\text{magnetite}} P_{O_2}}{[Fe_2O_3]^6_{\text{hematite}}} = \frac{\Delta H}{RT} + \text{constant} \qquad (10.6)$$

in which the square brackets indicate the molar fraction of the species in question in each of the solid solutions containing hematite and magnetite, and P_{O_2} is oxygen pressure. Relationships like this can be used either for estimating temperature if the redox state of the system is known, or vice versa for measuring oxygen pressure if temperature is known. The fugacity of oxygen in many natural rocks is distributed around the famous QFM (quartz–fayalite–magnetite) buffer:

$$3Fe_2SiO_4 + O_2 \Leftrightarrow 3SiO_2 + 2Fe_3O_4$$
$$\text{(fayalite)} \quad\quad \text{(quartz)} \quad \text{(magnetite)} \qquad (10.7)$$

This equation does not mean that these minerals are present in the rocks, but that the electron balance of the actual mineral assemblage is on average close to that of this buffer.

10.1 Early diagenesis

We have already introduced this issue in Chapter 8. Let us go back to our observers on the sea floor watching the sediment sinking beneath their feet (Fig. 10.1) and draw up the balance sheet of matter for one species, say sulfate, between two levels at depth Δz one above the other (Fig. 10.1).

The sediment porosity (proportion of interstitial water) is φ and the sulfate concentration in the water is $C(z)$. In the stationary state, i.e. after "some" time, we can write that the difference between the influx of sulfate at z less the outflux at $z + \Delta z$ is equal to the quantity of sulfate destroyed by biological activity. For our observer, the advective flux of sulfate is equal to $\varphi v C(z)$, where v is the sedimentation rate, corrected where necessary for compaction, which progressively expels water from the sediment. Likewise, the diffusive flux of sulfate is $-\varphi D \, dC(z)/dz$, where D is the diffusion coefficient of sulfate in salt water. We can then write the diagenesis equation, standardizing it to a unit area:

$$\left[\varphi v C - \varphi D \frac{dC}{dz} \right]_z - \left[\varphi v C - \varphi D \frac{dC}{dz} \right]_{z+\Delta z} = -P(z)\Delta z \qquad (10.8)$$

where $P(z)$ is the rate of destruction of sulfate by microbial activity. The rate $P(z)$ will be expressed in a suitable form, probably by a first-order kinetic with an appropriate stoichiometric coefficient (see Chapter 5) involving the availability of organic carbon. Solutions to the diagenetic equation are usually combinations of exponentials. By examining (10.8) above, we can infer that the concentration of sulfate in pore water, like the abundance of organic particles, must decrease with depth: Fig. 10.2 shows the concentrations of sulfate ion in the interstitial water of sediments collected from the Saanich River fjord (west coast of Canada). It could be shown that (10.8) implies exponential decrease in sulfates from the initial sulfate concentration of seawater, which Fig. 10.2 confirms. The reduced sulfur is in

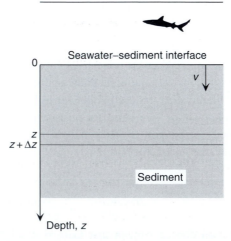

Figure 10.1   The diagenesis reference frame. Sediment sinks beneath the ocean floor at the rate *v*.

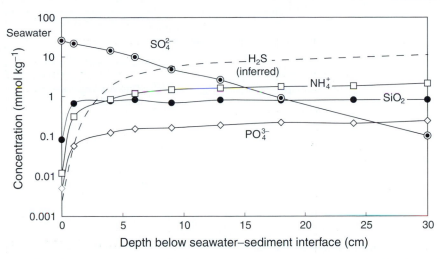

Figure 10.2 Concentration of various species dissolved in interstitial water of the Saanich River fjord (western Canada). With burial, sulfate is reduced to sulfide, organic nitrogen is reduced to ammonia (denitrification), and phosphate and silica of organic origin are remobilized by diagenesis. Data from Murray *et al.*, 1978.

the form H_2S, which eventually causes pyrite to precipitate, but in this core the sulfur data are missing. Laboratory measurement of diffusion coefficients and estimates of sedimentation rate by radiochronometric methods allow us to determine, by resolving the diagenesis equation, the rate at which sulfate is converted to sulfide and equally the rate of oxidation of the organic matter. Similar behavior is observed for nitrates in seawater. Dissolution of organic matter is reflected by a large increase in ammonia produced by the reduction of nitrogen in proteins and other organic compounds (denitrification). The increase in remobilized phosphate from phospho-organic compounds of soft matter and hard parts of certain organisms (fish teeth, ichthyoliths) is responsible for diagenetic reprecipitation of calcium phosphate (apatite), which may be intense enough to produce phosphate deposits of economic value. The silica remobilized by dissolution of diatom and radiolarian tests is locally reprecipitated as flint and the enclosing rocks can be further silicified as cherts. As seawater is highly undersaturated in SiO_2, the expulsion of interstitial water also provides an important source of silica for the ocean.

Until complete, these bacteria-controlled reactions are accompanied by substantial isotopic fractionation, of the order of 25 per mil for carbon and 20 per mil for sulfur, invariably with organic matter and sulfide having a preference for the lighter isotope.

10.2 Hydrothermal reactions

This term is reserved for all medium temperature reactions, typically from 100 to 500 °C, between aqueous solutions and the rock through which they circulate. These are largely, but not exclusively, hydration reactions. Hydrothermal solutions and ores are the sources of many metals of economic significance. The familiar thermal springs at the surface of

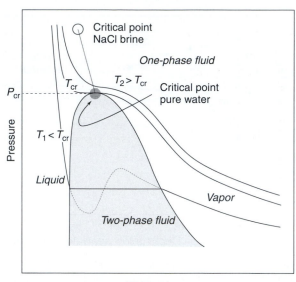

Molar volume

Figure 10.3 The critical behavior of aqueous solutions. At $T_1 = 100\,°C$, boiling of pure water upon decompression proceeds by phase separation (the dotted line shows a metastable trajectory). The horizontal segment connects the two points representing the molar volumes of the liquid and the vapor. At the critical temperature T_{cr}, these molar volumes become equal and transition from liquid to vapor takes place in a continuous manner. The state referred to as one-phase fluid is a continuum between the liquid state and the gas state. The lowest temperature and pressure T_{cr} and P_{cr} at which this can happen define the critical point. Note that adding salt to water significantly increases both T_{cr} and P_{cr}.

the continents give an imperfect sampling of hydrothermal solutions: because the boiling point of hydrous solutions increases from $100\,°C$ at ambient pressure to more than $350\,°C$ at a few kilometers below the surface, the compositions and temperatures of thermal springs do not faithfully represent the properties of the solutions in contact with deep rock. Most hydrothermal springs therefore have been affected by intense boiling, to which they are subjected as they rise to the surface. In contrast, the emergence temperatures of the solutions spouted by the black smokers along mid-ocean ridges can reach $400\,°C$ because of the high pressure prevailing at the ocean floor (200–450 atmospheres), and better represent the original hydrothermal solutions. Many geologically (and economically) important hydrothermal reactions take place at even higher temperatures (400–600 °C). High-temperature solutions are often trapped in some minerals, such as quartz, in the form of fluid inclusions.

In order to understand hydrothermal reactions, we must examine phase changes and critical phenomena in high-temperature hydrous solutions. Let us first take some liquid water at $100\,°C$ in an autoclave pressurized by a neutral gas such as air and progressively release the pressure to the ambient value (Fig.10.3): boiling goes on until all the liquid has changed into vapor. For the time it takes to evaporate the liquid, we have a two-phase fluid with liquid and vapor separated by a visible interface. If we measure the molar volume, it slightly increases upon decompression in the liquid domain, remains fixed during boiling, and increases again, but much more rapidly, in the vapor domain. Let's now repeat the experiment at higher pressure and therefore at higher temperature: a similar sequence of events is obtained, but the difference between the molar volumes of the vapor and the liquid decreases with pressure, until we reach the critical pressure of 22 MPa. Then the volumes of both phases are identical and we pass continuously from the liquid field into the vapor field without seeing any interface. The temperature of the critical transition also is fixed at 374 °C. In nature, water is rarely pure and salts are dissolved, sometimes in large proportions (highly saline solutions are called brines). Adding salt to the solution increases both the temperature and pressure of the critical point. For example, a brine with 10 weight percent NaCl has a critical point at 700 °C and 120 MPa. Brine boiling produces rather light and dilute vapors and other brines, dense and even more concentrated in salt than the original solution. Highly concentrated brines with several tens of percent salt are important because they can dissolve enormous amounts of metals such as transition elements and rare-earth elements which, upon cooling, produce ore bodies of occasionally economical value. A hydrothermal fluid rich in Cl^-, most likely because it derives from seawater, must contain a heavy load of cations simply to equilibrate the negative charges. Another reason why hot brine chemistry is different from that of low-temperature solution is the low dielectric constant of water which enhances the recombination of ionic species, such as Na^+, H^+, and Cl^-, into molecular species, such as NaCl and HCl. Such molecular species are true complexes and increase the apparent solubility of metals in hydrothermal solutions. Hydrothermal deposits may form, however, from more diluted solutions at relatively low temperatures: this is the case of the Pb–Zn (base metal) Mississippi Valley-type ore deposits.

Balancing hydrothermal reactions requires a combination of cation-exchange reactions between the rock and the solution and the principle of electrical neutrality. First we need to draw up an inventory of proton/cation exchange reactions such as (7.22), which can be simply re-written:

$$2NaAlSi_3O_8 + 2H^+ + H_2O \Leftrightarrow Al_2Si_2O_5(OH)_4 + 4SiO_2 + 2Na^+$$
$$\text{(albite)} \quad \text{(solution)} \quad\quad \text{(kaolinite)} \quad \text{(silica)} \quad \text{(solution)}$$

For pure mineral phases, this equation leads to:

$$\ln \left(\frac{[Na^+]}{[H^+]} \right)_{\text{solution}} = \frac{\Delta H}{RT} + \text{constant} \tag{10.9}$$

At a given temperature, this ratio is constant. The enthalpy ΔH of the reaction varies from one reaction to another. A solution rich in H^+ therefore displaces cations from the rock more efficiently than a high-pH solution. This phenomenon is heavily dependent on temperature. Moreover, the positive charges must balance the negative charges, which are very often dominated by Cl^-, an anion scarcely involved in any mineral reactions and whose concentration remains virtually constant. The chemistry of the solution and therefore its capacity to modify the chemistry of the rock depends essentially on all the thermodynamic properties of exchange reactions and on temperature.

Among all the parameters of hydrothermal activity that need to be understood, temperature is, along with Cl concentration, the most important. For thermometry, it is

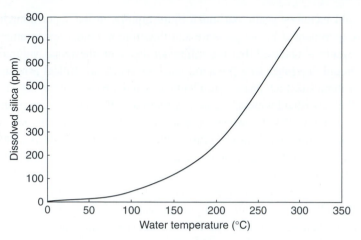

Solubility of silica in water. Silica contents of thermal waters can be used as a thermometer to a temperature of about 220 °C, and thus indicate the depth of equilibrium. Above this temperature, amorphous silica precipitates as the fluid rises.

commonplace to utilize analysis of the water–rock equilibria for hydrothermal reaction thermometry, particularly K/Na fractionation induced by the exchange reaction between feldspar and the solution:

$$K^+ + Na^+ \Leftrightarrow K^+ + Na^+ \tag{10.10}$$
$$\text{(solution)} \quad \text{(feldspar)} \quad \text{(feldspar)} \quad \text{(solution)}$$

Assuming that feldspars in the reservoir zone of the solution have the compositions of common feldspars from granites and metamorphic rocks, the mass action law yields the equation:

$$\log_{10}\left(\frac{[Na^+]}{[K^+]}\right)_{solution} = \frac{908}{T} - 0.70 \tag{10.11}$$

The sodium and potassium contents of a hydrothermal solution therefore allow an equilibrium temperature with feldspar to be deduced. Solutions can also dissolve minerals without any reaction product: this is the case for silica, the content of which in spa water is used as a thermometer by writing the thermodynamic law of saturation (Appendix C):

$$\frac{d\ln[SiO_2]_{solution}}{d(1/T)} = \frac{\Delta H}{R} \tag{10.12}$$

where ΔH is the heat of dissolution of silica in water. By means of a few approximations and by introducing experimental values, the silica thermometer equation is:

$$\log_{10}[SiO_2]_{solution} = -\frac{1306}{T} + 0.38 \tag{10.13}$$

(note the logarithms with different bases). Assuming that there is surplus silica, which is true for most continental rocks, the measurement of the silica content of thermal water gives, with this equation, the equilibration temperature of the solutions with the deep rocks (Fig. 10.4).

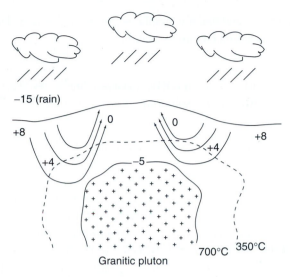

Figure 10.5 Geothermal system controlled by the emplacement of a granite pluton: rain water infiltrates and when heated in proximity to the intrusion becomes less dense and percolates back to the surface. Values shown (in ‰) are those of $\delta^{18}O$ of rock (and of precipitation). Rain water is strongly depleted in ^{18}O and changes the values of rocks, initially here at +8‰, to lower values of $\delta^{18}O$, depending on temperature and the water/rock ratio. Crystallization of the granite magmas may release magmatic waters.

From these solubility relationships, we gain some understanding of the mineral associations present in the fractures observed in granitic and metamorphic basements. Pegmatites, characterized by large crystals of albite and K-feldspars, precipitate from relatively high-temperature hydrous fluids (400–600 °C), while the quartz veins, which are commonly associated with ore deposits, form at lower temperatures (200 –400 °C).

Water–rock interactions may also leave a profound mark on the isotopic compositions of the geothermal systems associated with magmatic provinces, such as Wairakei in New Zealand, Larderello in Italy, and the Geysers of California. Felsic intrusions emplaced at a few kilometers below the surface act as deep heat engines; they promote convection of the pore fluids and isotopic exchange between meteoric water and the rock matrix (Fig. 10.5). Rain water being particularly depleted in the heavy isotopes of oxygen and hydrogen, the $\delta^{18}O$ and δD of the magmatic intrusions themselves and of their country rocks are significantly shifted towards lower values upon exchange with meteoric water.

The markedly marine signature of isotopic compositions of oxygen and hydrogen in solutions from black smokers indicates that these formed by seawater infiltrating the edges of the ridges and by reaction at depth with the still hot basalt. Hydrothermal reactions at the mid-ocean ridges play an important role in the magnesium cycle and in controlling the alkalinity of the ocean. It has been observed that water from the black smokers of the ocean ridges is particularly acidic, i.e. its pH is much lower (typically 3) than that of deep ocean water (7.6), and that it is totally devoid of magnesium. This can be explained by reactions of seawater with common basalt minerals:

$$\underset{\text{(olivine)}}{Mg_2SiO_4} + \underset{\text{(solution)}}{Si(OH)_4} + H_2O + \underset{\text{(solution)}}{Mg^{2+}} \Leftrightarrow \underset{\text{(serpentine)}}{Mg_3Si_2O_5(OH)_4} + \underset{\text{(solution)}}{2H^+}$$

whose effect is to exchange the abundant Mg^{2+} in seawater for protons and therefore to reduce its alkalinity. In such reactions, the silica in the solution comes from dissolution of various enclosing igneous rocks (basalt, gabbro).

The redox reactions between seawater and hot basaltic rock first explains why hydrogen, methane, and ammonia continuously seep out from mid-ocean ridges and in aquifers located in ultramafic rocks, notably those of ophiolitic environments. The most important reactions are probably the oxidation of the ferrous component of Fe-olivine (fayalite) by water and its complement, the serpentinization of Mg-olivine (forsterite), which we discussed in the previous chapter; and the reaction of hydrogen with CO_2 and N_2 to produce methane and ammonia.

As the state of oxidation of the oceanic crust at the time it is subducted also dictates the long-term evolution of the redox state of the mantle, similar reactions involving iron and sulfur are also important. The two following half-reactions of oxidation–reduction:

$$Fe^{2+} \Leftrightarrow Fe^{3+} + e^- \tag{10.14}$$

$$S^{2-} + 4H_2O \Leftrightarrow SO_4^{2-} + 8H^+ + 8e^- \tag{10.15}$$

can be represented schematically by oxidation in an acidic medium of the ferrous iron of magmatic rocks offset by reduction of marine sulfate into sulfide ions (see above for the origin of H^+ ions). By multiplying the first equation by eight and subtracting the second, we obtain:

$$\underset{\text{(seawater)}}{SO_4^{2-}} + \underset{\text{(seawater)}}{8H^+} + \underset{\text{(basalt)}}{8Fe^{2+}} \Leftrightarrow \underset{\text{(solution)}}{S^{2-}} + 4H_2O + \underset{\text{(solution)}}{8Fe^{3+}}$$

Ferric ion precipitates as hydroxide $Fe(OH)_3$ and S^{2-} as sulfide minerals of iron and copper (chalcopyrite), and zinc (sphalerite) forming the mineralized submarine chimneys of the black smokers. Sulfur isotopes indicate, however, that, within the mid-ocean ridge hydrothermal systems, sulfur derived from seawater through the previous reaction is dominated by sulfur leached from the basaltic rocks. The sulfate-reduction reaction also transforms part of the ferrous iron of the oceanic crust into ferric iron stored as oxides or silicates (epidote). Subduction of the oceanic crust therefore controls not only mantle hydration but also its relative proportions of ferric and ferrous iron, and its electrical conductivity. The Earth's mantle, into which enormous amounts of oxidized crust have been recycled over the Earth's history, is thus far more oxidized than the Moon's mantle.

10.3 Metamorphism

Metamorphic transformations affect all rocks drawn down deep into the Earth by subduction. They are particularly perceptible where continents collide, i.e. when mountain ranges form. In contrast to hydrothermal reactions, these are largely, although not exclusively, dehydration reactions under the effect of temperature or excess CO_2. Metamorphic facies correspond to given temperature and pressure ranges and these are usually bounded by specific mineralogical reactions: the greenschist facies (250–450 °C), the amphibolite facies (450–700 °C), and the granulite facies (> 700 °C) correspond to increasing temperatures at usual pressures; at higher pressures (> 30 km) we speak of blueschist facies, and at higher temperatures of eclogite facies.

The nomenclature of metamorphic rocks, based on their mineralogy, is fairly straightforward. Gneiss contains feldspar and quartz with variable proportions of other minerals and is frequently similar to granite in chemical composition. Schist contains little or no feldspar and is typically composed of quartz and mica; it is of similar composition to claystones. Amphibolite contains amphibole, with or without plagioclase feldspar; it is similar in composition to basalt. Most dehydrating metamorphic reactions could be described by dehydration reactions of the type described by (10.2) and represented by straight lines in a plot of $\ln P_{H_2O}$ vs. $1/T$ K, but it has become customary to represent metamorphic equilibria as curves on a simple pressure–temperature graph (T °C, P_{H_2O}). A metamorphic grid of this sort is commonly used to determine the temperature and pressure conditions prevailing in ancient metamorphic environments. Oxidation–reduction reactions are also used to determine temperature and oxygen pressure. Other reactions, finally, do not involve any fluid, as for example the polymorphic transformation of aluminum silicates:

$$Al_2SiO_5 \Leftrightarrow Al_2SiO_5 \Leftrightarrow Al_2SiO_5 \tag{10.16}$$
$$\text{(andalusite)} \quad \text{(kyanite)} \quad \text{(sillimanite)}$$

Such reactions, not usually directly involving fluids, are represented by straight lines in pressure–temperature diagrams (Fig. 10.6). The position of the Clapeyron curve of these different reactions in pressure–temperature space must be carefully calibrated by experiment or obtained by thermodynamics (see Appendix C).

At this stage we should look more closely at the relationships between fluid pressure – the fluid for simplicity we take to be pure water – and the pressure of the surrounding rock. As water is about three times less dense than rock, the weight of a column of water is three times less than the weight of a column of rock of the same height. Near the surface, the pores are interconnected and interstitial fluid pressure is therefore equal to hydrostatic pressure. The weight of the water column being about one-third the weight of the rock column, pressure in the rock matrix (lithostatic pressure) is higher than that of interstitial water. With depth, the rock is compacted and the pores tend to close, progressively isolating the pore water a few kilometers beneath the surface. Fluid pressure and rock pressure are then in equilibrium. When the temperature of this unit is raised by a metamorphic event, as the water has greater thermal expansivity than rock, it is at higher pressure than the rock

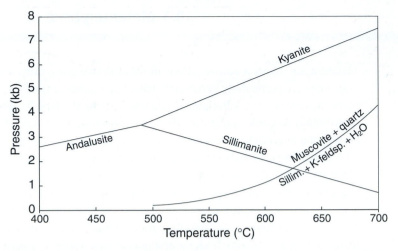

Figure 10.6 Metamorphic grid showing dehydration reaction curves that can be described by (10.2) and three solid–solid reactions for which the slopes are virtually constant.

matrix. As the tensile strength of the rock is generally low, this pressure contrast causes hydro-fracturing and water escapes.

Metamorphic dehydration reactions entail a loss of water but also of dissolved solids, and metamorphic rocks must be considered as open systems where very large quantities of water may have circulated. During this process, the chemical composition of the initial rock is transformed. As metamorphic reactions proceed, the distribution of trace elements and chronometric systems are severely disrupted but, because the system is open to fluid circulation, this rarely happens in a predictable way. Since, at the temperatures at which metamorphism occurs, ^{18}O tends to preferentially fractionate into the fluid (contrary to what happens at ordinary temperatures), the isotopic composition of the oxygen of rocks drifts toward lower $\delta^{18}O$ levels closer to those of the mantle. At temperatures in excess of 500 °C, the oxygen isotope fractionation coefficients tend to unity, there is little water left to exchange oxygen with, and isotopic changes become less important.

Extreme metamorphic conditions cause rocks to melt when they have a high water content; the hydrated melting of common crustal metamorphic rocks is referred to as anatectic melting. When water is absent or the dominant fluid is CO_2, granulite facies conditions pertain. Circulation of CO_2 promotes migration and the loss of elements that are normally inert under hydrated conditions: it is known that granulite facies rocks, which are so common at the base of the continental crust, have lost much of their uranium and some of their thorium. The production of heat in granulites of the deep part of the continental crust by the radioactive elements U, Th, and K is therefore usually very low, with, as a consequence, a distinctive isotopic composition of lead, which is in general unradiogenic.

Metamorphic assemblages and, at depth, the metasomatic assemblages derived by reaction between mantle rocks and percolating fluids and magmas, are commonly used to derive equilibration temperatures and depth (pressure)–temperature–time (through geochronology) pathways in the mantle and the crust. Mineral assemblages can only be considered as equilibrium assemblages (they are then referred to as parageneses) and can

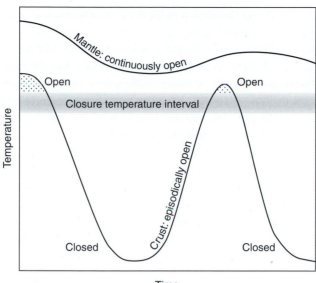

The common thermal pathways of rock samples from the mantle (top) and crust (bottom). The shaded zone represents the blocking temperature of a mineral reaction used to provide temperature and pressure estimates or the closure temperature of a chronometer such as garnet ^{176}Lu–^{176}Hf ages. Temperatures of mantle rocks typically exceed 900 °C: their minerals therefore remain continuously open to the exchange of elements and isotopes. In contrast, after original cooling to rather low temperatures, the *PT*–age estimates derived from crustal rocks may be episodically reset by geological events.

therefore be used as thermometers or barometers if they cool rapidly enough to freeze mineral compositions under the conditions prevailing at high temperatures. In Chapter 5, we saw that, in the most general case, diffusion may maintain chemical exchange among minerals and between minerals and surrounding fluids. It is important that such exchanges do not affect a large fraction of mineral elements, in other words that cooling is fast. In the case of moderate cooling rates, metamorphic assemblages reflect not equilibrium PT–age conditions but the blocking (closure) temperature of chemical exchanges. This is a common situation of high-temperature assemblages, such as granulites and eclogites, in slowly cooling metamorphic terranes: the ubiquitous temperature range of 700–850 °C observed for these rocks corresponds to the closure temperatures of many minerals, notably pyroxene and garnet.

Two additional complexities arise with pressure–temperature estimates in metamorphic and metasomatic assemblages (Fig. 10.7). First, it must be checked that the values observed reflect the temperature and the age of the original cooling and not those of a resetting during a subsequent reheating event above the closure temperature of the different mineral reactions and chronometric systems. Migration of magmas or hot fluids on top of rift or subduction zones are prime potential culprits for such resetting events. Second, most rocks from the mantle and the deep crust never cooled below these blocking temperatures: they remained essentially open systems since they formed. If such an open system behavior is nearly perfect, the thermobarometric estimates can be used to derive modern geotherms, e.g. the P–T relationship in the local mantle. Such a behavior is best assessed by checking that the chronometric ages correspond to the time these rocks rose to the surface (e.g. the eruption age of ultramafic inclusions in volcanic rocks). If, however, the chronometric ages of these rocks are substantially older than the time of their final ascent, the P–T observations should be treated with utmost care. In the absence of zircons in the mantle, garnet ^{176}Lu–^{176}Hf ages have proved to be particularly helpful in this context. It is therefore worth stressing that most rocks from the mantle should be a priori treated as open systems and that attempts to interpret the major and trace element concentrations and the isotopic properties of their minerals as reflecting those prevailing at the time they formed are probably bound to be erroneous. The lack of equilibration among minerals of mantle rocks is therefore not a sufficient indication of a metasomatic event. In contrast, most of the crustal rocks have cooled through the closure temperature and some of them have been reworked over different tectonic and magmatic events.

10.4 Water/rock ratios

The motivations for finding out how much water a given rock sample has "seen" during diagenetic, hydrothermal, or metamorphic processes are varied: we may want to assess the reserves of a particular element of economic value remaining in the source rock of a particular ore deposit, we may need to know how far water circulation affected the thermal regime of a given area, or we may worry that too much circulating water disturbed the initial geochemical properties of a rock, a mineral, or fossil remains. In order to provide a quick estimate of the extent of water–rock interaction, geochemists often refer to the concept of a water/rock ratio. Given a set of geochemical observations on an altered rock or a diagenetic/hydrothermal solution, geology may often give some hints at what the untransformed rock may have looked like (a basalt, a granite), while other constraints may be good enough to let us infer the geochemistry of the reacting fluid (meteoritic water at a given latitude, for instance). A number of geochemical properties can be used as well, concentrations and isotopes, but probably the most popular are δ^{18}O and ^{87}Sr/^{86}Sr. Let us suppose, for instance, that we measured the δ^{18}O$_{HR}$ of a hydrothermally altered basalt sample to be $-2‰$, while examination of thin sections suggests that the rock was hydrothermally altered at 350 °C in the greenschist facies. Paleogeography also suggests that when this particular basalt was erupted, the latitude was such that meteoric water had a δ^{18}O$_{MW}$ of $-10‰$. We will assume that the δ^{18}O$_{FR}$ of the fresh basalt was in the range of mantle-derived magmas, say $+5.5‰$. Can we find out how much water interacted with this particular basaltic sample before it turned into a metabasalt?

We will note R and W the mass of rock and water, respectively, reacting with each other, and neglect the fact that water and rock contain slightly different oxygen concentrations. We can write the isotopic mass balance during the reaction as

$$R\delta^{18}O_{FR} + W\delta^{18}O_{MW} = R\delta^{18}O_{HR} + W\delta^{18}O_{HW} \tag{10.17}$$

where HW refers to the hydrothermal solution, which, unfortunately, can no longer be sampled and analyzed. Rearranging, we get the water/rock (W/R) ratio as:

$$\frac{W}{R} = \frac{\delta^{18}O_{FR} - \delta^{18}O_{HR}}{\delta^{18}O_{HW} - \delta^{18}O_{MW}} \tag{10.18}$$

It is often observed that, for mass balance reasons, the $\delta^{18}O$ value of igneous whole-rocks is similar to that of their feldspar. This property allows us to estimate the $\delta^{18}O$ of the hydrothermal water from the $\delta^{18}O$ of the altered rock and the fractionation factor of $^{18}O/^{16}O$ between feldspar and water at 350 °C, i.e:

$$\delta^{18}O_{feldspar} = \delta^{18}O_{HW} + 4 \tag{10.19}$$

Rearranging, we get the water/rock (W/R) ratio as:

$$\frac{W}{R} = \frac{5.5 - (-2)}{(-2) - 4 - (-10)} = 1.875 \tag{10.20}$$

Water/rock ratios in the range 1–5 are very common. These values are in general minimum values and much higher values (up to several hundred) are not exceptional, especially during diagenesis. Examples based on Sr isotopes may be constructed in a similar way.

Exercises

1. Show that if the changes of enthalpy $\Delta H(T, P)$, entropy $\Delta S(T, P)$, and volume $\Delta V(T, P)$ of a reaction are approximately constant, (C.5) in Appendix C can be integrated into

$$\Delta G(T, P) \approx \Delta H_0 - T\Delta S_0 + \Delta V_0(P - 1) \tag{10.21}$$

   where $\Delta G(T, P)$ is the free enthalpy of the reaction at T and P and the subscript 0 denotes a reference state (usually 298 K and 1 bar). Show that at equilibrium, this relation provides an equation for the reaction curve. Use the previous result to draw the reaction curve defining the onset of the eclogite facies:

$$NaAlSi_3O_8 \Leftrightarrow NaAl_2SiO_6 + SiO_2$$
$$\text{(albite)} \qquad \text{(jadeite)} \qquad \text{(quartz)}$$

   in which the sodic plagioclase breaks down into a sodic pyroxene, which dissolves into ambient clinopyroxene and quartz. Use the following values: $\Delta H_0 = -2115$ J mol^{-1}, $\Delta S_0 = -32.25$ J mol^{-1} K^{-1}, and $\Delta V_0 = -17.0 \, 10^{-6}$ m^3.

2. Aragonite is a denser polymorph of calcite $CaCO_3$ and is therefore more stable at high pressure than calcite. Using the following values $\Delta H_0 = +1132$ J mol^{-1}, $\Delta S_0 = -0.146$ J mol^{-1} K^{-1}, and $\Delta V_0 = -2.78 \, 10^{-6}$ m^3 for the calcite $\Leftrightarrow$ aragonite reaction, calculate the pressure of the transition at 500 °C.

3. Show that the constant in (10.2) is equal to $\Delta S_0(T, P)/R$, the entropy of the reaction in standard conditions (hint: refer to Appendix C and use the definition of the free enthalpy to show that $\Delta G_0(T, P) = \Delta H_0(T, P) - T\Delta S_0(T, P)$). When the solids

are maintained at a pressure > 1 bar, in particular when solid and gaseous phases are identical, a small correction for mineral expansivity is necessary but is usually very small.

4. The reaction described by Eq. 10.1 has a ΔH_0 (298.1) of 111,090 J mol^{-1} and a ΔS_0 (298.1) of 188.9 J mol^{-1} K^{-1}. Assume that these two values remain constant, use the result of the previous exercise, and calculate P_{H_2O} for temperature range of 450–650 °C.

5. The reaction described by (10.7) has a ΔH_0 (298.1) of 528,600 J mol^{-1} and a ΔS_0 (298.1) of 235.8 J mol^{-1} K^{-1}. Assume that these two values remain constant and calculate P_{O_2} for range temperature of 800–1200 °C.

6. Magnesite (MgCO$_3$) reacts with quartz to produce forsterite according to the reaction:

$$2MgCO_3 \ + \ SiO_2 \ \Leftrightarrow \ Mg_2SiO_4 + 2CO_2$$
$$\text{(magnesite)} \quad \text{(quartz)} \quad \text{(forsterite)}$$

The ΔH_0 (298.1) of this reaction is 173,000 J mol^{-1} and the ΔS_0 (298.1) 350.2 J mol^{-1} K^{-1}. Assume that these two values remain constant and calculate P_{CO_2} for temperature range of 350–550°C.

7. Calculate the P_{CO_2} as in the previous exercise for a carbonate with equal molar proportions of calcite (CaCO$_3$) and magnesite (MgCO$_3$).

8. Use (10.13) to retrieve the ΔH_0 and ΔS_0 values for quartz dissolution in water at atmospheric pressure. From thermodynamic data tables, we find that ΔV_0 between the molar volume of dissolved silica and quartz is -9.1×10^{-6} m^3. Calculate the solubility of silica at 300 °C at the surface and 0.1 GPa. Discuss the potential implications for the interpretation of the chemistry of hydrothermal solutions.

9. Olivine is a solid solution of forsterite (Mg$_2$SiO$_4$) and fayalite (Fe$_2$SiO$_4$). Write the simultaneous equations of weathering by pure water (hydrolysis) reactions of fayalite to magnetite (Fe$_3$O$_4$) and of forsterite to serpentine Mg$_3$Si$_2$O$_5$(OH)$_4$ and explain why low-temperature alteration of peridotites is a source of hydrogen. How many moles of mantle olivine fo$_{90}$ (with 90 mole percent forsterite and 10 percent fayalite) does it take to produce one mole of H$_2$? Assuming that a mass equivalent to 10 percent (≈ 2 km^3 y^{-1}) of the annual production of oceanic crust is serpentinized, what is the annual flux of hydrogen from the mid-ocean ridges to the ocean and the atmosphere? What do you think its fate is?

10. Fresh mid-ocean ridge basalts (MORB) contain about 10 wt% FeO. Upon reaction with sulfate from seawater infiltrated into the young oceanic crust, some of the Fe^{2+} is oxidized to Fe^{3+}. How many grams of seawater with a concentration [SO$_4^{2-}$] = 28.9 mmol kg^{-1} of sulfate would it take to oxidize all the Fe^{2+} contained in one hundred grams of MORB?

11. Hydrothermal solutions spouted by black smokers at mid-ocean ridges result from interaction between seawater and fresh basalts at temperatures of 300–400 °C. The ^{87}Sr/^{86}Sr ratio of the hydrothermal solutions is 0.7040 and contrasts with the value of 0.7091 observed for seawater and 0.7025 for fresh MORB. The Sr content of the solution does not seem to be greatly affected by the hydrothermal process (8 ppm for both seawater and hydrothermal solutions). Fresh MORB contains about 120 ppm Sr.

Assuming isotopic equilibrium between hydrothermal solutions and the host basalt, calculate the water/rock ratio controlling the hydrothermal process. Many ophiolites show $^{87}Sr/^{86}Sr$ ratios of about 0.708. What is the apparent water/rock ratio of the hydrothermal processes leading to the formation of these rocks?

12. List potential electron acceptors (oxidizing substances), both ions dissolved in interstitial solutions and sedimentary minerals, that will eventually be used by organisms to oxidize organic carbon in sediments. Discuss how they affect the mineral composition and oxidation state of common sediments at the surface of the Earth.

Reference

Murray, J. W., Grundmanis, V., and Smethie, W. M. (1978) Interstitial water chemistry in the sediments of Sannich Inlet. *Geochim. Cosmochim. Acta*, **42**, 1011–1026.

Before discussing the formation of the major geological units of the solid Earth, we should review the internal structure of our planet as described by seismic wave studies (Fig. 11.1). The most important discontinuities observed by seismologists are:

- The base of the crust (called the Mohorovičić discontinuity or Moho), 40 km below the continents, but only 5–7 km beneath the oceans.
- The base of the lithosphere, on average 80 km below the oceans, and deeper still beneath the continents. Rather than a discontinuity, this is a rather diffuse transition. This is the lower boundary of the rigid tectonic plates. The softer part of the upper mantle underneath the lithosphere is called the asthenosphere.
- The 410 km discontinuity corresponding to a change in olivine structure (spinel or ringwoodite phase).
- The 660 km discontinuity corresponding to the transformation of all minerals into perovskite and minor Fe–Mg oxide (magnesio-wüstite). This is the base of the upper mantle.
- The mantle–core boundary at about 2900 km. Above this boundary is a seismically abnormal layer some 200 km thick, known as the D'' layer.
- The outer core–inner core boundary at about 5150 km. The core is composed mostly of metallic iron and nickel. The motion of the fluid outer core generates the Earth's magnetic field.

Plate tectonics is a powerful theory that unifies the geological expression of crustal and upper-mantle geodynamics (Fig. 11.2). The Earth's surface is covered with rigid lithospheric plates that may or may not carry continents. These plates are about 80 km

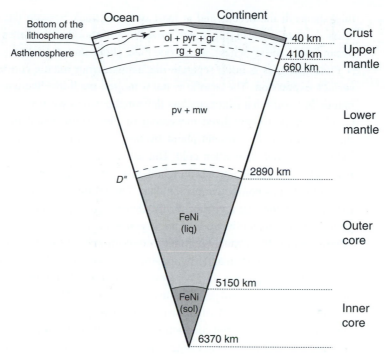

Figure 11.1 The internal structure of the Earth determined by the propagation of seismic waves from earthquakes. Essential minerals of the mantle are garnet (gr), magnesio-wüstite (mw), olivine (ol), perovskite (pv), pyroxene (pyr), and ringwoodite (rg).

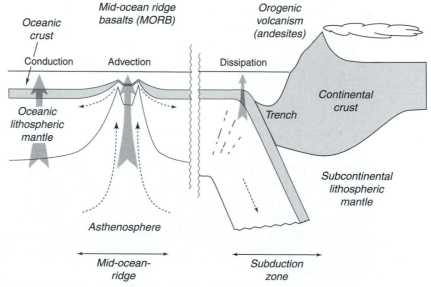

Figure 11.2 The essential components of plate tectonics: mid-ocean ridges, subduction zones, and rigid lithospheric plates. Oceanic crust is in gray. The three mechanisms of heat loss through the surface are shown: conduction through the lithosphere, advection of magmas at mid-ocean ridges, and dissipation of the potential energy of sinking plates at subduction zones, where plates bend.

thick beneath the oceans and at least twice as thick beneath the continents. They move apart along mid-ocean ridges and converge along subduction zones (marked by the deepest ocean floors, the trenches), where one plate slides under another.

Plate tectonics is not a separate mechanism from mantle convection, but is merely its surface expression. The mantle is not a molten medium: like ice flowing in glaciers, the mantle is a solid that is nonetheless deformable when worked over geological time scales. Convection is a generalized movement of the mantle maintained by density inversions (heavy above light) brought about by thermal contrasts within the Earth: a material is lighter when hot than when cold. But also, hot material deforms more easily than cold material, so that radioactive heating not only makes the mantle hotter and lighter but softer, which facilitates convective movements. The temperature dependence of viscosity therefore acts as a thermal regulator for the thermal regime of the mantle. Hot mantle material pours out from under the mid-ocean ridges, while cold lithospheric plates are injected deep into the mantle by subduction. Mantle convection therefore has the effect of extracting heat by replacing deep hot mantle by material that was cooled next to the surface. Heat is lost through the surface (Fig. 11.2) by conduction through the lithosphere and by advection of magmas at mid-ocean ridges and other volcanic sites. The potential energy of sinking plates is converted into heat (dissipation) at subduction zones, where the oceanic lithospheric plates bend. Since thermal conduction through the lithospheric plates is a rather inefficient cooling mechanism, it is not exaggerated to say that the mantle cools from below thanks to the subduction of young cold plates.

Density inversions are maintained by heat from radioactive elements (U, Th, K) contained in the mantle but also by the heat released from the core. In this case, fluid mechanics teaches us that convection in a medium heated from the bottom is unstable, with irregular spurts of hot, less dense material rising rapidly to the surface: these are hot spots or volcanic plumes that begin catastrophically at the core–mantle boundary. The heads of these instabilities produce gigantic eruptions covering areas larger than Texas with lava in comparatively short geological time spans (about a million years) and may break up whole continents. The separation of North America and Europe or of Antarctica and Australia is attributed to this process. The residual plume tail may remain active for tens of millions of years and, as plates move across the surface, it may form strings of volcanoes thousands of kilometers long, such as those of Hawaii. Where these plates are not continental, cooling makes them progressively denser and they sink into the mantle as planar structures at subduction zones. The cold parts constantly drag the remainder of the plate down into the mantle where it tears apart at its thinnest points. This sinking is offset by the formation of young lithosphere along the scars, the mid-ocean ridges, rather like a tablecloth slipping over the edge of the table, from the center of which it is being constantly woven and fed out. The subduction zones are the downgoing limbs of mantle convection, whereas material rises from the deep mostly as part of the general circulation or as narrow upwellings known as plumes.

The continents are formed by relatively light felsic material and behave like corks fixed on top of the lithospheric plates. Mountain ranges form when two continental "corks" collide: a fraction of the continental rocks may then be carried down to very great depths (up to 200 km). There, they are subjected to very high temperatures and pressures and their mineral assemblage becomes eclogitic. This is the origin of high-pressure metamorphism.

The most significant chemical fractionation processes within the Earth are the formation of oceanic lithosphere, hot spots, and continental crust. The fractionation processes attending melting and crystallization – the essential mechanisms by which magmatic rocks originate – are responsible for the diversity of chemical composition of such rocks. They also determine indirectly the geochemical variability of metamorphic rocks and clastic sedimentary rocks (clays, sands). The essential mixing processes are associated with hybridization (or contamination) of magmas and mantle convection. The current state of the mantle and the continental crust is the outcome of competition among all of these processes.

11.1 The geochemical variability of magmas

Magma, the common term for a molten rock, may contain crystals in suspension, usually called phenocrysts. Molten magma reaching the surface is known as lava. If the magma crystallizes completely, most commonly as an effect of slow cooling, the rock produced is said to be intrusive or plutonic. If cooling is too fast for crystallization to be completed, for example in a submarine eruption, the liquid is quenched to a glass. After cooling, it forms an effusive or volcanic rock.

The two most abundant types of magmatic rocks are basalt in ocean areas and granite in continental areas. A basalt reaches the surface at a temperature of 1150–1250 °C, and a granitic liquid (rhyolite) at about 1000 °C. Table 11.1 gives the characteristic major element compositions of magmatic rocks in their standard form, i.e. by oxide weight. Basalt is rich in iron, magnesium, and calcium, whereas granite is rich in silica and alkaline elements. We will now look at the two mechanisms responsible for the variability of magmatic rocks, melting of the mantle and crust, and magmatic differentiation.

11.1.1 Melting of the mantle and crust

To the best of our current knowledge, melting is localized in the upper 100 km of the mantle, where it produces basalts, and in the crust where it produces granites. There is some discussion about deeper zones of melting, notably at the bottom of the mantle where some low seismic velocity anomalies are present, but evidence is still fragmental and ambiguous. We will leave aside for now the issue of the early Earth and the magma ocean. Melting occurs on top of mantle upwellings, both under mid-ocean ridges and at hot spots because (i) the low density of melts with respect to their residue gives the melting curve (solidus) a positive dP/dT Clapeyron slope, and (ii) silicates are thermal insulators, which reduces the cooling path to become adiabatic, very nearly isothermal. A primary melt did not lose mineral phases and opposes a differentiated melt: the process by which crystals are lost is known as fractional crystallization or magmatic differentiation and will be discussed in

Table 11.1 Chemical composition of a few representative magmatic rocks: tholeiitic and alkali basalts, andesite, and granites (in weight percent oxide)

Locality	Rock	SiO_2	TiO_2	Al_2O_3	FeO	MgO	CaO	K_2O	Na_2O	P_2O_5
Mid-ocean ridge	Tholeiitic basalt	50.9	1.2	15.2	10.3	7.7	11.8	0.1	2.3	0.1
Hawaii	Alkali basalt	45.9	4.0	14.6	13.3	6.3	10.3	0.8	3.0	0.4
Hawaii	Tholeiitic basalt	49.3	2.8	13.4	11.5	7.7	11.0	0.4	2.3	0.3
Siberia (trap)	Tholeiitic basalt	48.7	1.4	14.8	10.3	7.3	9.7	0.9	2.4	0.1
Cascades	Andesite	60.4	0.9	17.5	6.4	2.8	6.2	1.2	4.3	0.2
Australia	Granite (I)[a]	69.5	0.4	14.2	3.2	1.4	3.1	3.2	3.5	0.1
Australia	Granite (S)[b]	70.9	0.4	14.0	3.2	1.2	1.9	2.5	4.1	0.2

[a,b] See pp. 224–225 for a definition of I- and S-type granites.

the next section. The factors of chemical variability of the primary melts extracted from the mantle and the crust are multiple and not always independent. Let us consider the most important of these.

1. *The nature of the source rock.* Partial melting of the peridotitic upper mantle, composed mainly of olivine, pyroxene, and aluminous minerals (feldspar, spinel, and garnet) yields basalts. Melting of the continental crust, formed by wet sediments accumulated by erosion, and by their metamorphic equivalents (schist and gneiss) produces granitic liquids.

2. *Pressure.* The pressure P at any depth of melting z is the weight of the overlying rocks; P and z are related through the hydrostatic condition $dP = \rho g dz$ (Fig. 11.3), where ρ is the rock density. Different planets have different gravities and therefore different pressure–depth relationships! At shallow depths, melting of the mantle produces tholeiitic basalts, i.e. basalts that are somewhat richer in Si and poorer in Mg than other types of basalts. At greater depths the trend is reversed and magma is richer in Mg and Fe and poorer in Si; these are alkali basalts.

3. *Temperature.* For a system of a given composition at a given pressure, temperature is directly related to the *degree of melting*: the major elements that enter the melt most readily (Si, Al, Ca, K, Na, Ti) and incompatible trace elements (Rb, Zr, Ba, rare-earths, Th, U, etc.) are relatively abundant in the first melts. As melting proceeds, the liquid becomes richer in refractory elements (Mg, Cr), fresh melt dilutes incompatible elements, and basalt gives way to picrite. Melt temperature and composition are not independent. Close to the minimum eutectic temperature, the melt fraction is small and the liquid composition is buffered by the residual minerals. This is the case not only for mantle melting at high pressure (Fig. 11.3), but also for the crust. At higher temperatures, melting progressively consumes residual minerals, and the liquid composition

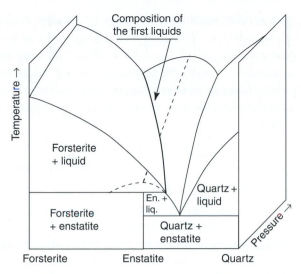

Composition of the first liquids

Temperature →

Pressure →

Forsterite + liquid

Forsterite + enstatite

En. + liq.

Quartz + liquid

Quartz + enstatite

Forsterite   Enstatite   Quartz

Figure 11.3 The effect of pressure on the composition of the first melt (bold line). With depth, peritectic reactional melting, producing basalts with a fairly high silica content (tholeiites), gives way to eutectic melting forming more olivine-rich basalts (after McBirney, 1992).

drifts toward that of the initial solid. In the mantle, an increase in temperature and degree of melting produces Mg-rich rocks. Komatiites are high-magnesian lavas found only in Archean terranes (except for the small Tertiary flows on Gorgona Island). They are explained by very intense melting of the mantle of the order of 50% and are thus evidence of the very high temperature of some parts of the mantle during the earliest ages of the Earth.

4. *Water and carbon dioxide content of the source.* The effect of water is similar to that of elements with low melting points (e.g. alkali elements). Water may form hydrous minerals such as mica and amphibole in the upper mantle and a number of unnamed minerals ("alphabet" phases known by letters) at depth, but it is also dissolved in the lattice of the so-called "nominally anhydrous minerals" such as olivine and pyroxene and greatly reduces their viscosity. Water is also very soluble in melts because it breaks the Si–O–Si double bond into two Si–OH single bonds. The presence of such "impurities" lowers the melting point of the rocks (Appendix C, (C.32)) in proportion of their *molar* fraction: since water has quite a small molar weight (18), its effect per *weight* percent on the melting point is much larger that that of other minor incompatible elements such as Ti or P. As long as a gas phase is absent, water behavior is otherwise similar to that of an ordinary incompatible element. It is noteworthy that the H_2O/Ce ratio is practically constant at 300 in basaltic magmas from different environments (mid-ocean ridges, oceanic islands, subduction zones) when measurements are made on melt inclusions within crystals, i.e. from non-degassed samples. As water is nonetheless the most abundant impurity, it soon reaches saturation and forms a vapor phase as the magma rises and separates out from the melt. From this point, if the vapor phase is dominated by water, the mechanical equilibrium of vapor and the surrounding rock requires the water pressure to be equal to the pressure of the surrounding rock and

the water concentration of the magma is primarily controlled by pressure. The presence of abundant water also changes the composition of the melt: a water-rich source rock produces a magma that is richer in silica than a dry rock (Fig. 11.3). Carbon dioxide at pressures < 3 GPa (90 km) has a behavior similar to water, except that its effect on melting is smaller due to its larger molar weight (44) and its lower solubility in melts. At higher pressures, CO_2 reacts with silicates to form carbonates according to reactions such as:

$$Mg_2SiO_4 + CO_2 \Leftrightarrow MgSiO_3 + MgCO_3 \qquad (11.1)$$
$$\text{(olivine)} \quad \text{(vapor)} \quad \text{pyroxene} \quad \text{magnesite}$$

Melting of carbonated mantle at high pressure tends to produce melts with a high carbonate content that may eventually form low-silica carbonate-rich melts (nephelinites) or even carbonate melts (carbonatites).

5. *Fertility*. This is the abundance of minerals with a low melting temperature, and it affects the capacity of the rock to produce magma (melt productivity). At a given temperature, a mantle rock rich in pyroxene and aluminous minerals will produce more liquid than a peridotite with abundant refractory olivine.

Geochemical observations of lavas, notably of trace elements, can be interpreted quantitatively with equilibrium melting or fractional melting models – see (2.17) and (2.31). More sophisticated models, mentioned earlier, draw on the percolation of fluids through rock pores or the compaction of molten rock. These provide a more realistic picture of the geochemistry of melt products. The mantle melting mechanisms themselves are still poorly understood. The liquid first appears at the grain boundaries. Because it is less dense than the solid mantle, buoyancy tends to expel the melt upwards as would a sponge full of water under its own weight. So long as the porous matrix can be deformed, liquids percolate, producing chromatographic fractionation effects. When they penetrate the colder, more rigid parts of the mantle, they collect in fractures and are channeled quickly toward the surface forming volcanic systems.

Melting in the continental crust, most often in the presence of water, produces granitic liquid: the composition of the melt is buffered by a mineralogical assemblage of quartz and feldspar (eutectic melting), so that the variability of granite composition is fairly narrow. The presence in the granite source of sedimentary material or material weathered at low temperature is attested to by many indicators, the most obvious being the commonly high level of their $\delta^{18}O$ value (8–14‰). During orogenic periods, substantial crustal melting may occur: if we imagine that a granite intrusion, some 150 km long, 50 km wide and 6 km thick (a common enough dimension in many shield areas), represents a partial melt rate of 30%, there must have been an enormous molten layer of nearly 20 km of crust at the time these granites formed. Crustal melting is therefore a large-scale process. Narrow swarms of granitic intrusions, occasionally elongated over thousands of kilometers (batholiths), often mark the edge of subduction/collision zones: hybrid magmas formed by mixing of mantle-derived basalts and andesites with felsic anatectic melts dominate these structures.

Some granites, known as S-type granites, form primarily when sediments and their metamorphic equivalents melt deep in the continental crust. Their high $\delta^{18}O$ values (9–15‰) imply that their source rock formed at low temperature in the presence of water. The abundance of muscovite and the low Fe^{3+}/Fe^{2+} ratio of the S-granites indicate that this source contained large proportions of oxygen-starved clay minerals. Their high $^{87}Sr/^{86}Sr$ and low $^{143}Nd/^{144}Nd$ ratios reflect the fact that this source has evolved over hundreds of millions, even billions of years, with a high Rb/Sr ratio typical of metasedimentary rocks. In contrast, other granites, known as I-type granites, have low $\delta^{18}O$ values (6–9‰) indicative of a source dominated by igneous rocks that were only slightly affected by low-temperature alteration (typically older plutonic rocks). Ubiquitous hornblende reflects the abundance of Na and Ca in the source rock. More than 25% of the iron is oxidized, while Sr is less radiogenic and Nd more radiogenic than in S-granites: all these features point to the melting of deeper levels of the continental crust.

Granitic magmatism is almost exclusively limited to the Earth's continents: a sample of lunar granite has been described, a few lavas of granitic composition are known in Iceland and the islands of Réunion and Ascension, and on mid-ocean ridges, but their isotopic characteristics (O, Sr, Nd) make them more like basalts than continental granites. These granitic rocks must have formed by remelting of the basaltic pile, although, in rare cases, extreme differentiation of basaltic magmas may also leave felsic lavas as the end-product.

11.1.2 Differentiation of magmatic series

We have seen that one of the causes of magma chemical variability is differentiation, already referred to at the end of Chapter 2. As magmatic liquids are less dense, at least at upper mantle pressure, they are subjected to the buoyancy forces of the surrounding rocks and therefore rise, initially by percolation when the matrix is deformable and then, in the lithosphere, along fractures. As they rise, they cool by contact with the walls and the solubility limit is reached for each mineral in turn (Fig. 11.4) in rather straightforward sequences. In basalts at high pressure, olivine saturation is reached first, followed by that of clinopyroxene and, finally, plagioclase. If cooling occurs at low pressure ($< 15\,km$), as under the mid-ocean ridges, the order of appearance of plagioclase and pyroxene is reversed. In some cases, an intermediate reservoir may temporarily accommodate the magma. This is the mythical magma chamber, a concept that pervades petrological literature. Such magma chambers are actually uncommon in island volcanoes, but are ubiquitous beneath ridges. This contrasting pressure-dependent behavior is clearly visible in the evolution of the major element concentrations in various types of basalts (Figs. 11.5, 11.6). Since Mg is much more compatible than Fe in mafic minerals (olivine and pyroxene), the FeO/MgO ratio (a convenient differentiation index) decreases when the degree of crystallization increases. Most ocean island basalts (except from those islands that lie on the ridge axis, such as Iceland, or adjacent to it, such as the Galapagos) differentiate at medium pressure and remain undersaturated in plagioclase: when plotted against FeO/MgO, the Al_2O_3, or TiO_2 concentrations attest to the essentially incompatible behavior of Al and Ti and increase in differentiated melts. In contrast, mid-ocean ridge basalts fractionate plagioclase: Al_2O_3 is compatible and decreases, while FeO and TiO_2, rejected by plagioclase, remain incompatible and increase in residual

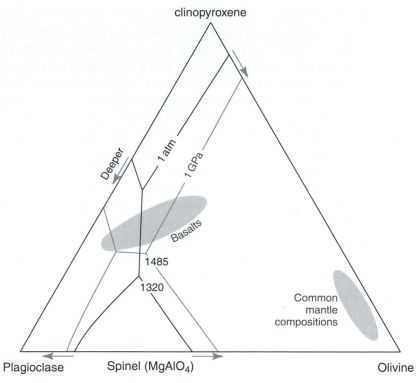

Change in the stability field of different minerals in basaltic melts with pressure. At ambient pressure, the minimum melt of mantle rocks (triple point at 1320 °C) is saturated in olivine, spinel, and plagioclase. Clinopyroxene therefore dissolves into the liquid. In contrast, at 1 GPa ($z \approx 30$ km), the minimum melt (1485 °C) is saturated in olivine, spinel and clinopyroxene. Plagioclase now dissolves into the liquid. After Presnall *et al.* (1978).

melts. In arc (orogenic) magmas, the relatively high water content suppresses again early plagioclase saturation: Al, Ti, and Fe are controlled by clinopyroxene and olivine fractionation.

The fate of the different species of iron Fe^{2+} and Fe^{3+} is germane to what petrologists refer to as changes in oxygen fugacity. Since silicates are electrical insulators, changes of the Fe^{3+}/Fe^{2+} ratio cannot be effected by simple electron transfer, but require incorporation of one or other species into minerals and are therefore controlled by the crystallizing mineral assemblages. Ferrous Fe^{2+} is compatible with olivine and pyroxene structure, whereas Fe^{3+} is incompatible until saturation of magnetite Fe_3O_4 is attained. We can divide the familiar Rayleigh distillation equation (2.29) for Fe^{3+} by that for Fe^{2+} in residual magmas (res) to obtain the Fe^{3+}/Fe^{2+} ratio as:

$$\left(\frac{Fe^{3+}}{Fe^{2+}}\right)_{res} = \left(\frac{Fe^{3+}}{Fe^{2+}}\right)_0 f^{D_{Fe3+} - D_{Fe2+}} \tag{11.2}$$

in which the 0 subscript refers to the undifferentiated magma, f is the fraction of residual liquid (res), and D stands for bulk solid–liquid partition coefficients. In all cases, $D_{Fe^{3+}}$ is

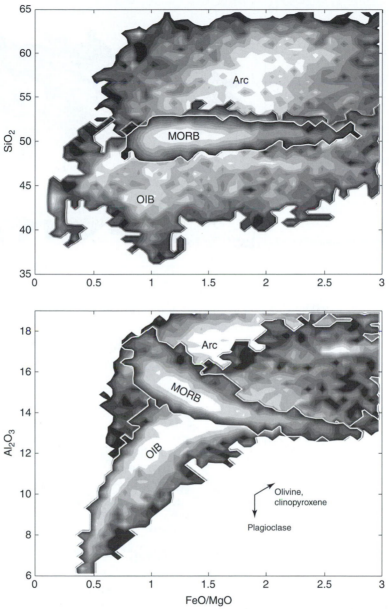

Figure 11.5 Changes in the behavior of some major elements in basalts from different tectonic settings as a function of plagioclase saturation. The arrows represent the effect of removing a particular mineral on the major element content of the residual melt. As shown by the evolution of SiO_2, the x-axis coordinate (FeO/MgO) represents the extent of differentiation: lavas become more differentiated from left to right. Plagioclase saturation in MORB as a result of low-pressure (< 5 kb) differentiation makes Al compatible and Fe and Ti more incompatible. Plagioclase undersaturation as a result of high pressure fractionation (OIB) or high water pressure (arc magmas) has the opposite effect: Al becomes incompatible in contrast to Fe and Ti, which become more compatible.

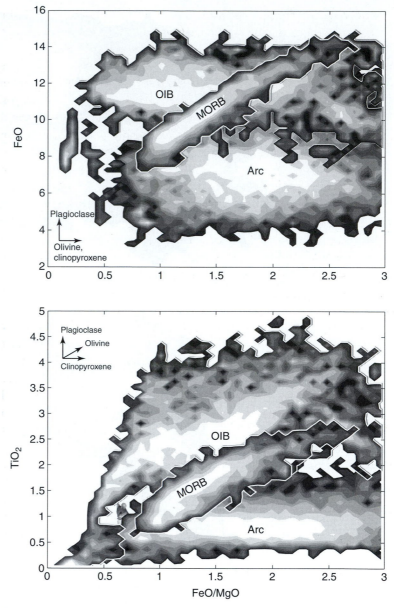

Figure 11.6 Continued from Fig. 11.5.

nearly zero until magnetite reaches saturation. In contrast with MORB, cumulates of both ocean island and arc magmas are plagioclase-free and therefore:

$$\left(D_{Fe^{2+}}\right)_{MORB} < \left(D_{Fe^{2+}}\right)_{OIB+arc} \qquad (11.3)$$

The Fe^{3+}/Fe^{2+} ratios must therefore be smaller in MORB than in any ocean island and arc magma of similar f. This behavior is seen by petrologists as indicating an evolution towards more oxidizing conditions (higher oxygen fugacity) in the latter. It is actually a simple consequence of ferric and ferrous iron being unable to trade electrons; Fe^{3+} and Fe^{2+} behave as different elements and their relative proportions are essentially controlled by their respective compatibility during magma differentiation.

The minerals thus formed may be denser (olivine and pyroxene) or less dense (plagioclase) than the magma from which they crystallize. The denser ones tend to sink and sediment out, while the buoyant plagioclase floats, with cumulates forming in both cases. Such crystallization followed by the separation of minerals selectively removes elements from the parent magma: this is magmatic differentiation by fractional crystallization. This process is responsible for the formation of magma suites of very variable compositions, for example the basalt–phonolite–trachyte suite, fine examples of which can be seen in composite volcanoes such as in the Canary or the Society islands. The chemical evolution of these suites is generally well understood. It is represented by situations of fractional crystallization (Rayleigh's law), possibly complicated by the assimilation of surrounding rocks (AFC = assimilation and fractional crystallization), or recurrent episodes of reservoir replenishment. Compatible elements (Ni, Cr, Sc, Sr) will be good tracers of fractional crystallization as they vary greatly for a low degree of fractionation. Nickel indicates the precipitation of olivine, Cr and Sc that of clinopyroxene, Eu and Sr that of plagioclase. It can be calculated, for example, from (2.30), that a modest 10% precipitation of olivine, for a partition coefficient of Ni between olivine and liquid of 15, lowers the parent magma content of this element by 77%. Excess Eu with respect to the neighboring rare-earth elements Sm and Gd will signal that a mafic rock is a plagioclase cumulate (gabbro), while a deficit requires plagioclase removal. In contrast, incompatible elements will be uniformly enriched in the residual liquid. Their relative distribution carries little or no information about magmatic differentiation: fractionation of 10% olivine enriches the magma in both Th and La (zero partition coefficient) by 11%. In contrast, the variability in incompatible elements induced by partial melting is very high compared with that induced by fractional crystallization. As already alluded to, incompatible elements are therefore particularly useful for the study of partial melting. In the more viscous granitic liquids, gravity separation of minerals is less efficient than in the normally much more fluid basalts. Granitic rocks are generally far more difficult to interpret than basalts in terms of geochemistry as they are plutonic rocks and fully crystallized: it is practically impossible to assert from examination of a sample of granite whether it is a sample of congealed liquid magma, the cumulates derived from it, or some intermediate combination of both.

11.2 Magmatism of the different tectonic sites

Figure 11.7 schematically shows the geographical distribution of various magma types in relation to their geodynamic site.

The temperature at which rock melts increases with pressure and decreases sharply in the presence of water. The low water content of the mantle (about 500 ppm) limits its effect on upper-mantle melting. Mechanical effects of trace amounts of water, however, stand out:

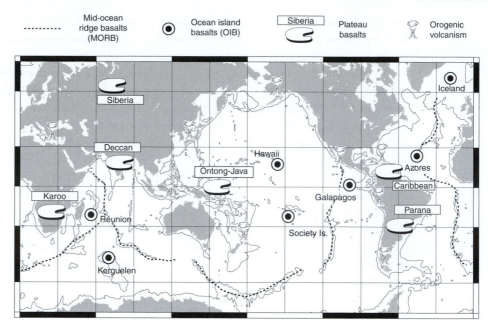

Mid-ocean ridge basalts (MORB)   Ocean island basalts (OIB)   Siberia   Plateau basalts   Orogenic volcanism

Figure 11.7 The geographical distribution of different types of volcanism. To bring out the shallower parts of the ocean (ridges and plateaus) the −3000 m depth contour is shown (drawn using GMT).

dry olivine is stiffer than wet olivine and melting leaves behind a fairly rigid residue. The production of lava is associated with decompression of the mantle in plumes and beneath oceanic ridges but also with dehydration of wet material (weathered sediments and basalts) sinking in subduction zones. These three sites produce basalts, i.e. lavas rich in Mg, Ca, and Fe and poor in Si. Other types of magma are associated with continental collision zones: these are felsic magmas rich in Si and poor in Mg, whose characteristic product is common granite. Continental margins are often associated with mixed magmatism, which goes by the name of orogenic magmatism. Examples are the deadly eruptions of Mount Pelée, Mount Pinatubo, or El Chichon. A common type of lava erupted by these volcanoes is andesite.

The nature of the mantle, which produces different types of basalt, is an active area of research. The greatest progress has been accomplished recently for the ocean island basalts (OIB) that form chains of volcanoes associated with hot spots, and the mid-ocean ridge basalts (MORB) that form the ocean floor. Production of MORB is connected with decompression of the upper layers of mantle, forming the asthenosphere, which rise beneath the mid-ocean ridges.

Mid-ocean ridge basalts are extremely depleted in highly incompatible elements such as Ba, Th, K, and light rare-earths compared with what might be expected of the melting of unprocessed mantle peridotite: the asthenospheric mantle had already lost a large proportion of its most readily fusible elements before the present-day MORB formed. This can be shown easily by comparing two trace elements with different degrees of compatibility, e.g. two elements of the rare-earth family, ytterbium (Yb) and lanthanum (La). If F is the

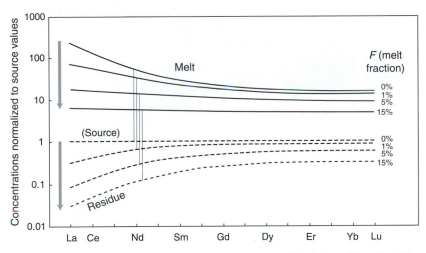

Figure 11.8 Distribution of rare-earth elements in melts from the mantle and in their residues for different degrees *F* of melting. Concentrations were calculated using equilibrium batch melting equations similar to (11.4) and then normalized to the values in the mantle source. The residues were assumed to contain 15% clinopyroxene with the following mineral–liquid partition coefficients: La: 0.03, Ce: 0.085, Nd: 0.19, Sm: 0.29, Dy–Yb: 0.44 and 85% of rare-earth free olivine and orthopyroxene. Note that the highly incompatible elements (La, Ce) are extracted much more efficiently from the residues than the more compatible elements (Yb, Lu). The tie-lines connect, for Nd, liquid and solid pairs for a same *F*. Since MORBs are depleted in La, Ce with respect to Yb, Lu (see Fig. 2.7), their sources are necessarily even more depleted, which demonstrates that they went through previous melt extraction events. This is why the mantle under the mid-ocean ridges is usually referred to as the depleted mantle (DM).

fraction of melt prevailing during formation of MORB and $D_{s/l}^{Yb}$ the partition coefficient of Yb between solid residue and melt, (2.17) can be written:

$$C_{MORB}^{Yb} = \frac{C_{source}^{Yb}}{F + D_{s/l}^{Yb}(1 - F)} \qquad (11.4)$$

As a first approximation, La is almost completely incompatible and its solid residue/melt partition coefficient $D_{s/l}^{La}$ is so small that it can be taken as zero. We can then approximate:

$$C_{MORB}^{La} = \frac{C_{source}^{La}}{F} \qquad (11.5)$$

Extracting *F* from (11.5) and replacing it in (11.4) gives:

$$\left(\frac{C^{La}}{C^{Yb}}\right)_{MORB} = \left(\frac{C^{La}}{C^{Yb}}\right)_{source} \left(1 + D_{s/l}^{Yb}\frac{1 - F}{F}\right) > \left(\frac{C^{La}}{C^{Yb}}\right)_{source} \qquad (11.6)$$

Because the term in the parentheses of the right-hand side is greater than unity, this equation shows that the La/Yb ratio must be greater in the magmatic liquid (MORB) than in the source mantle. This approach is valid, of course, for all trace elements. Figure 11.8 shows the distribution of rare-earth elements in melts from the mantle and in their residues for different degrees *F* of melting. The enrichment of incompatible elements in the melts with

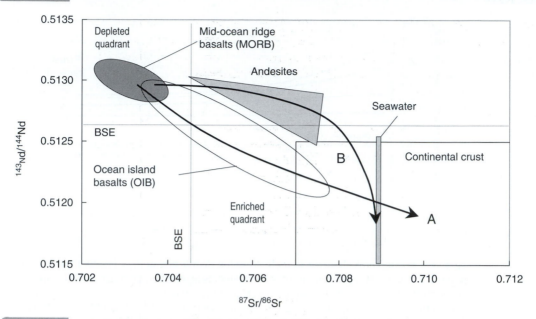

Figure 11.9 Relationship between $^{87}Sr/^{86}Sr$ and $^{143}Nd/^{144}Nd$ in the sources of oceanic basalts and andesites. The more incompatible character of Rb (compared with Sr) and of Nd (compared with Sm) during melting of mantle rocks explains the negative correlation of these ratios. The MORB originate from a source that is more depleted in incompatible elements by successive melting of the mantle than the OIB source. This depletion is ancient (several billion years), as indicated by the composition of slowly accumulating radiogenic elements. Andesites derive from either hybridization of the mantle by the continental crust (A), or melting of the mantle contaminated by dehydration fluids rich in seawater (B).

respect to their mantle source is visibly much greater for the highly incompatible elements (La, Ce) than for the less incompatible elements (Yb, Lu). The inescapable conclusion is that MORB depletion in highly incompatible elements, such as La, Ce, but also Th and Ba, compared with more compatible elements such as Sr, Yb, and Ti (see Fig. 2.7), implies that their source is depleted in these highly incompatible elements to an even greater extent.

The range of radiogenic isotope compositions show that depletion of the mantle under the mid-ocean ridges in incompatible elements is a very ancient phenomenon. The highly incompatible radioactive isotope ^{87}Rb is greatly diminished compared with the more compatible radiogenic isotope ^{87}Sr, and in the course of geological time a relatively low $^{87}Sr/^{86}Sr$ ratio of the asthenospheric mantle results (0.7025). By contrast, the parent isotope ^{147}Sm is more compatible than the daughter isotope ^{143}Nd. It is therefore enriched in residual mantle by comparison with the melt. In the course of time, the asthenospheric mantle acquires a relatively high $^{143}Nd/^{144}Nd$ ratio (see (4.33)) of the order of 0.5131. The $^{187}Os/^{188}Os$ ratio evolves like $^{143}Nd/^{144}Nd$, but in a more extreme way.

The values of the isotopic ratios of present-day magmatic rocks are usually plotted in a series of binary diagrams, the most popular being undoubtedly the plot of $^{143}Nd/^{144}Nd$ versus $^{87}Sr/^{86}Sr$ (Fig. 11.9). The geochemical jargon "depleted" simply refers to a deficit of incompatible (or fusible) elements with respect to compatible (or refractory) elements against an ideal average mantle model (Bulk Silicate Earth or BSE). The opposite of "depleted" is of course "enriched." Depletion can be determined for the time immediately before melting using the analysis of trace elements: for example, less-than-chondritic La/Sm ratios signal a refractory source depleted in incompatible elements. It may also be determined for long periods that precede melting by analyzing isotopic compositions of elements containing a radiogenic isotope (^{87}Sr, ^{143}Nd, ^{176}Hf, ^{206}Pb, ^{207}Pb, etc.) In MORB, Nd more radiogenic (^{143}Nd/^{144}Nd = 0.5132 or ε_{Nd} = +11) than in BSE (^{143}Nd/^{144}Nd = 0.51265 or ε_{Nd} = 0) indicates that the asthenospheric mantle was depleted with respect to a chondritic mantle at least 1–2 billion years ago.

The MORB source is therefore depleted in the more incompatible elements both in the short and long terms (Figs. 2.7 and 11.9). The oceanic lithospheric plate is isotopically indistinguishable from the asthenosphere, and so must be most of the upper mantle. Since granites, a major constituent of continental crust, are enriched in both the long and short terms, a core concept has emerged according to which depletion of the upper mantle reflects the extraction of continental crust throughout geological time. In contrast, the OIB source is generally depleted in the long term (as indicated by radiogenic isotopes) but enriched in the short term (as shown by their incompatible element distributions). A popular idea is that the OIB source has been enriched before melting by fluid and magma percolation reacting with the rock matrix: this process is known as metasomatism. In most cases, however, modern models of melting make this ad hoc interpretation rather unnecessary.

The orogenic magmas arising from the subduction zones, such as andesites, have a more complex history. If they are produced at a continental margin, such as the Andes, they clearly exhibit interaction between mafic magmas from the mantle and anatectic melts from the continental crust. If they are produced in oceanic convergence zones, the result is more surprising: the isotopic composition of their Nd is very similar to that of the Nd of descending lithospheric plates. By contrast, their ^{87}Sr/^{86}Sr ratio is particularly radiogenic, suggesting that marine Sr is present in the mantle source of these magmas. This is often taken as evidence that dehydration of sediments and altered basalts of the downgoing plate produces fluids which trigger melting in the overlying mantle wedge. Andesites are often seen as the resulting melts. This interpretation is confirmed by higher δ^{18}O values (> 7‰) than for the average of the ordinary mantle (≈ 5.2‰), therefore requiring the presence in their source material of rock components that have been subjected to weathering or alteration at low temperature. Andesites are richer in Si than basalts, which is consistent with the petrological concept stated earlier that water increases the Si content of melts. A number of trace elements exhibit abundance levels typical of such an environment: in particular the extreme depletion in Nb and Ta, which are thought to be locked in highly refractory minerals such as titanium oxide, in contrast to many incompatible elements (Rb, K, Ba, Sr, Pb, La, Th, etc.), which are returned toward the surface by dehydration of the plates. Petrology indicates that the origin of andesites cannot be ascribed to simple melting of the subducted crust as this process would produce lava far too rich in silica. This liquid is a special kind of granitic melt known as dacite. It can therefore be imagined that these liquids released by melting of the hydrated ocean crust react with the mantle located above the plate. Reaction between the mantle and the felsic liquids generated by hydrated fusion of the plate modifies the mantle overlying subduction zones: andesites are thought to

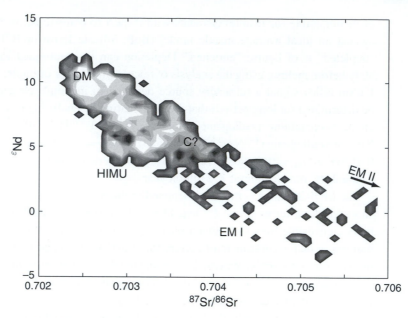

Two-dimensional histogram of Sr and Nd isotope compositions of oceanic basalts and the distribution of the four major mantle components (end-members): depleted mantle (DM, prevalent in MORB), high-^{238}U/^{204}Pb (HIMU) mantle (probably recycled oceanic crust), enriched mantle (EM) of type I (uncertain), and type II (recycled terrigenous sediments); C stands for the "common" component, which probably represents an average composition of the deep mantle.

derive from the melting of this anomalous mantle. The most popular interpretation is that andesites are melts formed upon invasion of the mantle wedge overlying the subduction zone by dehydration fluids. A critical question is the assessment of the respective contribution of the subducting plate and the mantle wedge to the inventory of each element in orogenic magmas.

Ocean island basalts pose an enormous challenge to geochemists. The complexity of their isotopic characteristics has caused lasting confusion, which is only now beginning to clear. First, OIB are isotopically heterogeneous. In the range of radiogenic isotopic ratios, e.g. for ^{87}Sr/^{86}Sr, ^{143}Nd/^{144}Nd, ^{177}Hf/^{176}Hf, ^{206}Pb/^{204}Pb, etc., these heterogeneities can only be accounted for by the mixture of a minimum number of isotopic "components." These are thought to correspond to different mantle sources, which each represent a different geodynamic object. The components are normally referred to by their acronyms: DM (depleted mantle) stands for a depleted source similar to the MORB source and to the asthenosphere found beneath the mid-ocean ridges; the HIMU component is characterized by particularly radiogenic lead (high μ, with $\mu = {}^{238}$U/^{204}Pb), a character that may be inherited from remelting of ancient ocean crust; EM (enriched mantle) stands for a mantle enriched in incompatible elements, either by incorporation of sediments (EM II) or by invasion of the mantle by mineralized deep fluids (EM I). A ubiquitous component, known as FOZO (for focus zone) or C (for common component), probably embodies the average composition of the deep mantle. Oceanic basalts normally occupy a specific range of isotope ratios, involving different radiogenic isotopes such as ^{87}Sr, ^{206}Pb, ^{207}Pb, ^{208}Pb, etc. (Figs. 11.10 and 11.11). Such chemical complexity is also found in the relative distribution

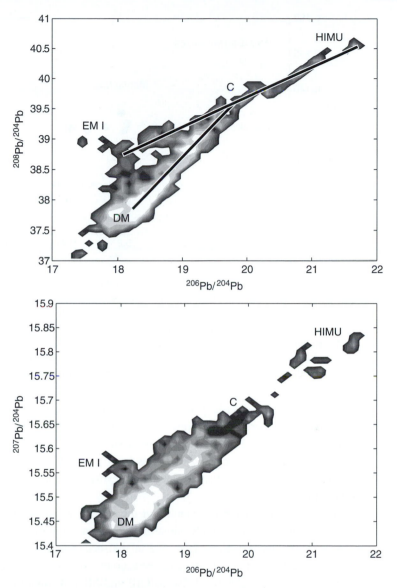

Figure 11.11 Two-dimensional histogram of Pb isotope compositions of oceanic basalts showing the high $^{238}U/^{204}Pb$ ratio of the mantle source of HIMU basalts (St Helena, Mangaia).

of trace elements, the most incompatible of which are normally greatly enriched in our understanding of oceanic basalt genesis.

With the new emphasis of some isotopic systems since the mid 1990s, notably the oxygen isotopes and the $^{176}Lu–^{176}Hf$ and $^{187}Re–^{187}Os$ systems, the complex geochemical landscape of the source of OIB is coming into sharper focus. Isotopic variations of oxygen indicate that mantle material that had experienced low-temperature sea-floor alteration is involved in the source of the Hawaiian basalts. Isotopic compositions of osmium and hafnium show that we find all the components of the ancient oceanic lithosphere recycled,

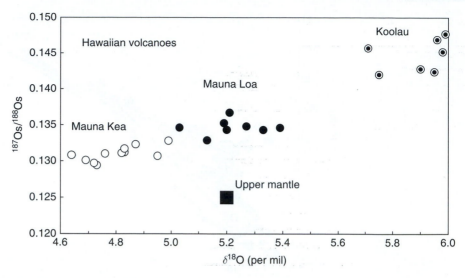

Figure 11.12 Relationship between the $^{187}Os/^{188}Os$ ratio and $\delta^{18}O$ of lava samples from three Hawaiian volcanoes. The variation in abundance in ^{18}O means that the source of these lavas contains a constituent that was processed under low-temperature surface conditions. Isotopic fractionation of Os requires the presence of a rock source that was once composed of ancient magmatic liquids and cumulates. The combination of these two characteristics suggests that these basalts remobilize ancient continental crust buried deep within the mantle. Other criteria preclude contamination by the present-day lithosphere (after Lassiter and Hauri, 1998).

from pelagic sediments (deep-sea clays), with high $^{176}Lu/^{177}Hf$ ratios, through the basaltic part of the weathered basalt crust, with high $\delta^{18}O$ and $^{187}Re/^{188}Os$, to cumulate rocks (gabbro) of the lower oceanic crust with their low $\delta^{18}O$ and $^{187}Re/^{188}Os$ (Fig. 11.12). It is becoming ever clearer that the process forming most of the OIB source is recycling of old oceanic crust, possibly after melting or devolatilization of fluids in the course of plate subduction.

Overall, our current understanding of the isotope geochemistry of oceanic basalts can be boiled down to three simple but strong statements: (i) it is created by two independent groups of processes taking place, one at mid-ocean ridges and the other at subduction zones; (ii) extraction of continental crust accounts for the overall depleted character of the mantle; and (iii) the contrast between OIB and MORB cannot be created in a short time interval, e.g. upon ascent and melting of a particular batch of mantle, but requires multiple cycles of convection, upwelling, and subduction.

The interpretation of the isotope compositions of the inert gases in the various types of basalt remains, however, not agreed upon. Typical ridge basalts have $^3He/^4He$ ratios four times lower than basalt of ocean islands such as Hawaii or Iceland. Let us refer back to (4.37), where we show the closed-system evolution of $^3He/^4He$. We can sketch the evolution of the $^3He/^4He$ ratios in the mantle source of ocean island and mid-ocean ridge basalts by assuming constant parent/daughter ratios (Fig. 11.13). Helium is distinctly more radiogenic in MORB than in OIB. The apparent $^{238}U/^3He$ ratio integrated over the Earth's history is therefore higher in the MORB source than in the OIB source. Given the much more volatile character of He compared with U and Th, it is inferred that parts of the OIB

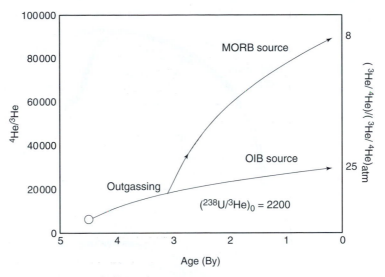

Helium isotope evolution of the mid-ocean ridge and ocean island basalt mantle sources, showing different $^{238}U/^{3}He$ ratios. Outgassing of helium upon melting produces mantle residue with higher $^{238}U/^{3}He$ ratios and therefore potentially more radiogenic helium, i.e. higher $^{4}He/^{3}He$ ratios, at a later stage. The more conventional notation of He isotopic compositions using atmospheric-He-normalized $^{3}He/^{4}He$ ratios is shown on the right-hand side.

source have never been as intensely degassed as the MORB source, and so have never passed close to the surface. The presence of neon, whose isotopic composition resembles solar neon in Hawaiian basalts, also supports this argument. Rare-gas isotope compositions had long been considered a solid proof of the primitive character of the mantle source of hot spots until overwhelming evidence that the Nd, Hf, Pb, and O isotope compositions are inconsistent with undifferentiated mantle made this hypothesis untenable. However, it is probable that, deep in the mantle, there are still expanses that have remained intact since the Earth's primordial differentiation. The lower mantle therefore is to some extent an intimate mixture of primordial and recycled material. Such a statement is not as blunt as it sounds and can be justified with the residence-time theory of Chapter 6: a residence time of 1 Ga in the mantle would require that a proportion of $e^{-4.5/1.0} \approx 1\%$ of the initial ^{3}He at the origin of the planet is still buried in the mantle. A residence time of 4.5 Ga would bring this proportion to 37%. Certain refractory rocks (dunite, harzburgite representing old lithospheric debris) may act as reservoirs for this "primitive" gas whose isotopic signature is transferred by early diffusion of the primordial component into it.

11.3 Mantle convection

Few problems have divided the Earth science community as much as the question of mantle dynamics. While all concur that convection does really take place, the consensus about the

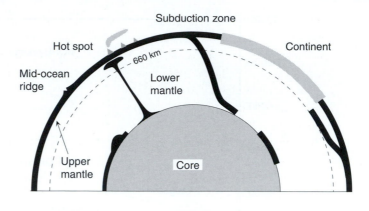

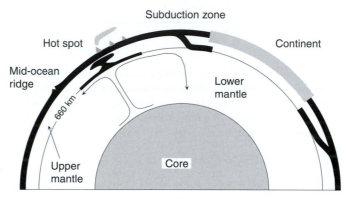

Figure 11.14 Two models of convection within the mantle. Top: whole-mantle convection. Subduction of lithospheric plates and rise of plumes form the convective system. Bottom: two-layer convection separated by the 660 km transition zone (notice the option of thermal coupling in the transition zone: upper mantle plumes are localized above the hot ascending zones of the lower mantle).

number of layers existing within the mantle and the way they contribute to its overall motion took decades to emerge.

The idea that signs of recycling of ancient subducted tectonic plates could be found at the surface in the source of MORB or OIB arose in the late 1970s (Fig. 11.14, top) under the pen of Al Hofmann and Bill White. It is now envisaged that ancient oceanic crust, transformed into pyroxenite by pressure, supplies a large part of the fusible material that gives rise to OIB. New OIB can be made out of old MORB! In the model of whole-mantle convection, the plates sink to the base of the mantle. The intense flow of heat at the core–mantle interface causes gravitational instabilities and after one or two billion years the deep material rises rapidly in the form of plumes (blobs). Isotope and trace element data were persuasive evidence for this concept in the early 1980s, but four factors soon came to oppose the idea of whole-mantle convection based on simple export of lithospheric plates to the lower mantle:

- The rate at which material is exchanged within the mantle at the present-day rate of plate tectonics would eradicate any differences between the lower and upper mantle in a very short time. This can be shown by replacing in (5.17) the values for residence time obtained from the mass of the two parts of the mantle and the flow of plates across the transition zone, to give a chemical relaxation time of 200–400 million years. Under these conditions, how could the geochemical identity and, above all, the isotopic identity of MORB be conserved relative to OIB?

- It was virtually impossible for a long time to find incontrovertible evidence of deep seismic activity that could indicate plate penetration into the lower mantle.
- With a mantle homogenized by convection, the radioactive element contents of the total mantle are equal to those of the upper mantle. So we can calculate the ratio between the heat produced by the Earth's radioactivity and that (43 TW) given off from the Earth's surface. This ratio, the Urey ratio, is then much smaller than unity (≈ 0.4). This would imply that the Earth still contains a considerable amount of the initial heat derived from the gravitational energy of accretion and core segregation locked at depth. If this condition is calculated back to the time our planet formed, it leads to an absolutely infernal temperature of formation, much higher than the vaporization temperatures of silicates.
- The argument from radiogenic argon has played a very significant role. The atmosphere contains a large quantity of radiogenic argon ^{40}Ar. If the total quantity of ^{40}Ar produced by decay of terrestrial ^{40}K is estimated, and if allowance is made for the very low levels of this gas in the degassed upper mantle, an inventory of it for the planet can be drawn up. We quickly reach the conclusion that nearly 50% of the mantle was never degassed and so has never risen near the surface beneath ridges.

These powerful arguments were the basis for the canonical two-layer convection model (lower panel of Fig. 11.14), which still pervades the literature. The upper and lower mantle are separated by the transition zone at 660 km and convect separately. Plates are recycled within the upper mantle and hot spots and their basalts (OIB) rise from the transition zone. The lower mantle is of virtually primeval composition and has very little involvement, supplying only the argon and helium found in OIB.

A number of observations cloud this ideal picture. Except for rare gases, there is little evidence that large expanses of primitive mantle contribute to the genesis of MORB or OIB. The second is that the terrestrial balance of some elements or isotopic systems is not at all consistent with the idea of a primitive lower mantle: Nb and Ta are depleted in both the continental crust and the MORB source with respect to similarly incompatible elements, the U–Pb, Re–Os, and Lu–Hf isotopic systems are not complete when known reservoirs are added up. The third observation was decisive. Modern high-resolution seismic tomography frequently shows lithospheric plates crossing the boundary between the upper and lower mantle.

All the arguments against whole-mantle convection have also been severely weakened by recent observations and concepts. The canonical two-layer convection model of the 1980s with separation at the 660 km transition zone seems no longer viable. This does not mean that the discontinuity does not hamper, and even locally stop transfer of material. What is the current situation? As we just discussed, it is probable that convection has not fully homogenized the initial mantle of the planet and that streaks of rocks inherited from the earliest time, have survived convective stirring at very great depths (1500 km)

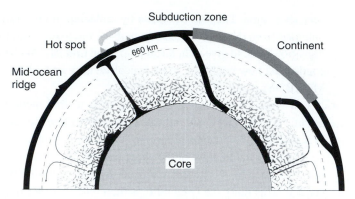

Figure 11.15 A compromise model of mantle convection (modified from Albarède and van der Hilst, 2002). The 660 km discontinuity has some holes which let large plates and plume heads through. Because of trench roll-back, some plates flatten and rest on the discontinuity for a few tens of Ma before foundering across the transition zone. The lower mantle is an intimate mixture of primordial and recycled material. The latter comprises both the refractory residues of melting at mid-ocean ridges and subducted oceanic crust severely outgassed at subduction zones. Helium with high $^3He/^4He$ values and solar neon soak most of the lower mantle.

(Fig. 11.15). A combination of these two effects would justify the presence of U, Th, and K at depth and a reasonable Urey ratio while at the same time leaving room for some whiffs of ancient rare gases to be felt in rocks that otherwise carry all the signs of recycled material in their source. We have seen that a primitive signal in basalts is not necessarily equivalent to a primitive source. One way of explaining the Ar paradox is to change the terrestrial K/U ratio by 30 percent. Large holes, megaplumes and avalanches through the transition zone may actually account for most of the traffic between the upper and the lower mantle. If the 660 km transition zone is not taken to be an effective barrier, subduction zones still act as filters. As the material enriched in lithophile elements, which makes up the continental crust, is extracted at subduction zones, the material penetrating into the lower mantle must be greatly depleted but still rich in basaltic components. Moreover, the subducted oceanic crust transforms under pressure into eclogite, a very dense rock compared with the remainder of the oceanic lithosphere. It will tend to become delaminated and to accumulate by gravity at the base of the mantle and to build up an important reserve of fusible elements. The emerging picture is therefore that of a zoned more than a fully layered mantle. The lower part is both the plate graveyard of the one-layer model and the repository of primordial material of the two-layer model. It is soaked in primordial rare gases and may locally contain some remains of the early planetary differentiation (see the discussion of 142 anomalies in the next chapter). Plumes initiate at the bottom of the mantle because this is where temperatures are the highest and therefore densities are the lowest. The particularly strong buoyancy allows mantle upwellings to rise all the way to the surface. The upper part of the mantle may contain essentially the same components, which is why tenuous OIB flavor is occasionally present in MORB and continental basalts, but the proportion of identifiable ancient material is much smaller, probably because of efficient mixing.

Geochemistry is walking down the last stretch of a long and difficult road. Most of the observations that once were controversial are now largely piecing together into a unified model. Some issues remain, but, yes, our understanding of the deep Earth looks certainly much better than a few decades ago.

11.4 The growth of continental crust

The extraction of the substance of continents from the mantle is an essential process in the evolution of the planet. It concentrates major elements, such as Si, Al, Na, P, and incompatible trace elements, such as Rb, Ba, the light rare-earths, and the radioactive elements, U, Th, and K, almost irreversibly near the surface. These elements share the characteristic of occurring in greater concentrations in melts than compatible elements. Proof that the continents are being permanently extracted from the mantle and that continental material is also being injected back into the mantle comes from isotopes. If we plot the evolution of certain isotopic ratios, and especially the ^{143}Nd/^{144}Nd ratio, against the age of formation of rocks, we generally obtain a graph such as in Fig. 11.16. If the continents and the mantle had separated out since the beginning of geological time and had remained isolated from one another, the slope of the straight lines $p_c \approx 0.14$ and $p_m \approx 0.22$ would reflect

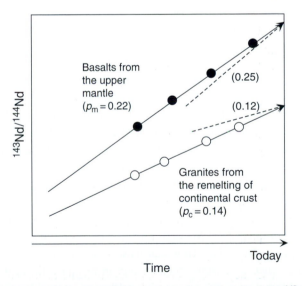

Figure 11.16 Isotopic evolution of Nd over time. The solid lines show the evolution of the ^{143}Nd/^{144}Nd ratio in mantle (basalt) and crust (granite) samples of different ages, the ratio being evaluated for the time the rocks were formed. The slopes of these two straight lines yield apparent ^{147}Sm/^{144}Nd ratios of the two geologic reservoirs. Conversely, the dashed lines show the changing ^{143}Nd/^{144}Nd ratio recalculated over time using ^{147}Sm/^{144}Nd and ^{143}Nd/^{144}Nd ratios observed in samples of present-day mantle and crust. The difference in the slopes of the two curves over time is proof that the mantle and crust exchange material. It reflects the crustal growth phenomenon from the mantle and the recycling of continental crust back into the mantle.

the ^{147}Sm/^{144}Nd ratios of these reservoirs. This is not so. Typical ^{147}Sm/^{144}Nd ratios measured for continental rocks and the mantle are 0.12 and 0.25, respectively. This indicates that reservoirs corresponding to mantle and continents are open to mutual exchanges and have the effect of raising the apparent ^{147}Sm/^{144}Nd ratio of crustal rocks and lowering that of mantle rocks. Both the growth of continents from the mantle and the recycling of continental crust by mantle convection have been caught red-handed!

Although there is apparent consensus that recycling of continental crust corresponds essentially to subduction of detrital sediments resting at the top of lithospheric plates, the inverse process of extraction of crustal material is not fully understood. As with convection, we can consider that there has been a canonical model of crustal growth based on the addition of andesite at converging plate boundaries. The Andean volcanoes, Mount Pinatubo or Mont Pelée, seem ideal agents for supplying mantle to the continents. This seems to be confirmed by the similarity of the chemical composition of lava from these volcanoes (including the famous Nb deficit anomaly) and the isotopic compositions of many elements, especially O, Sr, and Nd, to the composition of continental rocks.

Once again, superficial similarities mislead observers. A first alarming indication is the histogram of the age of formation of continental crust. No matter which method was considered, the first histograms of rocks ages produced in the 1960s (e.g. Gastil, 1960) and the 1970s were already extremely clear: continents form well-defined age provinces (Hurley *et al.*, 1962) and age histograms define very sharp peaks. The advent of dating using ion probes and laser-ablation ICP MS now allows the rapid production of large numbers of reliable zircon ages. Bedload sediments represent excellent samples of the crust from river watersheds and the detrital zircons they contain make great samples for the formation ages of the local crust. Hundreds of zircons from the bedload of four major rivers of the world (the Yangtze, the Mississippi, the Congo, and the Amazon) have been analyzed by Iizuka *et al.* (submitted). The concordant U–Pb (^{207}Pb*/^{206}Pb*) ages define very sharp peaks corresponding to well-known orogenic episodes. These are "magic" ages of very rapid crustal growth around 300, 600, 1100, 1400, 1800, 2100, and 2700 million years ago (Fig. 11.17). Such well-defined histograms are very unlikely to be the result of differential erosion: by any means, global processes such as erosion and sedimentation can be suspected to even out frequency peaks, not to create them. Geochronology also indicates that the time taken for these new continental expanses to form can be comparatively short. Millions of square kilometers of continent arose to form West Africa between 2170 and 2090 million years ago. Similarly, huge swaths of continental crust formed within a few tens of millions of years around 2700 Ma ago in the Superior Province of Canada with little crustal growth between the spikes. How can these characteristics be attributed to subduction, which is in essence a constant phenomenon in response to the formation of new oceanic crust, and therefore to the progressive outflow of internal heat? We can, of course, imagine that cold oceanic lithosphere occasionally collapses into the deep mantle during brutal events referred to as an avalanche. The thermal effects and mechanical consequences of such a process appear too violent, though, compared with the relatively constant character of the geological record since the earliest times (lava chemistry, sediment mineralogy).

Steady subduction in itself appears insufficient to account for crustal growth. Most subduction zones are associated with rather small amounts of lavas and most of the andesitic

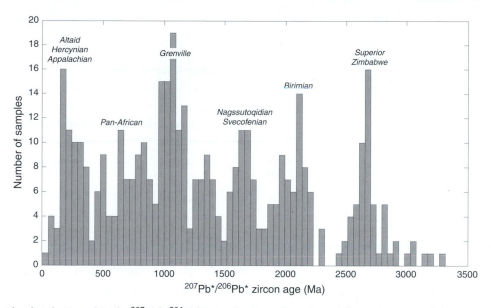

Figure 11.17 The distribution of U–Pb (^{207}Pb*/^{206}Pb*) ages in zircons from four of the major rivers of the world, the Yangtze, the Mississippi, the Congo, and the Amazon (courtesy Tsuyoshi Iizuka). Only concordant zircons are plotted. These data clearly show that granitic rocks, the primary carriers of zircons, and therefore representative of continental crust, formed during very well-defined pulses. Only a few local names of orogenic events are shown.

volcanoes show isotopic properties indicative that they are largely dominated by melts from the underlying continental crust rather than creating it. The crustal volume of the intra-oceanic arcs, such as the Western Aleutians, the Mariannas, or the South Sandwich Islands is actually quite insignificant. The reason why regular subduction produces so little crust is that modern oceanic crust is too cold to melt before it disappears into the deep mantle. Likewise, avalanches would not allow enough time for the sinking lithosphere to melt. A popular idea is that Archean subduction was hot enough to allow subducted oceanic crust to melt, but such a process is very unusual in modern times.

There is, nonetheless, a process of recurrent instabilities that can account for the episodic character of crustal growth. This is the eruption of plume heads, such as those at the origin of large igneous provinces (LIP) forming the traps of the Deccan in India or those of Ethiopia, and whose observable after-effects (plume stems) are the alignments of islands anchored to the hot spots of Hawaii or Réunion, among others. Other sudden submarine eruptions of very large size are known, such as the Ontong–Java plateau in the Pacific and the Caribbean plateau in the Atlantic. These plateaus currently form more than 10% of the ocean floor. Basalts from juvenile provinces (crustal segments that show a very short crustal history of the material at the the time they formed), such as the 0.6 Ga old Pan-African from Egypt and Arabia, the 2.1 Ga old Birimian of Africa, the 2.7 Ga old Superior of Canada, show remarkable geochemical similarity (Fig. 11.18) between continental protolith, i.e. the material extracted from the mantle and which coalesces to form continents and the ocean plateaus. Although andesitic volcanic rocks undoubtedly exist in

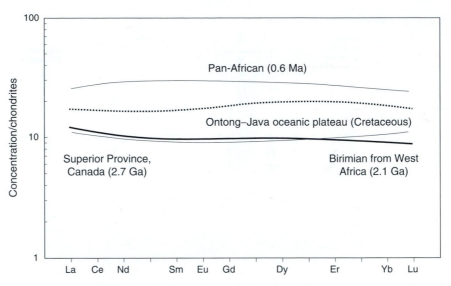

Figure 11.18 Distribution of concentration of rare-earths in the basalts of the Superior Province (Canada), the Birimian of West Africa, the Pan-African of the Near East, and the Cretaceous Ontong–Java ocean plateau. This similarity suggests that continents formed from volcanic material similar to that issuing from plume heads (after Abouchami *et al.*, 1990 and Stein and Goldstein, 1996).

such deposits, they invariably overlie basaltic plateau material and so were formed subsequently. The subduction mechanism therefore builds up continental material from existing oceanic plateau swept against and accreted to continents by plate tectonics. The normal fate of the oceanic crust and of most plateaus is to be removed from the surface by subduction, as it is clearly the case for the head of the Hawaiian plume under the Kamchatka Peninsula. However, when the volume of the plume head is large enough to make the lithosphere buoyant, a new continent is formed. This construction is usually brief, taking less than 100 million years.

From a strictly geochemical point of view, the composition of the crust is a compromise between magmatic input from the mantle (basalt and andesite) and output through erosion. For all practical purposes, interaction with seawater can be ignored because the oceanic residence times of all elements are much shorter than the age of the continents. This compromise can be evaluated by using the method in Chapter 6. At first sight, the composition of the material that erosion removes from the continent, as shown by the composition of rivers and their suspended loads, is very poor in silica, but rich in Mg and Ca. It is incorrect to assume that the composition of continents simply is identical to that of their protolith, i.e. the magmatic rock that may be their origin (Fig. 11.19). Continents must be seen as dynamic systems with both inputs and outputs. As hardly any significant differences in composition are observed between past and present crust, particularly in detrital sediments, it is considered that the system has reached a stationary state.

In fact, the assertion that present-day crust is indistinguishable from ancient crust applies only to its elemental content, whether of major elements, like Si and Mg, or of trace elements. When the isotopic compositions of Nd or Hf of basalts, either fresh or metamorphosed, at the time of formation are plotted, i.e. the composition of their source mantle,

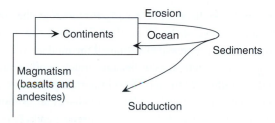

Figure 11.19 Crustal composition is the net result of an input – crustal growth out of magmas erupted from the mantle – and an ouput – erosion and subduction of sediments and oceanic crust. The composition of the continental crust does not therefore reflect that of any magmatism feeding it.

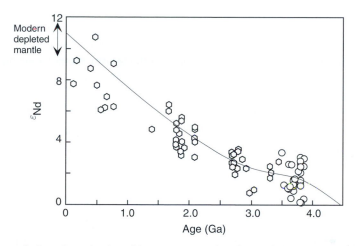

Figure 11.20 Isotopic evolution of neodymium (shown in units of ε_{Nd}) in rock from the mantle over geologic time. The slope of the curve of evolution is proportional to the Sm/Nd ratio. The existence of radiogenic Nd 3.8 Gy ago suggests that the Earth probably underwent intense fractionation of this ratio between 4.5 and 3.8 Gy ago (probably during its primordial differentiation). The Sm/Nd ratio remained low up to 2 Gy ago, then increased to its present-day value. Data compiled by Vervoort and Blichert-Toft (1999).

it can be seen that these compositions varied little, with slightly positive $\varepsilon_{Nd}(T)$ and $\varepsilon_{Hf}(T)$ values between 3.8 and 2.0 billion years (Fig. 11.20). This contrasts with virtually continuous growth of the radiogenic character of Nd and Hf from 2.0 billion years to the present. From the definition of the notation of ε given by (4.11) and of evolution given by (4.33), the plot of Fig. 11.20 is very similar to that of Fig. 4.12: for a closed system, the slope of the curve is proportional to the ^{147}Sm/^{144}Nd ratio of the source mantle. As the oldest known basaltic rocks (3.85 Ga from west Greenland) already contain radiogenic Nd, it can be deduced that the ^{147}Sm/^{144}Nd ratio of their source mantle at the very beginning of the Earth's history (>4 Ga or Hadean) was higher than the chondritic ratio. Rocks rich in incompatible elements were therefore extracted from the Earth's mantle very early in its history, probably forming a continental proto-crust. This proto-crust was since reworked into more recent crust or engulfed by subduction into the lower mantle.

During the Archean and the early Proterozoic, the slope of the isotopic evolution curve for basalt Nd, and so the $^{147}Sm/^{144}Nd$ ratio of the mantle, remained fairly low and then, at around 2 Ga, progressively attained the modern value. Interpretation of this observation is complex and controversial because much continental crust seems to have been formed around 2.7 billion years ago. Two important points are involved in the explanation of this observation: (1) the Earth underwent substantial primordial differentiation that plate tectonics took one or two billion years to erase; and (2) because of the cooling of the Earth, the significance of violent convective instabilities in the form of hot spots diminished in favor of slower and more regular heat loss through ridges. The fast cooling of the core also acted to stabilize mantle convection with respect to plume generation. We will show in the next chapter that the ^{142}Nd anomalies left by the extinct ^{146}Sm radioactivity in the 3.8 Ga old rocks from Isua, west Greenland require the existence of an early magma ocean and therefore support hypothesis (1). However, simply because heat production by radioactivity has declined by a factor of three over the last four billion years, the overall strength of mantle convection must be less today than in the Archean. This simple argument implies a reduction over the Earth's history of the juvenile magmatic contribution to crustal growth from hot spots. It is probable that if recycling of ancient crust is perceived as so important today, as for example in the Alps, the Himalayas, the Appalachians, or the Hercynian belt of Western Europe, it is because plumes have become much more infrequent.

References

Abouchami, W., Boher, M., Michard, A., and Albarède, F. (1990) A major 2.1 Ga event of mafic magmatism in West Africa on early stage of crustal accretion. *J. Geophys. Res.*, **95**, 17 605–17 629.

Albarède, F. and van der Hilst, R. (2002) Zoned mantle convection. *Phil. Trans. R. Soc.*, **A360**, 2569–2592.

♠ Allègre, C. J. (1982) Chemical geodynamics. *Tectonophys.*, **81**, 109–132.

♠ Gastil, G. (1960) The distribution of mineral dates in space and time. *Amer. J. Sci.*, **258**, 1–35.

♠ Hurley, P. M., Hughes, H., Faure, G., Fairbairn, H. W., and Pinson, W. H. (1962) Radiogenic strontium 87 model of continent formation. *J. Geophys. Res.*, **67**, 5315–5333.

♠ Hofmann, A. W. and White, W. M. (1982) Mantle plumes from ancient oceanic crust. *Earth Planet. Sci. Lett.*, **57**, 421–436.

Iizuka, T. *et al.* (2008) Detrital zircon evidence for Hf isotopic evolution of granitic crust and continental growth. Submitted.

Lassiter, J. C. and Hauri, E. H. (1998) Osmium-isotope variations in the Hawaiian lavas: evidence for recycled oceanic lithosphere in the Hawaiian plume. *Earth Planet. Sci. Lett.*, **164**, 483–496.

McBirney, A. R. (1992) *Igneous Petrology*. San Francisco, Freeman-Cooper.

Presnall, D. C. *et al.* (1978) Liquidus phase relations on the join diopside–forsterite–anorthite from 1 atm to 20 kbar: Their bearing on the generation and crystallization of basaltic magma. *Contrib. Mineral. Petrol.*, **66**, 203–220.

Stein, M. and Goldstein, S. L. (1996) From plume head to continental lithosphere in the Arabian–Nobian shield. *Nature*, **382**, 773–778.

Vervoort, J. D. and Blichert-Toft, J. (1999) Evolution of the depleted mantle: Hf isotope evidence from juvenile rocks through time. *Geochim. Cosmochim. Acta*, **63**, 533–556.

12 The Earth in the Solar System

The Earth is, together with Mercury, Venus, and Mars, one of the rocky planets of the inner Solar System. Our sampling of the Solar System is rather limited. Terrestrial samples are plentiful and the Apollo astronauts brought back hundreds of kilograms of lunar rocks. More than 30 meteorites from the Moon and 40 meteorites from Mars (the SNC meteorites) complete the well-identified planetary material. Achondrites are basaltic rocks which have been suggested to come from the asteroid Vesta in the asteroid belt, between Mars and Jupiter. Chondrites derive their name from the presence of abundant mm-sized molten blebs known as chondrules. They are fragments of much smaller, undifferentiated asteroids, and are divided into carbonaceous, ordinary, and enstatite meteorites. Carbonaceous meteorites are remarkable because they contain hydrous minerals, organic matter, refractory inclusions, and pre-solar grains (SiC, diamonds). Iron meteorites are actually Fe–Ni alloys: some of them represent the core of some small planetary bodies disrupted by impacts.

The composition and the conditions prevailing at the surface of each rocky planet are very different and distinct from those of the giant gas-rich planets, such as Jupiter and Saturn, of the outer Solar System. Astronomical observations indicate that the Sun and its companion planets formed by condensation of a cloud of gas and dust, called the solar nebula. The topic of this chapter is to describe the major processes that led from the early stages of the Universe to the formation of the elements, to the birth of the Solar System and finally to the formation of the Earth and its sister planets.

The speed at which the universe is expanding as measured by the "red shift" of light from the stars and the nuclear cosmic chronometers indicate that the Big Bang occurred about 14 billion years ago. Because our star, the Sun, is a mere 4.5 billion-year-old youngster, it is clear that the Solar System recycles material with a long history. The processes leading to the formation of elements are known collectively as nucleosynthesis. These elements form in stars by a combination of processes of thermonuclear fusion, neutron capture, and radioactive decay. Who would think that our warm sunshine is actually the output of the largest nuclear reactor for light years around?

12.1 The formation of elements

The most abundant element in the universe is hydrogen, an element formed by an electron orbiting around a single proton. Let us begin with the formation of a Sun-like star by gravitational collapse of an interstellar nebula, a cloud of gas and dust. The Sun itself, which makes up 98% of the total mass of the Solar System, consists of 71% H, 27% He, and 2% of heavier elements. The composition of the Solar System is given by spectroscopic analysis of sunlight, and it is this composition that will concern us here (Fig. 12.1). The accumulation of potential energy released by the collapse of the star's enormous mass raises the elements at its center to extremely high temperatures, typically several million K, until thermal agitation prevents further contraction. At these very high temperatures, the elements are totally stripped of their electrons, which leaves the Sun largely made of a dense gas of protons, electrons, and alpha particles. With the help of a tunnel effect, opposite to that referred to for the α decay process, the nuclei gain enough thermal energy to overcome the Coulomb repulsion barrier between them. The nuclei then fuse together and thermonuclear reactions begin, releasing gigantic energy reserves by conversion of mass. The star achieves steady state when the energy produced by nuclear reactions is balanced by the emission of neutrinos generated in the nuclear reactions and by electromagnetic radiation in outer space. At high pressure, however, the stellar gas is opaque. The highly energetic

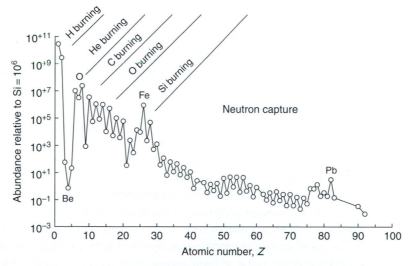

Figure 12.1 Chemical composition of the Solar System standardized to a million atoms of Si. The main nucleosynthetic stages are marked above the curve (after Pagel, 1997). Notice the stability peaks (e.g. Fe, Pb) and the lower abundance of elements with odd atomic numbers compared with their even-numbered neighbors, whose nuclei are more stable. The deficit in the light atoms Li, Be, and B results from destruction of these elements in the stellar interiors.

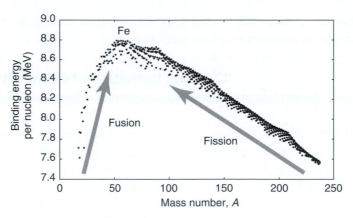

Figure 12.2   Binding energy per nucleon of different nuclides is maximum for iron (in millions of electron-volts). Fission of nuclides heavier than iron and fusion of nuclides lighter than iron are energy-producing reactions and, apart from their activation energy, occur spontaneously. Nucleosynthesis beyond iron will therefore consume energy.

radiation produced in the stellar interior is therefore absorbed and further converted into thermal energy: it can only escape at a wavelength corresponding to the surface temperature of the star. The surface temperature of the Sun (6000 K) corresponds to radiation in the window of visible wavelength.

Understanding these processes involves the concept of nucleon (protons and neutrons) bonding energy. In the fission of a uranium atom the total mass of the fragments, e.g. xenon and barium, is less than that of the initial atom. Conversely, in hydrogen atom fusion the mass of the resulting atom, e.g. lithium, is less than that of the initial hydrogen atoms. The extra mass is converted into thermal energy. Any reaction that releases energy is liable to occur spontaneously. If we plot the binding energy thus available per nucleon (Fig. 12.2) it can be seen that the maximum of the curve is located at the position of iron: if the activation barriers are crossed the fusion reactions of light elements and fission reactions of heavy elements will be spontaneous, while masses in the vicinity of iron will be stable.

Let us now examine the abundance distribution of elements in the Sun relative to their atomic number. A first observation is that elements with even atomic numbers are more abundant than those with odd atomic numbers, reflecting the greater stability of nuclei in which the nucleons are paired. This characteristic is one justification of element plots against a reference standard such as chondrites, which overcomes the odd–even effect. An irregular decline in abundance is observed with increasing mass number. In addition, abundance peaks occur at certain mass numbers such as that of iron (56), or for other nuclides with proton or neutron magic numbers of 2, 8, 20, 28, 50, 82, and 126. There is also a considerable apparent deficit in the low-mass elements Li, Be, and B.

From hydrogen to iron, nuclear fusion within stars is the predominant process. The fusion of hydrogen into helium is the dominant "low" temperature (10^7–10^8 K) process and the one going on in the outer part of our Sun right now. The helium produced is burned at greater depth within the star and therefore at a higher temperature (10^8 K) to form C, N, and O, which in turn are raised to higher temperatures (10^8–10^9 K) and produce Mg, Al, Ca, and Si. Fusion stops at iron because production of heavier nuclides requires energy input. Lithium, Be, and B, for their part, are burned in the stellar interiors to form heavier elements and the small quantities of these elements observed in planetary material were produced in the interstellar medium upon breakup of heavy nuclei by cosmic rays.

At the very high temperatures (several billion K) resulting from gravitational collapse of the core of the more massive stars, explosive silicon burning produces the abundant elements of the iron group (Fe, Ni, Cr). For the elements beyond iron, processes of a different type are at work that involve the absorption of neutrons by nuclei. Neutrons are in general easily absorbed (astrophysicists would refer to a large neutron absoption cross-section). They are not affected by the Coulomb electrostatic potential of the nucleus, and therefore can approach it to the distance at which nuclear forces become dominant. Three essential processes are the slow *s* process, the rapid *r* process and the *p* process, which stand for slow and rapid neutron absorption and for proton absorption, respectively. Most elements are produced by a combination of processes although some of them are essentially pure, such as Sn, Ba, Pb (*s*) or Xe, Eu, Pt (*r*). The relatively minor *p* process produces nuclei rich in protons located above the zone of stable nuclides (gray in Fig. 12.3) by reaction of the nuclides with high gamma radiation emitted when supernovae explode, and these will not be discussed here. There is some agreement that the *s* process must be associated with stars that have masses a few times that of our Sun from the so-called Asymptotic Giant Branch (AGB), which are known to lose enormous amounts of material continuously, disheveling themselves into molecular clouds which will later be used as the raw material for new stars. *r*-Process elements are formed in the violent explosion of some particular types of supernovae. There is no consensus, however, on how the more uncommon *p*-process elements form.

The *s* and *r* processes can be better understood by examining a segment of the chart of the nuclides (Fig. 12.3). It can be seen that the neutron-rich, unstable *s* and *r* nuclides (located to the right of the stability valley) decay by β^- emission, while the rare *p* unstable nuclides (left) decompose by electron capture or β^+ decay. This pattern of radioactive decay results in the preferential formation of nuclides in and around the valley of stability.

Neutrons are being produced permanently in stars by multiple reactions involving light elements. The *s* process starts beyond the peak of the iron group, i.e. beyond the top of the curve of the binding energy per nucleon (Fig. 12.2), with iron "seeds." It arises from high-temperature (300 million K) equilibrium between absorption of these neutrons by nuclei and radioactive β^- decay. The rate of change of the number n_i per unit volume of a stable nuclide containing i neutrons and which is bathed in a medium with a concentration of neutrons N_n agitated at the mean thermal velocity v_n is given by the equation:

$$\frac{dn_i}{dt} = -\sigma_i N_n v_n n_i + \sigma_{i-1} N_n v_n n_{i-1} \tag{12.1}$$

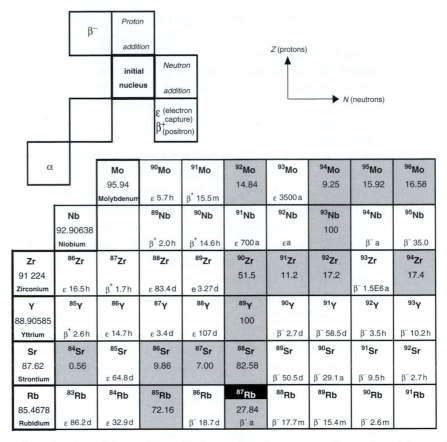

Figure 12.3 Excerpt from the chart of the nuclides with the number of neutrons on the x-axis and the number of protons on the y-axis. Stable nuclides are in gray. White on black are long-lived radioactive nuclides. White fields are short-lived radioactive nuclides. Each square shows the mass number, the isotopic abundance as a percentage, the decay process and the half-life. The atomic mass of the element is shown at the start of the row. Notice the difference in dominant nuclear processes on either side of the stability zone: β^+ and electron capture ε above the stability zone, and β^- below it.

where σ_i and σ_{i-1} are coefficients with the dimension of a surface area, called neutron-capture cross-sections, that can be determined by laboratory experiments. The reader will recognize in this expression the neutron flux $N_n v_n$ through surface areas σ_i and σ_{i-1}. As the excess of neutrons over protons increases, the σ values decrease and neutron capture becomes less and less efficient, while the probability and the rate of β^- decay increase (Fig. 12.4). When the nuclide with neutron number i is radioactive, the previous equation must be changed to:

$$\frac{dn_i}{dt} = -\sigma_i N_n v_n n_i + \sigma_{i-1} N_n v_n n_{i-1} - \lambda_i n_i \tag{12.2}$$

where λ_i is its β^- decay constant. For a given nuclide, a steady state is reached where the input by neutron absorption on the progenitor equals the output by radioactive decay plus

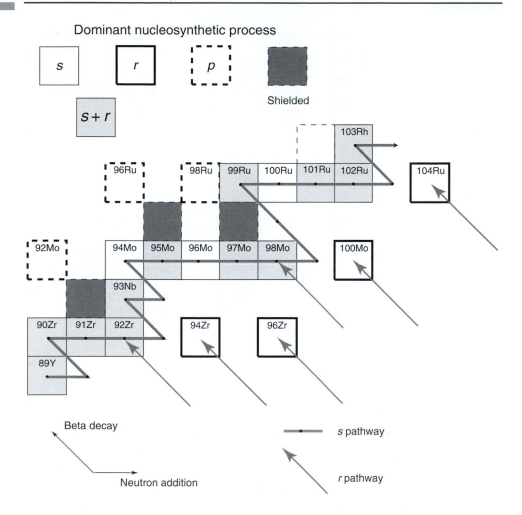

Figure 12.4 Example of nucleosynthesis by neutron capture for elements heavier than Fe. Slow *s* processes represent an equilibrium between neutron addition and β⁻ decay. Rapid (*r*) addition of neutrons in the expanding core of some stars produces neutron-rich isotopes by β⁻ decay. The *p* process yields proton-rich nuclei located above the valley of stability by reaction of the nuclides with high gamma radiation emitted by supernovae. Note the shielded isotopes between stable nuclides.

neutron absorption on the progeny, which, for stable isotopes, leads through (12.1) to the simple relationship:

$$\sigma_i n_i = \sigma_{i-1} n_{i-1} \tag{12.3}$$

An *s* pathway may be easily constructed, with its branchings and accumulation points, and the natural abundances of *s* nuclides can be successfully predicted as a function of the neutron flux. The odd nuclides have in general larger cross-sections and therefore are

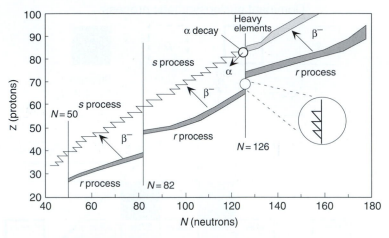

Figure 12.5 The origin of natural nuclides. The *s* process stops when α decay destroys the newly formed nuclei. When neutron absorption is too fast to allow loss of electrons (*r* process), it progresses to very stable nuclei (magic numbers) where it must wait for electron re-emission before it can go further (blown-up sawtooth). When the nuclei become too heavy, they are destroyed by fission.

less abundant than even nuclides. The abundance peaks (magic numbers) reported earlier correspond to extremely stable optimal filling of nuclear energy levels making the nuclei immune to neutron absorption. The *s* process stops at ^{209}Pb, when α decay becomes predominant and decomposes nuclei formed by the slow absorption of neutrons.

Other processes are required to explain many heavy nuclides, in particular those whose mass is greater than that of lead, such as U and Th. More highly energetic neutrons are added to nuclei when supernovae explode, faster than they are lost by β^- emission. The *r* process of rapid absorption of neutrons by nuclei follows an *r* pathway quite remote from the valley of stability to which nuclides eventually return by β^- emission (Fig. 12.5). This fast track is blocked at the level of particularly stable nuclei unamenable to the addition of neutrons (magic numbers 50, 82, 126). Further neutron absorption must wait at each notch for an electron to be lost by β^- emission before a new neutron can be absorbed and the nucleus can ratchet up the sawtooth of magic numbers. The process is stopped at the heavy masses when fission destroys nuclei faster than they can form.

The abundances of *p*-process nuclides is remarkably well predicted by stellar models and experimentally determined cross-sections and decay constants. The abundances of *r*-process neutron-rich nuclides is deduced by subtracting the contribution of the *s* process from the observed distribution of the elements in the universe. The names of Burbidge, Burbidge, Fowler, and Hoyle (usually referred to as B^{2}FH) is attached since 1957 to one of the most fascinating successes in the history of science. Overall, astrophysics has been very successful at predicting the abundances of elements and isotopes in the universe (Fig. 12.6). It has been less successful at identifying the stellar sites at which this complex alchemy takes place.

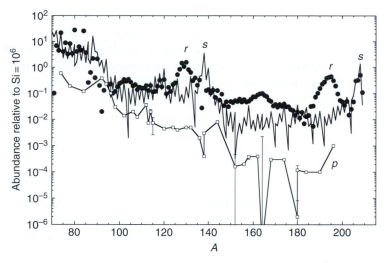

Figure 12.6 Breakdown of the solar abundances in components produced by the different nucleosynthetic processes (Arnould *et al.*, 2007).

12.2 The formation of the Solar System

The Solar System is composed of a medium-sized star – the Sun – and its retinue of planets. Some 99.9% of the mass of the Solar System is concentrated in the Sun and the dominant element is hydrogen. Our idea of how the Solar System formed goes back in its principle to Laplace. We start with a cloud of gas and dust formed from material lost by unstable large stars: grains of silicon carbide, diamonds, and various silicates presenting uncanny isotope compositions for all their elements are found in some chondritic meteorites and attest to the pre-solar history of the Solar System material. This is the solar nebula, which quickly assumes the shape of a disk. Material swirling around the center of mass (occupied by the nascent star), atoms, molecules, and particles fell to the equatorial plane of the disk. Thermal agitation caused countless collisions. Loss of momentum during these collisions and interaction of matter with the magnetic field of the Sun caused material to spiral inward toward the Sun. Although most of the mass ended up at the heart of the system, the orbiting part that was to form the planets condensed into rock fragments (planetesimals), which coalesced through successive collisions. The planets toward the inner part of the Solar System are more rocky because the high temperatures caused by intense radiation emanating from the Sun did not allow gases, such as water vapor and methane, to condense. It is commonly assumed that the "snow line," i.e. the imaginary limit of ice stability, was located somewhere near the asteroid belt between Mars and Jupiter. The outer planets such as Jupiter and Neptune formed in a colder environment and, in addition to their rocky cores, also collected gaseous species such as hydrogen and methane. All the observations of young stars and the astrophysical models concur that the residual gas was blown off in 3–5 million years by the solar wind, the intense electromagnetic radiation (ultra-violet,

X-rays, protons, ^{3}He) and particle fluxes emitted by the Sun during its early gravitational collapse (T-Tauri phase). The implication is that the accretion of Jupiter and the other giant planets from the Outer Solar System was by this time largely completed.

The numerous collisions between planetesimals increased the sizes of the resulting planetary bodies, reduced their number, and cleaned up their orbits. The gravitational pull of the massive outer planets perturbed their neighboring objects, which were either expelled from the Solar System or kept pouring down on the Sun and the inner planets. The outcome of these processes, which lasted less than 50 million years, was a small number of planets

The variability of oxygen isotope compositions in the Solar System

The stable isotope compositions of samples from individual planetary objects in the Solar System follow a mass-dependent fractionation line (Chapter 3), which indicates that they are derived from an isotopically homogeneous reservoir. Among the rare exceptions (Cr, Ti), oxygen isotopes are remarkably anomalous in defining two separate trends in the $\delta^{17}O$ vs. $\delta^{18}O$ plot of Fig. 12.7: (i) the regular mass-dependent fractionation trend with a slope of 0.5, such as the terrestrial fractionation line or the SNC (Martian meteorites) trend, and (ii) the 1:1 fractionation line of refractory inclusions and chondrules from carbonaceous chondrites such as the Allende CV3. The anomalies of the different samples are usually described by calculating a $\Delta^{17}O$ parameter, which is approximately equal to $\delta^{17}O - 0.5\,\delta^{18}O$. Martian meteorites and ordinary chondrites have positive $\Delta^{17}O$ values (they plot below the terrestrial fractionation line or TFL), whereas refractory inclusions and chondrules from carbonaceous chondrites have negative $\Delta^{17}O$ values (they plot below the TFL). By analyzing the solar wind implanted in metal grains from lunar soil and in traps carried by the Genesis mission it was found that the oxygen isotope composition of the Sun lies on this 1:1 line with a $\delta^{18}O$ of -40 to -60‰, i.e. with a very negative $\Delta^{17}O$. Remarkably, ozone produced by electrical discharge in molecular oxygen shows such fractionation along the 1:1 line (Thiemens and Heidenreich, 1983), a discovery that found important applications in atmospheric chemistry.

There is a large consensus on the variety of processes that may have formed the trend (i) such as vapor–solid fractionation and low-temperature alteration. In contrast, the nature of the mass-independent process (ii) is not agreed upon. Some scientists believe in the so-called "self-shielding" of CO dissociation by UV radiation from the young Sun (remember that CO is particularly abundant at high temperature): the wavelengths of UV absorption by $C^{16}O$, $C^{17}O$, and $C^{18}O$ are slightly shifted with respect to one another. Since the abundances of the isotopes are very different, it takes different distances for the nebular gas to absorb UV of a particular radiation: the wavelength of the most abundant $C^{16}O$ is smothered much faster than that of the minor isotopes. Another candidate is the abundant molecule SiO. This idea is opposed by those who believe that isotopic exchange between the co-existing O and CO is too fast to make self-shielding an efficient process. Alternative interpretations appeal to symmetry effects in the kinetics of some gas reactions (think of the differences between the two ozone molecules $^{16}O-^{16}O-^{18}O$ and $^{16}O-^{18}O-^{16}O$). It is remarkable that the processes that severely affected the isotope compositions of one of the most abundant elements in the Universe still remain unresolved.

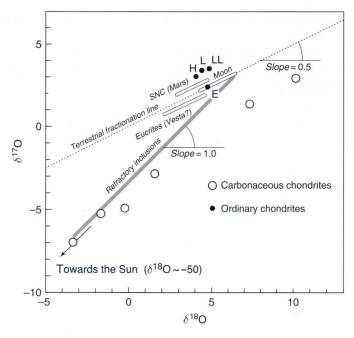

Figure 12.7 The $\delta^{17}O$ vs. $\delta^{18}O$ plot of different planetary bodies (in ‰). Basaltic achondrites are igneous rocks chipped off from an asteroid, which may be Vesta. Individual planetary bodies plot along mass-dependent fractionation lines parallel to the Earth–Moon fractionation line. Samples of SNC (Martian) meteorites and ordinary carbonaceous chondrites do not lie on this line. The letters LL, L, H, and E next to the symbols of ordinary chondrites refer to chondrites with different iron contents and oxidation states. The refractory inclusions are globular assemblages of Ca–Al-rich minerals, such as spinel, anorthite, and melilite. They plot on a 1:1 line (see box for discussion). Note that the Earth cannot be made up of ordinary chondrites except those from the E group (after Clayton *et al.*, 1991), but this goes against other geochemical evidence (see text).

of similar size: Mercury, Venus, Earth, and Mars. An exception is the asteroid belt occupied by many small rocky bodies. The meteorites continually falling to Earth come from this zone and are the debris of asteroids whose orbits were strongly perturbed by the huge mass of neighboring Jupiter. Some of the meteorite parent bodies have been identified, such as the asteroid Vesta. Our planet is not the only one to be hit by such fragments. We know that impacts of such objects on the Moon and Mars have thrown out fragments now found as meteorites on Earth, and a constant albeit tenuous exchange of material between the various planets has been the rule over geological time. A particularly significant terrestrial collision led to the formation of the Moon. The strongest geochemical argument – and probably the strongest argument in general – for the common origin of the Earth and the Moon is that the oxygen isotopes of rocks of both planets fall on the same mass fractionation line (Fig. 12.7), thereby indicating a common reservoir for this element. It has also recently been suggested that the giant impact resulted in large-scale isotopic homogenization of the vaporized material. It is imaginable, therefore, that a grazing impact of a body the size of Mars knocked off the material that now forms the Moon. At the time of impact, the Earth was still unfinished and its mass was only two-thirds of what it is today. The material of the impactor was highly refractory, greatly depleted in volatile substances, and intensely reduced.

The initial state of the terrestrial planets depended on their size and composition. The energy sources that may have contributed to heating the interior of these planets are well understood:

- Initial gravitational energy, which has markedly heated and almost certainly melted the external parts of the largest planets.
- Gravitational energy of separation and crystallization of the iron and nickel core. This energy is enough to raise the Earth's temperature instantaneously by 2000 K.
- Radioactive heating by decay of short half-life nuclides such as [26]Al whose descendant [26]Mg has been identified in several planetary objects.

We will see upon discussing the extinct radioactivity of [182]Hf that the Earth's core formed soon after the formation of the Earth itself. We will also see that the extinct radioactivity of [146]Sm requires that the upper mantle of the Earth, the Moon, and Mars went through a stage of wholesale melting, as what is termed a magma ocean. If so, it may be that the deep mantle still contains rocks formed at that time, an idea that still remains a challenge to geodynamicists.

12.3 Condensation of planetary material

The prediction of reasonable condensation sequences relevant to planetary accretion is one of the most successful contributions of thermodynamics to our understanding of the Solar System. The gravitational energy given off by the collapse of the solar nebula to form the Sun, and the radiative transfer of thermal energy thus produced, heated the proto-planetary material and vaporized it at temperatures in excess of 2000 K. As it cooled, the hot gas, initially dominated by molecular H_2, He, N_2, O_2, H_2O, CO, and SiO, recombined and recondensed to form the solids, liquids, and gases observed at ambient temperature. Astrophysical models provide temperature and pressure distributions throughout the nebula and their evolution over time. By applying elementary thermodynamic principles we should therefore be able to deduce the order of condensation of the various minerals involved in the formation of planetary bodies. Although it would be incorrect to assume that condensation is a well-understood equilibrium process, the insight into the diversity of planetary compositions and the mineralogy of planetary interiors provided by thermodynamic modeling is quite invaluable.

The method is very similar to that described in Section 7.3 for the calculation of speciation in solutions. A first task is to draw up an inventory of the number of moles N of components (O, Al, Si, etc.) and potential species (CH_4, Al_2O_3, H_2O) regardless of their state, allowing the mass balance to be written, e.g. for silicon:

$$N_{Si} = N_{Si}(g) + N_{SiO}(g) + N_{Mg_2SiO_4}(s) + N_{MgSiO_3}(s) + \cdots \qquad (12.4)$$

where g and s indicate whether the compounds are in a gaseous or solid state. As there are more unknowns than equations, the necessary number of mass action law equations must be completed. We can write gaseous equilibria of the type:

$$2H_2\,(g) + O_2\,(g) \Leftrightarrow 2H_2O\,(g) \tag{12.5}$$

and relate partial gas pressures by means of the mass action law:

$$\frac{P_{H_2O}^2}{P_{H_2}^2\,P_{O_2}} = K(T, P) \tag{12.6}$$

where $K(T, P)$ is a known function of temperature and pressure. The condensation of planetary material produces different solid phases, which makes equilibrium calculations a tedious task. Let us write, for example, that magnesian olivine (forsterite Mg_2SiO_4) appears by condensation of Mg vapor, SiO, and water:

$$2H_2(g) + O_2(g) \Leftrightarrow 2H_2O(g) \tag{12.7}$$

$$2Mg(g) + SiO(g) + 3H_2O(g) \Leftrightarrow Mg_2SiO_4(s) + 3H_2(g) \tag{12.8}$$

By considering that olivine is nearly pure forsterite, the mass action law can be written:

$$\frac{P_{H_2}^3}{P_{Mg}^2\,P_{SiO}\,P_{H_2O}^3} = K(T, P) \tag{12.9}$$

where $K(T, P)$ is the constant characteristic of this equilibrium. As most matter is initially formed from hydrogen it can be assumed that the partial hydrogen pressure P_{H_2} is equal to total pressure P_{tot} and that, applying Dalton's law, the pressure of each gaseous component is proportional to its molar proportion in the gas. Starting with a composition of solar gas, rather cumbersome calculations can resolve this complex system of equations with decreasing temperatures. The condensation sequence can be predicted from the most refractory minerals (such as melilite and perovskite of refractory inclusions of the famous Allende meteorite) through the common mantle minerals (olivine, pyroxene), the iron in the core, to the hydrated minerals (serpentine), and even water (Fig. 12.8). The mineralogy of each planet can be predicted fairly satisfactorily from its position in the proto-solar nebula and therefore from its temperature.

When the temperature of the solar nebula decreases, elements condense in groups: the refractory siderophile and lithophile elements (e.g., Ti, Si, Ca, Al, Mg, Fe from 1800 to 1300 K) come first, then the alkali elements (Na, K, Rb) and the group of high-temperature chalcophile elements (Zn, Cu, Ga from 1200 to 900 K), followed by the group of low-temperature chalcophile elements (Pb, Ti, Hg from 800 to 500 K), and finally by the volatile atmophile elements (N, H, C, rare gases below 300 K). How effectively volatile elements are incorporated varies from planet to planet and depends on the distance from the Sun, which determines the accretion temperature, and also on the early history of each planet, in particular the massive impacts between planetesimals. If we compare a volatile lithophile element such as potassium with a very refractory lithophile element such as uranium it can be seen that the K/U ratio, whose primordial value of 60 000 is given by carbonaceous

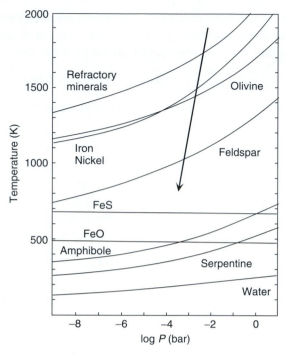

Figure 12.8 Condensation sequence of different minerals from the solar nebula calculated by adiabatic cooling (without heat exchange) of a gas of solar composition (after Lewis, 1995).

chondrites, falls to 10 000 for the Earth (probably 7000 for Mars), and to 3000 for the Moon. This is consistent with the dramatic difference in ^{87}Sr/^{86}Sr between chondrites on the one hand, and the Earth and Moon on the other hand; Rb is an alkali element similar to K, very volatile, whereas the alkaline-earth element Sr is quite refractory. Although the distribution of Sr isotopes in chondrites has been disturbed by the low-temperature alteration of their parent body, it is clear, as pointed out by Paul Gast in 1960, that the ^{87}Sr/^{86}Sr ratio of the Earth (≈ 0.709) is much less radiogenic than the mean ratio of chondrites or of the solar photosphere (≈ 0.745), which attests to a very low Rb/Sr ratio for our planet (Fig. 12.9). It is incontrovertible that the Earth, in addition to K, lost most of its Rb very early after accretion. The Earth and the Moon even more are therefore particularly depleted in volatile elements.

It is commonly said that the amount of water held by the terrestrial mantle (≈ 200 ppm) is similar to the mass of the ocean. An old idea pervading literature and based on the analysis of basalts is that water from the oceans and most atmospheric gases were outgassed from the mantle throughout geological time. But if such a large proportion was lost of elements (K, Rb) that only vaporize at temperatures in excess of 1000 K, how much gas and water, with a much lower condensation point, could have been preserved in the accreting material? Probably not much, so the Earth accreted essentially dry and water was added later from a different source. This is a conundrum that we will address in the next section.

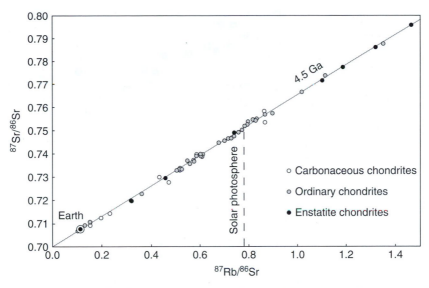

Figure 12.9 The loss of volatile Rb by the Earth. Comparison of the Earth's position with respect to chondrites in the $^{87}Rb/^{86}Sr$–$^{87}Sr/^{86}Sr$ isochron diagram. The $^{87}Sr/^{86}Sr$ ratio in the Earth is an average of mantle and crust values. In spite of the scatter created by low-temperature processes on the parent bodies of chondrites, terrestrial strontium is much less radiogenic than Sr in chondrites and in the solar photosphere. Such a substantial deficit of Rb with respect to the more refractory Sr is a strong indication that the Earth is strongly depleted in all volatiles.

12.4 The composition of the Earth and its core, and the origin of seawater

The very difficult question of the Earth's composition has puzzled the geological community for over 40 years. Assembling the elemental inventory of the continental crust, the mantle, and the core is not an easy task since we have no robust estimate for the composition of the lower mantle and the core. We will see later that the Earth is essentially contemporaneous with the other planetary bodies of the Solar System for which we have samples: the Moon, the parent bodies of meteorites, and Mars. The major deep-seated fractionation processes, particularly the segregation of iron and nickel to form the core, which enable our planet to have a magnetic field, or the transfer of magmas to the surface, produced a strongly layered solid Earth. Closer to the surface, weathering, sedimentation, and metamorphism are other remarkable processes of chemical differentiation. The intense processing and the geochemical variety of terrestrial samples accessible to observation together with the existence of inaccessible domains in the deep Earth, such as the lower mantle and core, do not allow us to build up a verifiable picture of the mean composition of the Earth.

It is thought that the composition of the bulk Earth must resemble that of the least differentiated bodies of the Solar System. The Sun and carbonaceous chondrites CI, of which one of the best known representatives is the Orgueil meteorite, have compositions that differ

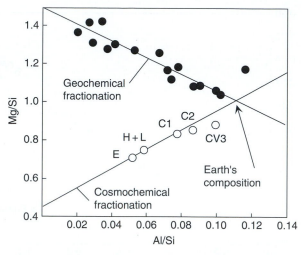

Evaluation of the composition of the primitive Earth by intersection of a trend for mantle rock and a trend for chondritic meteorites. The choice of refractory lithophile elements allows the share of the core and the crust to be ignored (after Jagoutz *et al.*, 1979). E: enstatite chondrites; H, L: high-Fe and low-Fe chondrites, respectively; C: carbonaceous chondrites, with 1–3 defining the degree of metamorphism.

only by their contents of highly volatile elements such as hydrogen and nitrogen. The most common ordinary chondrites are depleted in gaseous elements and are more reduced than carbonaceous chondrites. Since chondrites come from the asteroid belt, well beyond the orbit of Mars, they did not suffer the same heating as the Earth, which was substantially closer to the radiation inferno of the nascent Sun. In addition, the dynamical simulations of planetary accretion show that the planets from the Inner Solar System must have accreted material from between Jupiter and Mercury. There is therefore no reason why the Earth should be identical in its composition to any particular type of chondrite, with the probable exception of the refractory elements. Chondrites represent an unlikely parent material for the Earth. The oxygen isotope composition of most chondrite classes, either carbonaceous or ordinary chondrites, is different from that of the Earth–Moon system. A special category of meteorites, enstatite chondrites, could satisfy this criterion, but then we can infer from the Si content of the Earth's mantle that the core should contain 8–14 wt% Si, for which there is little supporting experimental evidence. The broadly chondritic composition of the Earth is often challenged today but, at least for the refractory elements, nevertheless remains a reasonable reference particularly in the face of no better alternative.

The composition of the primitive mantle (Bulk Silicate Earth, BSE), i.e. the mantle prior to crust extraction, whatever the misgivings about its very existence as an extant geodynamic entity, can be determined through simple mass balance. It can be considered that the separation of the core had as its only effect to enrich the non-siderophile or chalcophile elements by a constant factor (ignoring the crust) in the silicate fraction. The compositions of meteorites and of the Earth's mantle are plotted on the graph of Mg/Si vs. Al/Si (Fig. 12.10), along with that of fertile peridotites. The composition of these mantle samples

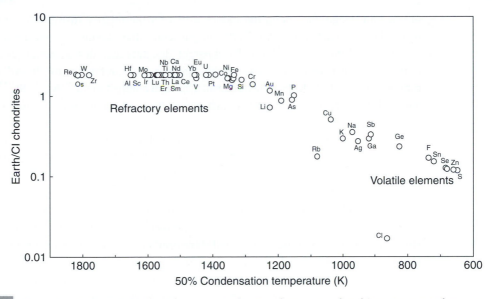

Figure 12.11 Depletion of the Bulk Earth with respect to the CI carbonaceous chondrites. For most elements except H and He, these chondrites have a composition similar to that of the Sun and therefore represent the most primitive solid material in the Solar System. The *x*-axis represents the temperature at which 50% of the initial element inventory has been taken up by solid phases in a standard condensation model. The Earth is strongly depleted in the most volatile elements such as K, Zn, S. Refractory elements such as Zr, La, Th, in contrast, have not been lost. They are slightly more abundant in the Earth than in CI because of volatile element depletion.

is close to the composition assumed for the primitive mantle. The mantle and chondrites each define a trend. It is proposed that one trend corresponds to planetary petrological differentiation and the other to cosmochemical differentiation (fractional condensation) in the proto-planetary nebula. If the primitive material of the Earth belongs to the chondrite family it should lie at the intersection of these two alignments. Of course, we have no evidence that the alignment of mantle compositions amenable to sampling passes through the primitive mantle. If a significant reservoir of one of these elements is not correctly identified, the evaluation is incorrect. This is probably the case for the core: its density, estimated from the speeds of elastic waves, implies that it contains, in addition to Fe and Ni, light elements such as C and S, and maybe even a substantial fraction of Si, which biases the compositional models based on mass balance. In spite of these limitations, this method is still the most reasonable to date.

Once the composition of the primitive mantle is obtained, the terrestrial concentration of the lithophile elements – which do not contribute to core composition – can be easily evaluated. It then becomes apparent (Fig. 12.11) that the Earth is depleted in the most volatile elements, such as K and Na. By comparison with ordinary chondrites, which also went through intense devolatilization, and using as a guide a scale of condensation temperatures, such as the temperatures obtained from the models described in the previous section, we can infer with some confidence the composition of our planet. To the best of our knowledge, the chondritic model works for the relative distribution of the refractory lithophile elements in the Earth. In particular, the chondritic concentrations of the ^{147}Sm–^{143}Nd and ^{176}Lu–^{176}Hf pairs of parent–daughter nuclides in chondrites are reliable indicators of planetary values and can be used as a robust reference for the Nd and Hf isotopic evolution of the Bulk Silicate Earth. This is in contrast to the ^{87}Rb–^{87}Sr, ^{238}U–^{204}Pb, and ^{187}Re–^{187}Os pairs in which at least one of the nuclides is either volatile (Rb, Pb) or siderophile (Os).

One aspect of the composition of the Earth and its mantle relates to the formation of its core. As indicated by the extinct radioactivity of ^{182}Hf (see below), the metallic core almost certainly formed within a few tens of My of the Earth's accretion. Core segregation is now seen as a runaway process releasing enormous amounts of gravitational energy, enough to melt the iron–nickel alloy and a substantial fraction of the mantle. It might be expected that during the segregation of the metallic core, highly siderophile elements strongly partitioned into liquid iron and were nearly completely removed from the mantle. Geochemical evidence tells us otherwise. Some moderately (Ni, Cr) to strongly (Pt, Re, Os) siderophile elements are far more abundant in the mantle than they would be if the mantle had been in contact with iron at the time it migrated into the core. Isotopic analyses for the parent–daughter ^{187}Re–^{187}Os pair indicate that the Re/Os abundance ratios in the mantle are chondritic, while the metal/silicate partition coefficients are very high ($>10\,000$) and different for both elements. A widely accepted explanation of this is that, shortly after the core segregated, the Earth was subjected to intense meteoritic bombardment, and the highly siderophile elements observed in the mantle reflect this chondritic input, termed the late veneer. This process replenished mantle Re, Os, Ni, etc., in relative abundances identical to those found in meteorites.

A good model of the elemental composition of the Earth can provide some idea of its core composition. The properties of the Earth's magnetic field indicate that the core is composed of iron. Seismic shear waves do not propagate in the upper part of the core, which must therefore be liquid. The planetary inventory of nickel, the presence of nickel in iron meteorites, and metallurgical data, all indicate that the core is essentially an iron–nickel alloy (8% Ni). The densities obtained from wave propagation speeds also point to the presence of light elements such as oxygen, sulfur, and less probably magnesium and silicon. To evaluate core composition, McDonough (1999) suggested comparing the relative abundances of lithophile elements (elements that are a priori absent from the core) and siderophile elements of equivalent volatility (measured by the condensation temperatures of these elements). For elements such as Fe, Ge, As, and S, the difference in abundance of lithophile and siderophile elements of equivalent volatility standardized to that of the Earth reveals a deficit represented by the elemental composition of the core. For all its elegance, this exercise depends on an assumption that, although reasonable, is difficult to test and leads to considerable uncertainties.

The issue of core composition is surprisingly germane to the question of where and when terrestrial seawater originated. It is reasonably well established by astrophysical models of star evolution that either the Earth's environment was too hot to allow any atmosphere to accrete or that any original atmosphere was rapidly blown off from the Earth's surface by the intense radiation emitted by the nascent Sun. It seems reasonable to assume that primordial terrestrial water is best represented by water outgassed from oceanic basalts. This water was depleted in deuterium by about 80 per mil or more with respect to SMOW (standard mean ocean water). Seawater therefore has a separate origin that some scientists do not hesitate to attribute to the outer Solar System. Comet showers are suspected to have been frequent during some particular geological periods and thus were considered first. Comets are celestial bodies originating in the outermost Solar System (the Kuiper belt) whose highly elliptical orbits bring them close to the Earth. They are made of abundant ice surrounding a small rocky nucleus. The measurement of the D/H ratio of Halley's comet by the Giotto probe and of the Wild2 comet minerals retrieved by the Stardust mission revealed a D/H ratio twice that of seawater, therefore limiting the input of cometary water to less than 15% of the ocean. Alternatively, water may have been added to the Earth as frost from the abundant planetary bodies that, right after the Solar System formed, were cruising in the outskirts of Jupiter. These planetesimals were ejected very early from their orbit by the gravitational pull of this planet, which acted just like a giant slingshot. In either case, we would expect the terrestrial abundances of heavy rare gases, such as xenon, to be higher than they actually are. A mixed source, mantle outgassing, plus a contribution from distant planetesimals and comets, would probably account reasonably well for these observations.

Convection in the core powered by gravitational energy and radioactive decay sustains the Earth's magnetic field. Geophysicists working on the dynamo theory would like, however, to see core convection powered by more radioactive decay than is currently accepted by geochemists. Experiments of elemental partitioning between pure molten iron and silicate have so far been unable to incorporate substantial amounts of K into the metal. The class of iron meteorites that are believed to best represent the core of proto-planets are devoid of potassium. The two points of view could be reconciled if the high abundance of certain light elements such as Si, S, or O, which the velocity of seismic waves in the core requires, increased K solubility in molten iron alloy compared with that of pure iron.

12.5 The early Solar System

Contrary to an often-heard assertion, the Earth itself has never been dated as there is no unmodified, inherited sample from the earliest times of our planet. In an atmosphere rife with religious prejudice, the age of the Earth was one of the great debates of the nineteenth and twentieth centuries between geologists, who claimed much time was required for sedimentary strata to be deposited and organisms to evolve, and physicists. Using the theory of heat conduction, Lord Kelvin calculated an age of the Earth of 94 million years (see box). One of the more imaginative attempts at dating the Earth was made in 1899 by the British astronomer Joly who calculated the time required for an ocean initially made of fresh water to become as saline as it is today if its only salt input was from rivers. This age τ can be calculated as the ratio of the mass of chlorine of seawater to the flux of chlorine from rivers, which using the data from Appendices A and G gives:

$$\tau = \frac{(1.4 \times 10^{21}) \times (1.89 \times 10^{-2})}{1.035 \times (3.6 \times 10^{16}) \times (7.8 \times 10^{-6}/1)} = 91 \times 10^6\ y$$

where allowance is made for the density of seawater ($1.035\ kg\,l^{-1}$) with respect to river water and of rivers. What we are really seeing here is the residence time of chlorine in seawater. In addition to the unsupported assumption of an initially fresh-water ocean, the theory described in Chapter 5 states that over the course of geological time the ocean has reached steady state for salt concentration. It therefore lost any memory of its origins and no significant age can be deduced by such a computation.

As ages of more than three billion years started appearing at the hands of the earliest geochronologists, namely those who utilized lead isotopes and the U–He technique, it became clear that Kelvin's calculations overlooked two fundamental processes, internal heating by radioactive decay of U, Th, and K, and mantle convection. It was only in 1954, however, that Claire Patterson produced the first meteorite isochron with chondrites and iron meteorites on it (Fig. 12.12). The age of 4.55 Ga indicated by this isochron has been challenged many times, but to no avail. Troilite (FeS) from one of these meteorites, Canyon Diablo, contains no detectable uranium and therefore yields the initial isotopic composition of lead in the Solar System. The $c.$ 4.54 Ga age of the Earth has been repeatedly confirmed by all known chronometers, which in itself is startling corroboration of the validity of the principles of radiometric dating.

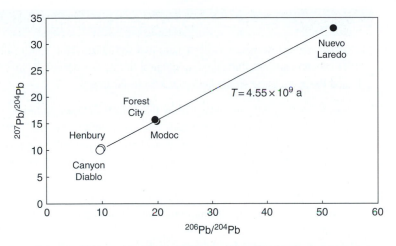

Figure 12.12 Patterson's (1956) geochron is in fact an isochron (^{206}Pb/^{204}Pb, ^{207}Pb/^{204}Pb) established from three chondritic meteorites (solid circles) and iron sulfide of two iron meteorites (open circles). The age established is therefore that of the Solar System.

The oldest crustal segment on Earth (Isua, Greenland) is 3.85 Ga old. Recently, detrital zircons dated between 3.9 and 4.4 Ga have been identified in an Australian sandstone at Jack Hills, which require the existence by that time of some granitic crust. However, most ages older than 4.3 Ga are mostly isolated patches in younger zircons and probably are to be taken with a grain of salt. Very ancient ages, of more than 4.4 Ga, have also been found for rocks and meteorites of the Moon and Mars.

Patterson's isochron has improperly been called the geochron, although almost all the measurements of compositions of terrestrial samples lie not on the isochron but to its right. The essential argument for dating the Earth is, in fact, an indirect one relying on what is known as extinct radioactivities, which provide a precise but relative chronology between planetary objects. It should be recalled therefore that the Earth has not been dated directly but that chondrites have, and with great precision (4.568 Ga with an uncertainty < 1 Ma). The question then becomes "how much younger than primitive meteorites could the Earth be?" and the answer has been provided by extinct radioactivities.

The first evidence of an extinct radioactivity was discovered as an excess of ^{129}Xe in the Richardton meteorite by John Reynolds in 1960. The parent nuclides of extinct radioactivities, such as ^{26}Al, ^{182}Hf, ^{129}I, have short half-lives (Table 4.1), and therefore large decay constants. The concept is simple: if the parent nuclide was still extant when a planetary body went through a major differentiation event such as accretion, core segregation, or magma ocean crystallization, associated parent/daughter fractionation should show up in the isotopic variability of the daughter isotope, even though today the radioactive parent is long gone. It is the physicist's way of dating by fossil remains! Depending on the half-life of the parent, this assumption gives us access to a range of time intervals, from a few millions of years for ^{26}Al to a few hundreds of millions of years for ^{146}Sm. Let us try to work out the isochron equations suitable for a system in which the parent nuclide is now extinct.

Are the decay constants truly constant?

One may wonder how we can be so sure that decay constants (probability of decay per unit time) remain constant throughout geological times. The answer is unambiguously that they do and there are three strong lines of evidence for this:

1. The decay constants are measured by different techniques, counting of decay events and accumulation of radiogenic isotopes, and give, within the errors of the measurements, values with a typical precision of a fraction of a per mil to a few percent depending on the nuclide. These values are very consistent for different laboratories throughout the world.
2. The ages of any particular object, typically a meteorite, a lunar rock, or an ancient volcanic sample are very consistent from one method to another, e.g. $^{238}U-^{206}Pb$, $^{40}K-^{40}Ar$, and $^{147}Sm-^{143}Nd$.
3. The interior of the Earth is kept hot by the decay of four radioactive nuclides, ^{238}U, ^{235}U, ^{232}Th, and ^{40}K. The energy liberated by the decay of each of these nuclides is dictated by the equivalence of mass and energy discovered by Einstein and used every day to produce energy in nuclear power plants. The decay energy is strictly a function of the mass difference between the original nuclides and the decay products. Today, the rate at which the Earth is heated by radioactivity is about 18 TW (1 TW $= 10^{12}$ W), while the Earth loses 42 TW through its surface (the rest is "fossil heat" inherited from accretion). If instead the same amount of radioactive energy was released over a few thousand years, the interior of the Earth would be hot enough for the planet to melt and even evaporate entirely. The situation would be far worse than the thermal regime created early on in the first millions of years of the Solar System history by the decay of ^{26}Al (see below).

Although the decay constants can still be refined for dating purposes, it is very unlikely that they could be very wrong or have substantially changed throughout geological time.

We assume that planetary material precipitates straight out of the nebular gas under conditions of isotopic equilibrium and that the isotopic properties of the solar nebula are those of the least differentiated material for which we have samples, the chondritic meteorites. For $t \gg \lambda^{-1}$, P becomes negligible and therefore the closed system condition reads:

$$D_{\text{today}} = P_t + D_t \tag{12.10}$$

at all t, where P and D are the parent and daughter nuclides, respectively. Let us write this equation for a sample (spl) and, as for the conventional isochron, let us divide it by the number of atoms of a stable isotope D' of D:

$$\left(\frac{D}{D'}\right)_{\text{today}}^{\text{spl}} = \left(\frac{D}{D'}\right)_{t}^{\text{spl}=\text{SN}} + \left(\frac{P}{P'}\right)_{t}^{\text{spl}=\text{SN}} \left(\frac{P'}{D'}\right)_{\text{today}}^{\text{spl}} \tag{12.11}$$

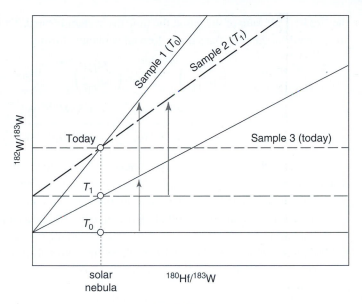

Figure 12.13 The evolution of isochrons in the system ^{182}Hf–^{182}W ($T_{1/2}$ = 8.9 Ma). Contrary to standard isochrons, the abscissa is the ratio of two stable isotopes. The solar nebula (open circle) is supposed to be an infinite reservoir, so its composition belongs in any isochron, even though it must be remembered that it changes through time. The isochrons represent three "samples" (planetary objects) formed at $T = T_0$, $T = T_1$, and today. After a few half-lives, there is no ^{182}Hf left and the ^{182}W/^{183}W does not evolve any more.

in which we emphasize that the sample was born in isotopic equilibrium with the solar nebula (SN) (P' is a stable isotope of the parent nuclide P). Using (12.10), this equation is equivalent to:

$$\left(\frac{D}{D'}\right)^{\mathrm{spl}}_{\mathrm{today}} = \left(\frac{D}{D'}\right)^{\mathrm{SN}}_{\mathrm{today}} + \left(\frac{P}{P'}\right)^{\mathrm{spl}=\mathrm{SN}}_{t}\left[\left(\frac{P'}{D'}\right)^{\mathrm{spl}}_{\mathrm{today}} - \left(\frac{P'}{D'}\right)^{\mathrm{SN}}_{\mathrm{today}}\right] \qquad (12.12)$$

which is the equation of an isochron in a plot of (D/D') vs. (P/P'). This equation shows that the point SN belongs at all times to the isochron (it is simply a non-fractionated sample of the solar nebula) and therefore that the isochrons revolve around this point as it steepens vertically through time (Fig. 12.13). The situation is reminiscent of the ^{238}U–^{230}Th isochron. For the ^{182}Hf–^{182}W chronometer ($T_{1/2}$ = 8.9 Ma), the last equation reads:

$$\left(\frac{^{182}\mathrm{W}}{^{183}\mathrm{W}}\right)^{\mathrm{spl}}_{\mathrm{today}} = \left(\frac{^{182}\mathrm{W}}{^{183}\mathrm{W}}\right)^{\mathrm{SN}}_{\mathrm{today}} + \left(\frac{^{182}\mathrm{Hf}}{^{180}\mathrm{Hf}}\right)^{\mathrm{spl}=\mathrm{SN}}_{t}\left[\left(\frac{^{180}\mathrm{Hf}}{^{183}\mathrm{W}}\right)^{\mathrm{spl}}_{\mathrm{today}} - \left(\frac{^{180}\mathrm{Hf}}{^{183}\mathrm{W}}\right)^{\mathrm{SN}}_{\mathrm{today}}\right]$$
$$(12.13)$$

This is the equation of the extinct radioactivity isochron (the conventional isochron cannot be used when $P = 0$). When ^{182}W/^{183}W is plotted against the ^{180}Hf/^{183}W (note that the two nuclides of the latter ratio are stable), samples formed at the same moment in isotopic equilibrium with the solar nebula form a straight line with a slope indicative of the ^{182}Hf/^{180}Hf ratio of the gas at the time the system formed. The ^{182}Hf/^{180}Hf ratio of the solar nebula at the time the sample formed is obtained from:

$$\left(\frac{^{182}\text{Hf}}{^{180}\text{Hf}}\right)_t^{\text{spl}\,=\,\text{SN}} = \left(\frac{^{182}\text{Hf}}{^{180}\text{Hf}}\right)_0^{\text{SN}} e^{-\lambda_{182\text{Hf}}t} = \tag{12.14}$$

$$\frac{\left(^{182}\text{W}/^{183}\text{W}\right)_{\text{today}}^{\text{spl}} - \left(^{182}\text{W}/^{183}\text{W}\right)_{\text{today}}^{\text{SN}}}{\left(^{180}\text{Hf}/^{183}\text{W}\right)_{\text{today}}^{\text{spl}} - \left(^{180}\text{Hf}/^{183}\text{W}\right)_{\text{today}}^{\text{SN}}} \tag{12.15}$$

If the ^{182}Hf/^{180}Hf ratio of the solar nebula at the reference time $t = 0$ is assumed, the result can be converted into an age. This age dates the time at which the dated sample shared the same ^{182}Hf/^{180}Hf ratio as the solar nebula. If no history of the ^{182}Hf/^{180}Hf ratio is assumed for the solar nebula, dividing this equation for one sample by the same equation for a second sample gives the age difference between the two samples.

Similar equations hold for a number of extinct short-lived chronometers, which each provide a time scale for different planetary phenomena:

1. The ^{26}Al–^{26}Mg system ($T_{1/2} = 0.75$ Ma) dates the Al/Mg fractionation events associated with the condensation of high-temperature minerals (oxides, aluminates, pyroxenes from the refractory inclusions) from the solar nebula.

2. The ^{182}Hf–^{182}W system ($T_{1/2} = 8.9$ Ma) dates the segregation of the siderophile element W from the lithophile element Hf, i.e. the formation of planetary cores, such as for the Earth–Moon system and for Mars, at ≈ 30 Ma after the formation of the Solar System (Fig. 12.14).

3. The ^{53}Mn–^{53}Cr system ($T_{1/2} = 3.7$ Ma) dates the mantle–core differentiation of planetary objects.

4. The ^{146}Sm–^{142}Nd system ($T_{1/2} = 103$ Ma): the 20 ppm difference in ^{142}Nd abundances between the terrestrial mantle–crust system and chondrites demonstrates the existence of a magma ocean very early in the history of the Earth. There is a small excess of ^{142}Nd in early Archean samples, such as those from the 3.85-Ga-old terranes from west Greenland, with respect to modern samples. Coupling these anomalies with the more conventional ^{147}Sm–^{143}Nd in much the same way as ^{238}U–^{206}Pb and ^{235}U–^{207}Pb are coupled dates the end of the magma ocean episode a few tens of Ma after the formation of the Solar System.

5. The ^{129}I–^{129}Xe system ($T_{1/2} = 15.7$ Ma) came historically first. Its low closure temperature makes it susceptible to metamorphic perturbations. The presence in MORB of excess ^{129}Xe with respect to the atmosphere indicates that the mantle was outgassed, i.e. the parent–daughter I/Xe reduced, while ^{129}I was still extant (< 75 Ma after the accretion of the Solar System).

6. The ^{60}Fe–^{60}Ni system ($T_{1/2} = 1.5$ Ma) is very particular since ^{60}Fe is the only undisputed nuclide inherited from supernovae. This controversial system therefore dates the injection of stellar material into the nascent Solar System.

To some extent, ^{235}U nearly qualifies as an extinct radioactivity!

Applications of these chronometers to the Earth and the early Solar System have recently produced startling results. Historically, differences in the isotopic abundances of ^{129}Xe

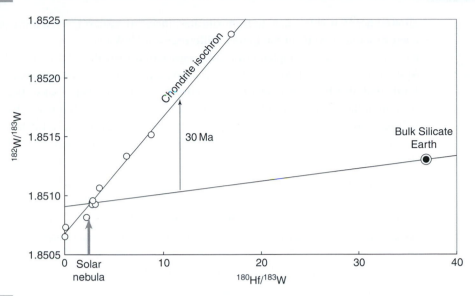

Figure 12.14 The age of the Earth's core (Yin *et al.*, 2002). Different mineral phases of chondrites make a ^{182}Hf–^{182}W isochron which defines the time at which the Solar System clock was set. At some later time, the Earth's core formed with most W going into the metal and most Hf being left in the silicate (mantle). The terrestrial two-point isochron goes through the point representing the mantle and the point of the chondrite isochron with the ^{180}Hf–^{182}W value of the solar nebula (chondrite whole-rock or solar photosphere). The difference in the slopes of the two isochrons assigns an age of ≈ 30 Ma to core formation.

between the atmosphere and gases from CO_2 wells and subsequently oceanic basalts proved that ^{129}I was still extant when the atmosphere formed. It was actually the first time the Earth could be dated to within a few tens of Ma of the meteorites!

As per today, a relatively clear chronology of the early Solar System and planetary differentiation is finally emerging. The Universe is about 14 Ga old, the age of the Big Bang. The U–Pb age of the oldest objects of the Solar System, the refractory inclusions known as CAI (for calcium–aluminum-rich inclusion) found in some chondrites (Allende) formed 4568 Ma ago. It is believed that this time represents the collapse of the solar nebula into a star, our Sun, and its planetary companions, a mixture of gases and grains inherited from molecular clouds and contaminated by exploding supernovae. The CAI inclusions are made of refractory minerals and represent the high-temperature residues of the violent irradiation by the T-Tauri Sun. Chondrules, these mysterious small blebs (≈ 1 mm) of molten silicates and which represent a large fraction of planetary material, formed 1–3 Ma after the CAI, probably in the wake of shock waves. It is probably at this stage that gas and dust were blown off by the UV and X-ray wind of the Sun; telescope observations show that the dark molecular clouds that show up across most young Sun-like stars disappear in 3–5 My. By then, large planetary objects must have been largely completed, first and foremost the giant planets (Jupiter, Saturn). The heat liberated by the decay of ^{26}Al and the release of gravitational energy by impacts ensured that many of these planets were largely molten (magma ocean stage). Melting further

272

The Earth in the Solar System

allowed heavy iron to separate from silicates, therefore releasing even more gravitational energy (about 2000 K for the Earth). Differences in ^{182}W abundances between chondrites and the Earth dates completion of core segregation to within the first 30 Ma of Solar System formation. Differences in ^{142}Nd abundances between meteorites and the silicate Earth, and also between modern rocks and the 3.85-Ga-old Isua rocks, date the crystallization of the magma ocean to the first tens of Ma of the planet's history. The period that still remains mysterious is the transition to plate tectonics: when did it happen? Which parameter (water?) made the Earth, and only the Earth, evolve towards plate tectonics? Why is the Earth endowed with a dynamo?

12.6 The Moon

The Moon's radius is 1738 km, its gravity a mere 17% of that of the Earth, and its density 3940 kg m^{-3} (compare with the Earth's 5515 kg m^{-3}). Thanks to the Apollo missions, we have several hundred kilograms of lunar samples from six localities on the visible side of the Moon and a wealth of observations made by the astronauts. Antarctic and desert meteorites have added more than 30 extra lunar samples, some of which are suspected to come from the far side. Although oxygen isotopes tell us that the Earth and the Moon formed from the same mass of debris left by the collision of a body the size of Mars with the proto-Earth, our satellite has a very different history and geology from that of our planet. A consequence of the small gravity is that the Moon does not retain most gases and therefore has no atmosphere. Gravimetric and altimetric data sent back by the satellite Clementine tell us that the Moon has a crust some 60–100 km thick, much thicker than the Earth's continental crust. A metal core was probably present during the early history of the Moon, but the absence of a magnetic field shows that it has now solidified. With respect to the Earth, the Moon is depleted in volatile elements and is extremely reduced, so much so that metal iron is a common mineral in magmatic rocks. It is also depleted in highly siderophile elements, notably Ni and the platinum group elements, which suggests that by the time of the giant impact, the supply of late veneer had already dried up. The Moon is a dead body whose last volcanic eruptions occurred 3.2 billion years ago. During the first billion years following accretion, the surface of the Moon was subjected to intense bombardment by planetary objects of different sizes, which gave our satellite its characteristic heavily cratered aspect. There is apparently no element of the original surface left untouched by cratering.

Lunar rocks are classified into three main groups: (1) intrusive rocks of the lunar highlands, especially anorthosite, with its high plagioclase content, and gabbro, which, in addition to plagioclase, contains olivine or pyroxene; (2) KREEP basalt, an acronym indicating that it is very rich in potassium, rare-earths, and phosphorus; and (3) lunar mare basalts. The intrusive rocks are 3.8–4.5 Ga old, KREEP basalts 3.6–3.8 Ga old, and mare basalts 3.2–3.9 Ga old. The feldspathic highlands dominate the far side. Very large impacts, such as the one that excavated the Mare Imbrium at around 3.8 billion years, form a nebular super-event known as the Late Heavy Bombardment (LHB). An idea popular at this time is

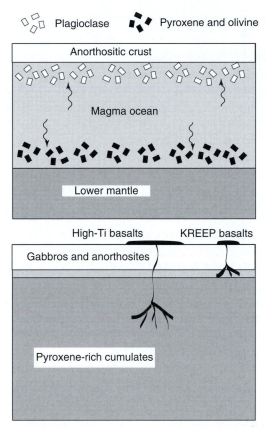

Figure 12.15 The magma ocean from which the lunar mantle and crust originated. Heat from the impact that separated the Earth and the Moon left the outer part of the Moon molten. Plagioclase feldspar floated to create the anorthositic crust (lunar highlands), while olivine and pyroxene settled out to form the mantle. The KREEP basalts derived from the last residual liquid accumulated beneath the crust. Basalts from lunar maria derive from the subsequent remelting of mantle cumulates.

that the LHB was an isolated event which resulted from the destabilization of small planetary bodies in the outer Solar System upon outward migration of the giant planets. The LHB produced a consolidated rubble known as breccia, which is particularly abundant in the samples brought back by the astronauts. Chemical maps produced by the satellites Clementine and Lunar Prospector show that the ejecta forming the wall of the Mare Imbrium are particularly rich in KREEP. The surface of the Moon is covered with regolith, a thick soil created by incessant meteoritic bombardment (gardening) of a body with no protective atmosphere.

The best documented and most popular model for the formation of lunar rocks is that of the magma ocean (Fig. 12.15). Observations on the first samples of breccia, brought back by Apollo 11, suggested that the relatively abundant anorthosite is an essential component of the surface, while lunar mare basalt chemistry indicated a form of basalt–anorthosite complementarity. It is thought that after the impact that formed the Moon some

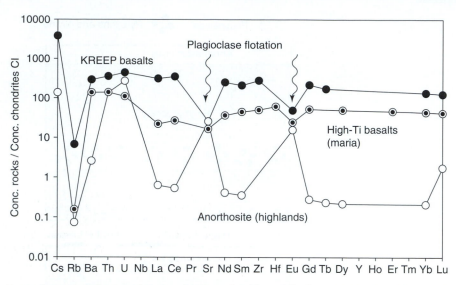

Figure 12.16 Trace elements in lunar rocks. Anorthosite and gabbro of the lunar highlands exhibit surplus Eu and Sr from the accumulation of plagioclase by buoyancy (cf. the partition coefficient of Fig. 2.5). The KREEP basalts exhibit negative anomalies complementary to those of anorthosite. Lunar maria basalt, being older, displays similar negative anomalies inherited from the source in the mantle and depleted in plagioclase.

4.56 billion years ago, its mantle was largely molten. As this enormous mass of magma cooled, which may have taken several hundred million years, all of its upper part was saturated in plagioclase, olivine, and pyroxene. Plagioclase, being lighter than the liquid, floated to form the highland crust, while the heavier olivine and pyroxene sedimented out, filling the magma ocean from the bottom upward. The last residual liquids, particularly rich in incompatible elements (including K, P, the rare-earths, and Ti), subsisted for several hundred million years beneath the anorthositic crust. What arguments are there to support this model? The presence of a strong positive europium anomaly in anorthosites and other intrusive rocks, while KREEP and other basalts display very marked negative anomalies (Fig. 12.16). Incorporation of Eu in plagioclase is very efficient in comparison with other rare-earths (Fig. 2.5). Such a large anomaly cannot be generated by plagioclase fractionation from the basalts themselves, and must already be present in the source of these magmas (see Fig. 2.10) and therefore be inherited from cumulates produced in a global melting event (the magma ocean). Efficient segregation of plagioclase by buoyancy from a magmatic liquid was at the origin of the intrusive rocks of the lunar highlands. The KREEP basalts are formed out of a mantle impregnated with residual liquids: the very high content of their mantle source in heat-producing elements (U, Th, K) may actually have kept this particular layer partially molten for billions of years. This model is supported by the contrasting unradiogenic character of the Nd of the anorthosites at the time they formed and the radiogenic Nd of the basalts (Fig. 12.17). The inference from this is of complementary Sm/Nd fractionation processes of the early lunar mantle, which is further evidence of magmatic fractionation.

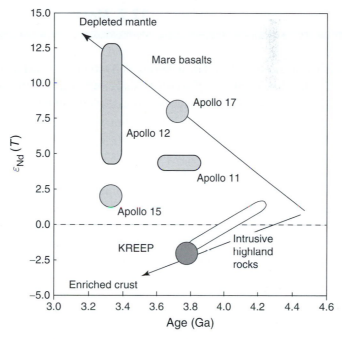

Figure 12.17 Evolution of the Sm–Nd isotope system on the Moon. Anorthosite of the lunar maria and KREEP basalts indicate geochemical "enrichment" (low Sm/Nd ratio and therefore unradiogenic Nd). By contrast, Nd of the source of lunar maria basalts is radiogenic, indicating that the mantle source is geochemically depleted, which is consistent with the hypothesis of pyroxene accumulation. Compare this with the evolution of the Earth's mantle and crust (Figs. 11.16 and 11.20).

How did the basalts of the lunar maria form in this process? Their relatively young ages exclude their being erupted out of the magma ocean system and their modest enrichment in incompatible elements indicates that they do not represent residual liquids. Nonetheless, these basalts exhibit a negative Eu anomaly indicating that their mantle source is a complement to the lunar highland crust. The most common interpretation is that they formed by remelting, at around 3.8–3.2 billion years, of pyroxene-rich cumulates present in great quantities in the mantle and produced by crystallization of the magma ocean and accumulated in the lowest points of the topography, the impact craters. The highly variable titanium contents of these basalts show that another mineral, ilmenite ($FeTiO_3$), was also present in very variable quantities at the time of melting. The sources of these lavas in the lunar mantle therefore represent cumulates of a magma ocean at relatively advanced stages of crystallization.

The essential features of lunar mantle differentiation are the saturation in plagioclase of the magma ocean and its lack of water. On Earth, basaltic magmas can precipitate plagioclase down to about 30 km. On the Moon, where gravity is six times less, the equivalent pressure is only reached at about 180 km. It can be seen therefore that potential plagioclase production by lunar magmas is much more substantial than on Earth and that the potential production of anorthositic crust is much greater on the Moon.

12.7 Mars

Mars has a radius of 3386 km, a density of 3940 kg m^{-3} and its gravity is 38% of that of the Earth. Mars is a heavy planet for its size, which is usually taken as a sign of a high iron content. Its modern magnetic field is essentially nil but the very ancient highlands are magnetized along broad stripes, which attests to the presence of a dynamo very early in Mars' history. The planet retains, admittedly somewhat imperfectly, an atmosphere, with a pressure equivalent to 0.7% of the Earth's atmospheric pressure and which is capable of generating violent winds. The composition of the atmosphere, reported by the Viking landers, is dominated by CO_2 (95.3%) with traces of N_2 (2.6%) and Ar (1.6%). The orbital satellite MOLA told us that the thickness of the Martian crust varies from 40 to 80 km. There is a marked contrast between the two hemispheres: the southern hemisphere is composed of elevated volcanic plateaus, which are heavily cratered and therefore very old. The surface of the southern hemisphere is eroded by what are unquestionably fluviatile systems. The northern hemisphere is a vast flat depression covered with eolian sands. Large old craters show in the basement through the sedimentary blanket of the northern plains. Orbiter spectroscopic observations of the sands show the ubiquitous presence of olivine, pyroxene, and ferric oxide, with occasional clay minerals. The rovers Spirit and Opportunity provided evidence of abundant iron sulfate in these sediments, whereas carbonates are missing. Whatever water masses have been present at the surface of the planet, they must have been too acid (pH < 6) for carbonate to precipitate. Extrapolating the lunar cratering chronology indicates that most of the southern hemisphere and the basement of the northern hemisphere are covered with very large craters older than ≈3.9 Ga, i.e. nearly contemporaneous with the Late Heavy Bombardment of the Moon. Cratering chronology for the sediments of the northern hemisphere supports an age in excess of 3.2 Ga. Some strato-volcanoes, such as those of the Tharsis plateau, have a rather fresh morphology: although their deeper layers are very old (>3.4 Ga), lava flows younger than 200 Ma are undoubtedly present. A polar ice cap composed mainly of carbonic ice forms and melts annually at high latitudes.

It is believed that Mars never went through a plate tectonic regime, probably because the lithosphere quickly became too rigid (too dry, too thick) ever to bend. The mantle is certainly convective, but beneath the thick lithosphere: this is the so-called stagnant lid regime, rather inefficient at extracting heat, and which probably also smothered the core dynamo very early on in the planet's history.

Martian stratigraphy is divided into three stratigraphic systems with boundaries tentatively set at 3.7 and 3.0 Ga by cratering chronology. The ancient cratered and rugged terranes of the southern hemisphere belong to the oldest Noachian, and terminated with the heavy bombardment period. This period was a time of generalized volcanic activity. The overlying Hesperian corresponds to the activity of the major strato-volcanoes, of the outflow channels (Valles Marineris), and to the infilling of the northern plains. The Amazonian is the youngest period, which saw very little volcanic and sedimentary activity.

Although there have not yet been any missions to recover samples from Mars, the rovers analyzed a variety of sedimentary and volcanic samples in place. It was concluded that

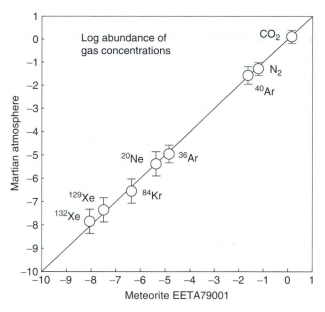

Figure 12.18 Gas in SNC (Martian) meteorites. The similarity of their gas content to that of the Martian atmosphere measured by the Viking probes (standardized to nitrogen) is a very forceful argument in favor of SNCs originating from Mars.

sulfate deposits are very abundant, clay minerals unusual, carbonate absent. Basalts and more differentiated lavas were also identified. In addition, the origin of more than 40 meteorites has been ascribed to Mars by powerful arguments. These meteorites are known as the SNC meteorites (from the three group-leading meteorites Shergotty, Nakhla, and Chassigny). The arguments that they are samples from Mars are: (1) these meteorites form a homogeneous group in the $(\delta^{17}O, \delta^{18}O)$ plot (Fig. 12.7); (2) the relative abundance of gases trapped in the Martian meteorite EETA79001 found in the Antarctic ice is identical to that of the atmosphere analyzed by the Viking probes (Fig. 12.18); and (3) the relatively young ages (160–600 million years) found for most SNCs require the existence of a planet with recent volcanic activity, while the ages of exposure to cosmic radiation determined from these meteorites indicate that they have not spent more than a few million years in interplanetary space and that they probably came from not very far away. Mars and its super-volcanoes, with their very well-preserved morphology, such as Olympus Mons and its caldera with vertical cliffs several kilometers high, is the only planet in the inner Solar System that could be a possible candidate.

The SNCs are cumulate rocks (gabbro, peridotite) affected by the impact that allowed them to escape Martian gravity to the extent that the plagioclase has lost its crystalline structure and become amorphous. Little is certain from this blind sampling except that the SNCs must represent the Martian lithosphere. The isotopic compositions of Sr, Nd, and Hf reveal a very clear contrast between rocks from a source enriched in incompatible elements and others from a depleted source: we can contrast samples from the Martian "mantle" and samples from a Martian "crust." The existence of isotopic anomalies left by the extinct radioactive nuclides ^{146}Sm and ^{182}Hf and their variation from one meteorite to another further indicate that the Martian lithosphere is nearly as old as the planet itself. Plate tectonics is therefore quite unlikely to have modeled the planet's surface. Recent Pb-Pb work suggests that shergottites are actually older than 4.1 Ga and that the younger ages inferred from mineral isochrons reflect later impact events. It is therefore a very long time since Mars experienced any intense internal activity, with the possible exception of the few volcanoes visible at its surface, and that even its metal core is frozen as it does not generate a magnetic field. Although some young lava flows are undoubtedly present, the dynamics of Mars' interior, just as that of the Moon's interior, was not strong enough to maintain a surface volcanic activity for more than about a billion years after planet formation.

However, it is generally accepted that Mars once had a denser atmosphere than today and possibly oceans. The Viking probes measured the composition of the Martian atmosphere: high D/H ratios with $\delta D = 4000$ and $^{15}N/^{14}N$ ratios 60 percent larger than that of the Earth indicate that hydrogen and nitrogen were quickly lost by the planet. The lack of a Martian magnetic field did not let a stable magnetosphere form that would protect it from the solar wind. Water photolysis by solar UV radiation and hydrogen loss may have been accompanied by oxygen production which was either lost to space or used by oxidation of ferrous iron. The presence of sulfates and of low-temperature carbonates in SNC meteorites are evidence of water circulation, corroborating satellite observations of hydrographic networks, probably of very ancient age. The current status of groundwater is still under discussion. The existence of biological activity – a hypothesis given much media coverage by NASA thanks to photographs of minute elongated calcitic objects in the Martian meteorite ALH84001 discovered in Antarctica – encounters a certain degree of disbelief among the scientific community. The wider implications of the claim for society are such that the possibility must, of course, be further investigated. If traces of biological activity are to be found, it will probably be deep in the sedimentary layers of the northern hemisphere, far from the lethal solar radiation.

12.8 Venus

Venus has a radius (6052 km), a surface gravity (8.87 m s^{-2}), and a density (5243 kg m^{-3}) very similar to the terrestrial values. There is not much detailed geochemistry we know for sure about Venus. The atmospheric pressure at the surface is 90 times the terrestrial pressure. The 460 °C surface temperature makes the planet an inferno of CO_2 (96.4%) and N_2 (3.4%), with clouds of sulfuric acid! Venus has essentially no magnetic field. The Venera Soviet probes survived hardly long enough to analyze the atmosphere and take a few pictures of a soil covered with volcanic fragments. The Magellan radar experiments showed a pronounced topography with highlands, volcanoes, and circular volcanic structures of unknown origin. Although the range of elevations (13 km) is comparable to that on Earth, the planet does not show the distinctive dichotomy of continental and oceanic reliefs that, on Earth, results from plate tectonics. The surface lacks the large and old craters that would attest to the presence of very ancient terrane. Adapting the cratering rates of the Moon to

Venus actually suggests that no surface is older than about 800 Ma. The nature of this young resurfacing is not well understood.

The composition of Venus' atmosphere is ascribed to a runaway greenhouse effect: the enormous amount of CO_2 led to a rapid increase of the temperature, to the complete vaporization of water, and to its photolysis by ultraviolet radiation in the upper atmosphere. The recent mission, Venus Express, did observe that oxygen and hydrogen are lost by the upper atmosphere in 1:2 atomic proportions.

12.9 Planetary atmospheres

A nagging issue of planetary sciences is why some planets, such as the Earth and Venus, are endowed with an atmosphere, while others such as Mars, Mercury, and the Moon have none or almost none. Some planets are too small for their gravity field to hold back gases with a light molecular weight, but this is not argument enough to explain why Mars has such a thin atmosphere (700 Pa), whereas Venus seems only to have preserved an enormous amount of carbon dioxide. The most important factor is the atmospheric temperature: if some atmospheric molecules travel faster than the escape ("Jeans") velocity (see below), they will be lost and the atmosphere will disappear. The answer is a little more complicated and the discussion can use a little bit of physics.

A first idea is to compare the kinetic energy $\frac{1}{2}M\overline{u^2}$ carried by a gas of molecular weight M at temperature T with mean squared velocity $\overline{u^2}$ with the potential energy of the gas in the gravity field of the planet (we assume that the atmosphere is thin and therefore that its distance to the center is equal to the radius R_E of the planet). We start from the law of gravitational attraction to express that the potential is minus the derivative of the attraction and get:

$$\frac{1}{2}M\overline{u}^2 = G\frac{M_E M}{R_E} \tag{12.16}$$

where M_E stands for the mass of the planet and G for the universal gravity constant. Newton's law of mechanics relates the gravitational force and acceleration of gravity through

$$G\frac{M_E M}{R_E^2} = Mg \tag{12.17}$$

and the escape velocity u_e is:

$$u_e = \sqrt{2g R_E} \tag{12.18}$$

This expression gives the value of 11 km s^{-1} for the Earth and 2 km s^{-1} for the Moon. We now relate the velocity to the temperature through

$$\frac{1}{2}Mu_e^2 = \frac{3}{2}RT \tag{12.19}$$

which gives the mass-dependent temperature T_e of escape:

$$T_e = \frac{2}{3} \frac{R_E M g}{R} \tag{12.20}$$

We can now conclude that all the gas of molecular weight $< M$ on a planet of radius R_E are lost when the atmospheric temperature exceeds T_e. Overall such a simple argument explains why there is no atmospheric gas on the Moon, very little on Mars, whereas the larger planets Venus and the Earth enjoy a thick atmosphere, not to mention Jupiter and Saturn. Although this calculation gives us a good first-order indication, the lack of helium in the terrestrial atmosphere and the lack of some gases on Venus and Mars, most notably N_2, indicate that the situation is more complicated. For example, the Earth has an acceleration of gravity of 9.81 m s^{-2} and a radius of 6.371×10^6 m. Helium-3 ($M = 3 \times 10^{-3}$ kg) should be lost at temperatures in excess of $(2/3) (6.371 \times 10^6) (3 \times 10^{-3})$ 9.81/8.31 = 15 000 K. A first obvious reason is that our calculation uses the mean velocity $\bar{u}$ of the gases: some molecules must dash faster than $\bar{u}$, while some others must be much slower and remain in the atmosphere. We therefore need a more sophisticated description of atmospheric velocities.

Let us stay for simplicity with our simple monatomic gas (e.g. Ar). If the atmosphere is thick enough, molecules, will collide with each other and exchange their energy, a process referred to as thermalization. An atom of molecular mass M traveling in the direction x with the velocity v_x transports a kinetic energy $\frac{1}{2} M v_x^2$. The proportion $f(v_x)\,dv_x$ of atoms with velocity between v_x and $v_x + dv_x$ will be given by the Boltzmann distribution:

$$f(v_x)\,dv_x = C \exp\left(-\frac{M v_x^2}{2RT}\right) \tag{12.21}$$

where C is a constant that will be determined later. We can write the same equation for the other two dimensions and multiply the three equations to obtain the distribution of velocity in any direction, but are mostly interested in the distribution of speeds $u = \sqrt{v_x^2 + v_y^2 + v_z^2}$ (the modulus of the velocity), which, with a bit of mathematics, is given by:

$$f(u)\,du = 4\pi \left(\frac{M}{2\pi RT}\right)^{3/2} u^2 \exp\left(-\frac{M u^2}{2RT}\right) \tag{12.22}$$

This is the famous Maxwell–Boltzmann velocity distribution (Fig. 12.19) giving the probability that an atom has a speed between u and $u + du$ and which shows that, to a good approximation, RT is the variance of the velocity distribution. The root-mean-square speed, which is the parameter we can use to describe the kinetic energy, is $\bar{u} = \sqrt{3RT/M}$ We should now be happy with the solution and contend that the fast tail of the Maxwell–Boltzmann distribution above the escape velocity should leave the planet. It shows that temperature is rather the equivalent of the variance of the speed than of its mean (molecules traveling together are cold regardless of their speed). Collision after collision, the slower molecules would progressively thermalize again and produce more of the fast molecules. Sooner or later, most of the light gases should therefore be gone. One more time, it does not quite work that way! The reason is that it is not enough that a molecule travels faster than the escape velocity, the way out must be clear of collisions. The important concept is therefore that of the mean free path between collisions: if a molecule traveling faster

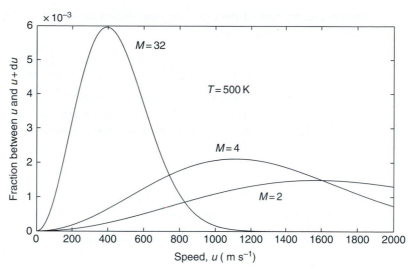

Figure 12.19 The Maxwell–Boltzmann distribution of speeds for hydrogen ($M = 2$), helium ($M = 4$), and oxygen ($M = 32$). On the Moon ($u = 2$ km s^{-1}), H_2 and He will quickly be lost while CO_2 will take longer to leave the gravity field. In contrast, only H_2 will rapidly be lost from the Earth's atmosphere.

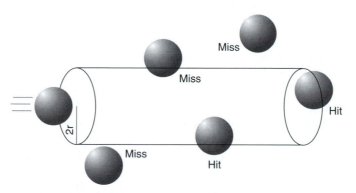

Figure 12.20 Collisions within gas. A molecule will sweep a cylinder of radius twice its radius. The mean free path is the mean distance between successive collisions.

than the escape velocity does not bump into any other molecule before being set free from gravity, it is lost. Otherwise, it is very likely to lose energy in collisions with the more abundant slow molecules and must wait until it gets a chance of another good push to be taken further.

What about the mean free path? Let us assume that our molecule is a sphere with radius r and a footprint (cross-section) of $4\pi r^2$ (Fig. 12.20). It will sweep free if there is no other molecule in a cylinder of twice its radius. The number of collisions dN is equal to the volume $4\pi r^2 \bar{u} dt$ of the cylinder swept during the time $dt \times$ the number density ρ of molecules in the unit volume:

$$dN = \left(\frac{1}{2}\right) 4\pi r^2 \rho \bar{u} dt = 2\pi r^2 \rho \sqrt{3\frac{RT}{M}} \tag{12.23}$$

where the factor 1/2 accounts for the fact that a collision involves two molecules. The mean distance between collisions (mean free path) ℓ is equal to the distance $v\, dt$ traveled by a molecule during the time dt divided by the number of collisions dN:

$$\ell = \frac{\bar{u} dt}{2\pi r^2 \rho \bar{u} dt} = \frac{1}{2\pi r^2 \rho} \tag{12.24}$$

At 1 bar (10^5 Pa) and 25 °C, the number density of nitrogen molecules ($r = 0.19 \times 10^{-9}$ m, $M = 28$) in a pure N_2 atmosphere is:

$$\rho = \frac{1}{V} = \frac{\mathcal{N}P}{RT} = \frac{(6.022 \times 10^{23})\,(10^5)}{8.31 \times 293} = 2.4 \times 10^{25}\,\mathrm{m}^{-3} \tag{12.25}$$

The mean speed is $\sqrt{3RT/M} = \sqrt{3 \times 8.31 \times 298/0.028} = 515$ m s^{-1}. The mean free path ℓ is $[(2)\,(3.14)\,(0.19 \times 10^{-9})^2\,(2.4 \times 10^{25})]^{-1} = 1.8 \times 10^{-7}$ m.

Such a very small distance between collisions shows that no gas molecule under the pressure conditions of the Earth's surface should ever leave the atmosphere. This is the standard tropospheric regime. It is only under the conditions of the upper atmosphere, where the gas molecules are rarefied, that gas loss may start becoming significant. Let us try to evaluate how high in the atmosphere we should go before we see this happen. We show in Appendix H that the variation of temperature in an adiabatic atmosphere in which conduction and radiation can be neglected is given by:

$$T = T_0 \left(1 - \frac{Mg}{C_P T_0} z\right) \tag{12.26}$$

where z stands for the altitude above the ground, M for the gas mean molar weight, g for the acceleration of gravity, C_P for the specific heat at constant pressure, and subscript 0 refers to conditions at the surface of the planet. The density is given by:

$$\rho = \rho_0 \left(1 - \frac{Mg}{C_P T_0} z\right)^{\frac{C_V}{R}} \tag{12.27}$$

where C_V stands for the specific heat at constant volume. For a diatomic gas such as N_2 or O_2, $C_V = \frac{5}{2}R$ and $C_P = \frac{7}{2}R$, so that the density of the atmosphere decreases with the elevation even faster than its temperature. The length scale ($C_P/Mg = (3.5 \times 8.31)/(0.028 \times 9.81) \approx 100$ m) is characteristic of a 1 degree drop in temperature.

We now understand that when the pressure drops enough to reduce the collisions, temperature has also decreased to the point where most molecules do not travel fast enough anymore to leave the gravity field! Heating by solar ultraviolet radiation, however, drastically affects the situation provided the appropriate gases are present at the right elevation. We have seen that diatomic molecules such as N_2, O_2, and H_2 do not absorb radiation. But if enough H_2O, CO_2, CH_4, and O_3 is present in the atmosphere and in particular at high altitude, the planet starts losing the high-velocity fraction of different gases. Absorption of solar ultra-violet radiation is strong enough to ionize the atmospheric gases partially and create a plasma (a mixture of positively charged ions and electrons): this is the ionosphere, which acts as a shield against most of the deadly solar wind and cosmic rays. The composition of the upper atmosphere and the temperature it creates therefore condition the preservation or the loss of atmospheric gases from planetary atmospheres and the well-being of life.

Loss may be amplified by hydrodynamic escape, an entrainment mechanism that cattle would describe as molecular stampede. Let us assume that the atmosphere of a small planet is essentially composed of hydrogen molecules and contains traces of N_2. On average, the hydrogen molecules travel faster upwards, down the pressure gradient, and keep pounding over and over against the slower nitrogen molecules. If the atmosphere is thin enough for the collision between nitrogen molecules to be infrequent, these remain in thermal disequilibrium and some of them will soon find themselves moving faster than their escape velocity and leave the gravity field. Hydrodynamic escape has been invoked in particular to explain why Venus' atmosphere is devoid of nitrogen. This effect can be felt for the isotopes of elements as heavy as xenon.

Loss of gases, such as oxygen and hydrogen, from the Earth, Mars, and Venus' ionosphere, has clearly shaped the evolution of their atmospheres. The internal magnetic field of our planet creates a protective shell, known as the magnetosphere, that protects the atmosphere from erosion by the solar wind; yet leakage through the polar ("auroral" zone) is still substantial. Today, the Earth loses tens to hundreds of moles of hydrogen and oxygen per second and these numbers fluctuate as a function of the strength of the terrestrial magnetic field and of the solar activity. The atmospheres of Mars and Venus, being devoid of an internal magnetic field, are less well protected than Earth. Since light isotopes are preferentially lost with respect to the heavy ones, this process is most clearly identified in the isotope compositions of light gaseous elements. The D/H and ^{15}N/^{14}N ratios of terrestrial planets is higher than those of the Sun or the giant planets and characterize the residual character of their atmospheres. They let us dream of the days Mars had an atmosphere not so different from that of our planet. If gravity had helped, another planet might also have received the gift of life.

Exercises

1. Figure 12.21 shows the portion of the chart of the nuclides in the Ce–Gd range.
   a. Draw the path of the s process in that range.
   b. Assign each stable nuclide to the most probable process (s, r, p or a mixture of these).
   c. Use pure s Sm nuclides and the data of Table 12.1 to calculate the relative contributions of the s and r processes to ^{147}Sm and ^{149}Sm (neglect the p contributions).
2. Show that for the ^{182}Hf–^{182}W chronometer ($T_{1/2} = 8.9$ My) the "isochron" diagram $x = (^{180}\text{Hf}/^{183}\text{W})_\infty$, $y = (^{182}\text{W}/^{183}\text{W})_\infty$ (where $_\infty$ stands for now) has a slope $s = (^{182}\text{Hf}/^{180}\text{Hf})_t$ and an intercept $i = (^{182}\text{W}/^{183}\text{W})_t$, where t is the time at which the last

Table 12.1 Element abundances (normalized to Si $= 10^6$), cross-sections σ (in mb = millibarn or 10^{-28} m^2), and isotopic abundances (in percent) of the nuclides in the Ce–Gd range.

	Abund.	σ (mb)	Isotopic abund.		Abund.	σ (mb)	Isotopic abund.
Ce	1.14			Sm	0.258		
^{140}Ce		11	88.48	^{144}Sm		92	3.1
^{141}Ce		76		^{147}Sm		973	17.5
^{142}Ce		28	11.08	^{148}Sm		241	11.3
Pr	0.167			^{149}Sm		1820	13.8
^{141}Pr		111.4	100	^{150}Sm		422	7.4
^{142}Pr		415		^{151}Sm		2710	
^{143}Pr		350		^{152}Sm		473	26.7
Nd	0.828			^{153}Sm		1095	
^{142}Nd		35	27.13	^{154}Sm		206	22.7
^{143}Nd		245	12.18	Eu	0.0973		
^{144}Nd		81.3	23.8	^{151}Eu		3775	47.8
^{145}Nd		425	8.3	^{152}Eu		7600	
^{146}Nd		91.2	17.19	^{153}Eu		2780	52.2
^{147}Nd		544		^{154}Eu		4420	
^{148}Nd		147	5.76	^{155}Eu		1320	
^{150}Nd		159	5.64	Gd	0.33		
Pm				^{152}Gd		1049	0.2
^{147}Pm		1290		^{153}Gd		4550	
^{148}Pm		2970		^{154}Gd		1028	21.8
^{149}Pm		2510					

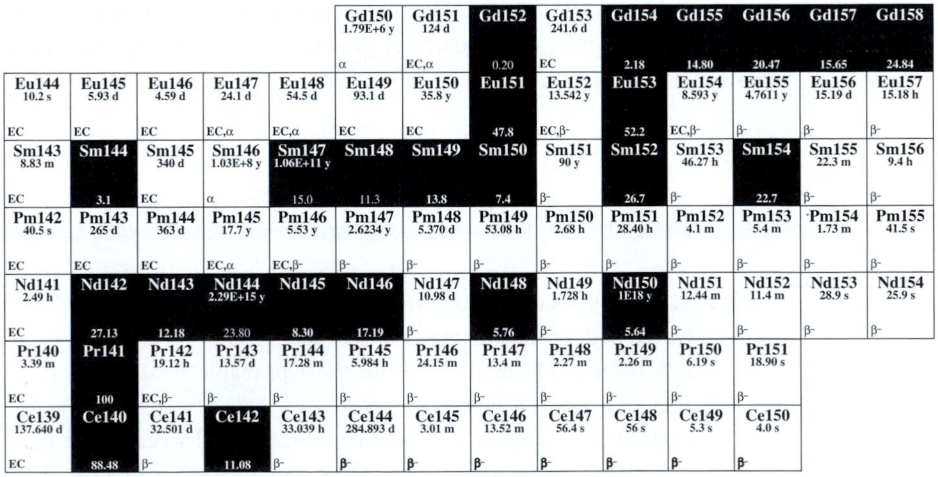

The chart of the nuclides in the Ce–Gd range (Figure 12.21). Each cell lists nuclide, half-life, and decay mode or (for stable/long-lived) isotopic abundance.

Element	139	140	141	142	143	144	145	146	147	148	149	150	151	152	153	154	155	156	157	158
Gd												Gd150 1.79E+6 y; α	Gd151 124 d; EC,α	Gd152; 0.20	Gd153 241.6 d; EC	Gd154; 2.18	Gd155; 14.80	Gd156; 20.47	Gd157; 15.65	Gd158; 24.84
Eu						Eu144 10.2 s; EC	Eu145 5.93 d; EC	Eu146 4.59 d; EC	Eu147 24.1 d; EC,α	Eu148 54.5 d; EC,α	Eu149 93.1 d; EC	Eu150 35.8 y; EC	Eu151; 47.8	Eu152 13.542 y; EC,β-	Eu153; 52.2	Eu154 8.593 y; EC,β-	Eu155 4.7611 y; β-	Eu156 15.19 d; β-	Eu157 15.18 h; β-	
Sm					Sm143 8.83 m; EC	Sm144; 3.1	Sm145 340 d; EC	Sm146 1.03E+8 y; α	Sm147 1.06E+11 y; 15.0	Sm148; 11.3	Sm149; 13.8	Sm150; 7.4	Sm151 90 y; β-	Sm152; 26.7	Sm153 46.27 h; β-	Sm154; 22.7	Sm155 22.3 m; β-	Sm156 9.4 h; β-		
Pm				Pm142 40.5 s; EC	Pm143 265 d; EC	Pm144 363 d; EC	Pm145 17.7 y; EC,α	Pm146 5.53 y; EC,β-	Pm147 2.6234 y; β-	Pm148 5.370 d; β-	Pm149 53.08 h; β-	Pm150 2.68 h; β-	Pm151 28.40 h; β-	Pm152 4.1 m; β-	Pm153 5.4 m; β-	Pm154 1.73 m; β-	Pm155 41.5 s; β-			
Nd			Nd141 2.49 h; EC	Nd142; 27.13	Nd143; 12.18	Nd144 2.29E+15 y; 23.80	Nd145; 8.30	Nd146; 17.19	Nd147 10.98 d; β-	Nd148; 5.76	Nd149 1.728 h; β-	Nd150 1E18 y; 5.64	Nd151 12.44 m; β-	Nd152 11.4 m; β-	Nd153 28.9 s; β-	Nd154 25.9 s; β-				
Pr		Pr140 3.39 m; EC	Pr141; 100	Pr142 19.12 h; EC,β-	Pr143 13.57 d; β-	Pr144 17.28 m; β-	Pr145 5.984 h; β-	Pr146 24.15 m; β-	Pr147 13.4 m; β-	Pr148 2.27 m; β-	Pr149 2.26 m; β-	Pr150 6.19 s; β-	Pr151 18.90 s; β-							
Ce	Ce139 137.640 d; EC	Ce140; 88.48	Ce141 32.501 d; β-	Ce142; 11.08	Ce143 33.039 h; β-	Ce144 284.893 d; β-	Ce145 3.01 m; β-	Ce146 13.52 m; β-	Ce147 56.4 s; β-	Ce148 56 s; β-	Ce149 5.3 s; β-	Ce150 4.0 s; β-								

Figure 12.21 The chart of the nuclides in the Ce–Gd range. The solid squares represent stable nuclides or nuclides with a long half-life. EC stands for electron capture.

		^{180}Hf/^{183}W	^{182}W/^{183}W
Table 12.2 ^{182}Hf–^{182}W results for two whole-rock chondrites and their silicate and metal phases (Yin *et al.*, 2002)			
CHUR		2.84	
Bulk Silicate Earth		36.87	1.8513
Moon		53.88	1.8515
Dhurmsala,	Silicate	16.96	1.8524
Dhurmsala,	Metal	0.10	1.8507
Dhurmsala,	Whole-rock	3.52	1.8511
Dalgety Downs	Silicate	6.30	1.8513
Dalgety Downs	Metal	0.02	1.8507
Dalgety Downs	Whole-rock	2.84	1.8510

equilibration of W isotopes took place while ^{182}Hf was still extant; ^{183}W and ^{180}Hf are stable non-radiogenic nuclides used for normalization.

3. The data listed in Table 12.2 give the ^{180}Hf/^{183}W and ^{182}W/^{183}W ratios measured in two whole-rock chondrites (Dhurmsala and Dalgety Downs) and their silicate and metal phases. Examination of the ^{180}Hf/^{183}W ratios shows that Hf is lithophile and W siderophile.

   a. From the slope of the isochron formed by the six points, calculate the $(^{182}Hf/^{180}Hf)_t$ ratio of the chondritic material at the time its constitutive phases last equilibrated with each other.

   b. Calculate the $(^{182}W/^{183}W)_t$ ratio of the mean chondritic reservoir (CHUR) at that time, given its mean ^{180}Hf/^{183}W listed in the table.

   c. Using these values as those of the parent reservoir, calculate the $(^{182}Hf/^{180}Hf)_t$ ratio of the Earth at the time the core and silicate material last equilibrated.

   d. Calculate the apparent age of core segregation from the Bulk Silicate Earth.

   e. Do the same calculation for the Moon.

4. Define similar isochron diagrams for other extinct radioactivities: ^{26}Al–^{26}Mg (normalize to ^{24}Mg and ^{27}Al), ^{53}Mn–^{53}Cr (normalize to ^{52}Cr and ^{55}Mn), ^{60}Fe–^{60}Ni (normalize to ^{58}Ni and ^{56}Fe), ^{146}Sm-^{142}Nd (normalize to ^{144}Nd and ^{144}Sm). Refer to Table 4.1 for the decay constants. From your knowledge of the geochemical properties of the elements of the parent and daughter isotopes and from the half-life of the radioactive nuclide, discuss some potential geochronological applications of each system.

5. Dating the Universe I.

   a. Explain why the radioactive nuclides ^{238}U and ^{232}Th must have been produced by the r process.

   b. The present-day ^{232}Th/^{238}U of the Solar System is 3.7. Calculate the value of this ratio 4.56 Gy ago. Refer to Table 3.1 for the decay constants.

   c. Let us assume that these two nuclides have been created at a constant rate since the formation of the Universe and call p_i their production rates ($i = {}^{238}$U or ^{232}Th).

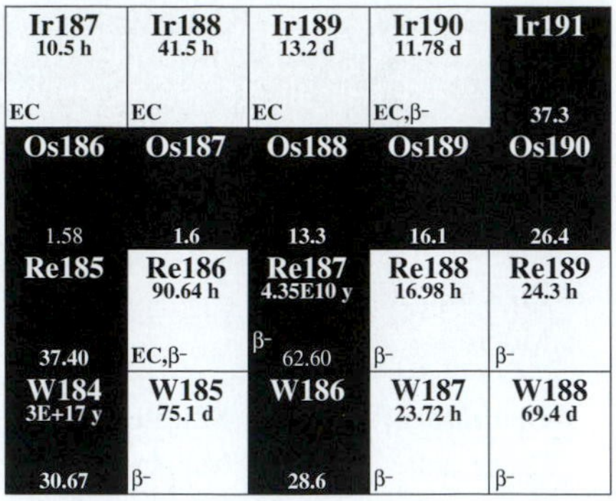

Ir187 10.5 h	Ir188 41.5 h	Ir189 13.2 d	Ir190 11.78 d	Ir191
EC	EC	EC	EC,β-	37.3
Os186	**Os187**	**Os188**	**Os189**	**Os190**
1.58	1.6	13.3	16.1	26.4
Re185	**Re186** 90.64 h	**Re187** 4.35E10 y	**Re188** 16.98 h	**Re189** 24.3 h
37.40	EC,β-	β- 62.60	β-	β-
W184 3E+17 y	**W185** 75.1 d	**W186**	**W187** 23.72 h	**W188** 69.4 d
30.67	β-	28.6	β-	β-

Figure 12.22 The chart of the nuclides in the W–Ir range. The solid squares represent stables nuclides or nuclides with a long half-life.

Show that the number of nuclides at time t after the Big Bang, the number N_i of nuclides in the Universe, is given by:

$$N_i = \frac{p_i}{\lambda_i} \left(1 - e^{-\lambda_i t}\right)$$

   d. From the ratio $p_{232\text{Th}}/p_{238\text{U}} = 1.65$, calculate the age of the Universe.

6. Dating the Universe II. We now use the slowly decaying ^{187}Re–^{187}Os chronometer. Figure 12.22 shows the portion of the chart of nuclides in the W–Ir range.

   a. We neglect the contribution of the p process to nucleosynthesis in this range. Explain why the ^{186}Os and ^{187}Os nuclides are pure s nuclides. Draw the s-process pathway on Fig. 12.22.

   b. Using the abundances shown on the figure and cross-sections of 422 (^{186}Os) and 896 (^{187}Os) millibarns (1 mb $= 10^{-28}$ m^2), find the abundance of radiogenic ^{187}Os due to the decay of ^{187}Re since the beginning of the Universe.

   c. Using the solar abundances of Re and Os (0.0517 and 0.675 atoms per 10^6 Si), calculate the age of the universe. Refer to Table 4.1 for the decay constants. You will need to consider that ^{187}Re decays so slowly that its abundances are not affected by radioactivity. Although other solutions are easily worked out (e.g. continuous production, see previous exercise), it is easier to assume that all the ^{187}Re was created in one single event just after the Big Bang.

7. We assume that core formation completely eliminated the very siderophile elements, in particular Ir and Os, from the terrestrial mantle. The modern mantle nevertheless contains 3.43 ppb of Os and 3.19 ppb of Ir. What is the proportion of late veneer in the form of chondrites with 520 ppb Os and 490 ppb Ir that has contributed to the modern mantle? Neglect the contribution of the continental crust. Assuming that the mantle is dry and that oceans were added by the late veneer, what is the water content of the incoming projectiles? Refer to Appendix F for the appropriate constants.

8. The acceleration of gravity g at the surface of a planet varies as $g = GM/r^2$ where $G = 6.67\ 10^{-11}$ m^3 kg^{-1} s^{-2} is the universal gravitational constant, and M and r the mass and radius of the planet, respectively. Calculate g at the surface of the planets listed

Table 12.3 Radius r and density ρ of different planets in the inner Solar System.

	Venus	Earth	Moon	Mars	Vesta
r (km)	6052	6371	1737	3390	520
ρ (g cm^{-3})	5.24	5.52	3.34	3.93	3.16

in Table 12.3. Let us consider a basaltic magma formed in the shallow mantle of these planets and assume a constant mantle density.

a. If the cross-over between plagioclase and clinopyroxene (i.e. the pressure at which the order of saturation is reversed) is at 0.5 GPa, at which depth will this cross-over be located? Discuss the implications in terms of the evolution of a molten planet.

b. In the terrestrial mantle, garnet is stable at pressures in excess of 2 GPa. What is the equivalent depth for the other planets? Discuss the presence of garnet in the mantle of each planet.

References

Arnould, M., Goriely, S., and Takahashi, K. (2007) The r-process of stellar nucleosynthesis: Astrophysics and nuclear physics achievements and mysteries. *Phys. Rep.*, **450**, 97–213.

♠ Burbidge, E. M., Burbidge, G. R., Fowler, W. A., and Hoyle, F. (1957) Synthesis of the elements in Stars. *Rev. Mod. Phys.*, **29**, 547–650.

♠ Clayton, R. N., Mayeda, T. K., Goswami, J. N., and Olsen, E. J. (1991) Oxygen isotope studies of ordinary chondrites. *Geochim. Cosmochim. Acta*, **55**, 2317– 2337.

Jagoutz, E., Palme, H., Baddenhausen, H., *et al.* (1979) The abundances of major, minor, and trace elements in the Earth's mantle as derived from primitive ultramafic nodules. *Proc. Lunar Planet. Sci. Conf.*, **10**, 2031–2050.

Lewis, J. S. (1995) *Physics and Chemistry of the Solar System*. San Diego, Academic Press.

McDonough, W. F. (1999) The Earth's core, in C. P. Marshall and R. F. Fairbridge (eds.) *The Encyclopedia of Geochemistry*. Amsterdam: Kluwer, pp. 151–156.

Pagel, B. E. J. (1997) *Nucleosynthesis and Chemical Evolution of Galaxies*. Cambridge: Cambridge University Press.

♠ Patterson, C. (1956) Age of meteorites and the Earth. *Geochim. Cosmochim. Acta*, **10**, 230–237.

♠ Reynolds, C. (1960) Determination of the age of the elements. *Phys. Rev. Lett.*, **4**, 8–10.

♠ Thiemens, M. H. and Heidenreich, J. E. (1983) The mass-independent fractionation of oxygen: A novel isotope effect and its possible cosmochemical implications. *Science*, **219**, 1073–1075.

Yin, Q., Jacobsen, S. B., Yamashita, K. *et al.* (2002) A short timescale for terrestrial planet formation from Hf–W chronometry of meteorites. *Nature*, **418**, 949–952.

13 The element barn

This chapter is intended to provide a geochemical overview of a number of important elements. The elements will be grouped according to mixed criteria, in particular their position in the periodic table and their geochemical properties. We will describe the major mineral phases that host these elements in the mantle and the crust, their properties in solution, and the processes by which they are transferred from any major reservoir (mantle, crust, ocean) to its neighbors. We will not reproduce here the terrestrial abundances, which can be found in Appendix A. We will nevertheless provide the reader with some important data. Condensation temperatures in the solar nebula (Wasson, 1985; Lodders, 2003) define the volatile versus refractory character of the element. The solubility and complexation data in surface waters (Morel and Hering, 1993) and the residence times in seawater (Broecker and Peng, 1982) constrain the concentration level and speciation in natural waters at low temperature. Different parts of geochemical cycles may receive uneven attention. This inhomogeneous treatment reflects the power of geochemistry: different elements are used to trace different processes. The atmophile elements, which essentially fractionate in to the atmosphere and the ocean are not considered in this chapter (N, O, H), while only the long-term aspects of the carbon cycle have been addressed.

13.1 Silicon

Most common form: Si^{4+}
Ionic radius: 0.26 Å
Stable isotopes: 28 (92.23%), 29 (4.67%), 30 (3.10%)
Atomic weight: 28.086
Condensation temperature: 1311 K

Dissociation of silicic acid in water:

$$H_2SiO_3 \Leftrightarrow HSiO_3^- + H^+ \ (-\log K = 9.86)$$

No significant complex in waters
Reactions limiting solubility in water:

$$SiO_2 + H_2O \Leftrightarrow H_2SiO_3 \ (-\log K = 4.0) \text{ (quartz)}$$
$$SiO_2 + H_2O \Leftrightarrow H_2SiO_3 \ (-\log K = 2.7) \text{ (amorphous silica)}$$

Residence time in seawater: 20 000 years

Silicon is, after oxygen and iron, the third most abundant element in the Earth. Although Si concentration in the core may reach several percent, this ubiquitous element is essentially lithophile and refractory. It is the most abundant cation among the silicates that constitute the mantle at shallow depth (olivine, pyroxene, plagioclase, garnet), at intermediate depth (ringwoodite, majorite), and in the deep mantle (perovskite). It also forms the major igneous minerals in the crust of both mafic igneous rocks (olivine, pyroxene, amphibole, plagioclase) and felsic igneous rocks (quartz, feldspars). Silicon is a major constituent of clastic sediments (quartz and clay minerals) and makes up a substantial fraction of metamorphic minerals.

At pressures corresponding to the surface, the crust, and the upper mantle, Si is tetrahedrally coordinated and occupies the center of a four-oxygen tetrahedron. At very high pressure, Si can be accommodated in octahedral sites. Since melts from the mantle are systematically enriched in silicon with respect to the residue, this element may be classified as slightly incompatible. It forms stoichiometrically fixed minerals: the various forms of silica SiO_2 are stable at different pressures: quartz and amorphous silica are stable in the crust and the upper mantle, while coesite and stishovite are stable at increasingly higher pressures. A basalt typically contains 45–50 wt % SiO_2 and granitic melts 70 wt % SiO_2. Quartz is unstable in the presence of Mg-rich olivine. Silica solubility in hydrous fluids increases with temperature and with pH (see Chapter 10).

High-temperature fluids transport large quantities of SiO_2, which, upon cooling, form the ubiquitous quartz veins seen in metamorphic basements. Warm diagenetic fluids carry silica, which precipitates in the upper sedimentary layers: this is the origin of the familiar flintstones that grow on any seed (fossil, pebble) occurring in limestones. When the replacement of the initial sediment is total, a siliceous rock known as chert is obtained.

As a result of the very low solubility of silica at ambient temperature, quartz is essentially left untouched by erosion and forms the familiar sands that, upon diagenetic cementation, become sandstones. For the same reason, silica concentration in seawater is very low and, as a result of biological activity, the ocean is actually undersaturated in silica. Silica is used by both phytoplankton (diatoms) and zooplankton (radiolara). It is therefore depleted in the upper layer of the oceans, while the falling debris redissolves at depth. Deep old waters are the most enriched in silica. Oceanic upwellings, such as the Atlantic Ocean off the coast of Morocco or the equatorial belt, bring up abundant sources of silica. Sediments from those areas are rich in silica deposited by falling organisms. Other localities, with warm hydrothermal springs on the sea floor, such as in the neighborhood of ridges and other volcanic areas, are also rich in siliceous sediments. The older waters from the southern oceans being rich in silica, it is not surprising to see a substantial fraction of the sea floor of the southern hemisphere covered with diatom-rich oozes.

13.2 Aluminum

Most common form: Al^{3+}
Ionic radius: 0.39 Å (tetrahedral) and 0.54 Å (octahedral)
Stable isotope: 27 (100%)
Atomic weight: 26.982
Condensation temperature: 1650 K
Complexes in water: hydroxides
Reactions limiting solubility in water:

$$\frac{1}{2}Al_2Si_2O_5(OH)_4 + \frac{7}{2}H_2O \Leftrightarrow H_4SiO_4 + Al(OH)_4^- + H^+ \ (-\log K = 19.5)$$

(kaolinite)

Residence time in seawater: 620 years

Aluminum is the sixth most abundant element in the Earth. It is a highly refractory lithophile element. The radioactive isotope ^{26}Al quickly decayed into ^{26}Mg in the first millions of years of the Solar System's evolution. It provided substantial heating to the early planetary bodies, and the isotopic composition of Mg is one of the most widely used extinct radioactivity chronometers. It is unlikely that Al enters in large proportions in the core, but it is a major constituent of many major minerals at any depth in the mantle and in the crust. In the mantle, it enters plagioclase up to pressures of about 1 GPa, spinel to 2 GPa, and garnet beyond. At these high pressures, Al also enters clinopyroxene in large proportions: garnet and clinopyroxene dissolve into each other to form majorite, an essential mineral phase of the mantle above the 660 km discontinuity. At higher pressure, Al is hosted in a perovskite structure, but its precise behavior is still largely unknown. The major mineral that hosts Al in igneous rocks is feldspar: only plagioclase occurs in basalts, while plagioclase and alkali feldspar may occur together in felsic rocks. Biotite mica may occur in both types of rocks but normally accounts for only a small part of the Al inventory. In sedimentary rocks, Al is hosted in clay minerals such as kaolinite and illite and, occasionally, in detrital feldspars. In metamorphic gneisses and schists, Al largely resides in feldspars and micas.

Aluminum can be tetrahedrally coordinated and in this coordination it replaces Si in the center of oxygen tetrahedra. It can also be octahedrally coordinated and form solid solutions with elements such as Ca, Mg, and Fe. During melting, the Al-rich minerals (feldspar, spinel, garnet) quickly dissolve into the melt and Al therefore behaves as a moderately incompatible element. During low-pressure fractionation of basalts, Al is removed by plagioclase precipitation. At higher pressure, plagioclase solubility in silicate melts increases, and this mineral does not precipitate until a late stage in the magmatic differentiation. Aluminum is therefore useful in assessing the depth of differentiation of basaltic series. In mid-ocean ridge and continental flood basalts, Al concentrations do not vary much with fractionation because they are buffered by plagioclase removal: these lavas are differentiated at low pressure in the plagioclase stability field. In contrast, Al concentrations increase steadily with fractionation of Hawaiian basalts: these rocks evolve at higher pressure in the absence of plagioclase. Typical concentrations of Al_2O_3 in basaltic and granitic melts average 15 wt %.

Aluminum solubility in hydrous fluids is low, except at very high temperature and high pH. This solubility is controlled by the stability of different complexes with the hydroxyl ion OH^- and the solubility of clay minerals in equilibrium with the solution, e.g. kaolinite. The minimum in solubility at pH $\approx$ 6 reflects the amphoteric character of this element. During weathering, Al solubility is controlled by kaolinite reactions of alkalinity production (7.22). The destruction of feldspars by CO_2-rich fresh water leaves Al in clays: this element is most efficiently transported by rivers to the sea in the suspended load. In seawater, Al solubility is controlled by the input of airborne clay particles transported from deserts, which gives Al distribution in the water column an unusual profile with concentration decreasing down the thermocline (Fig. 7.15).

13.3 Potassium

Most common form: K^+
Ionic radius: 1.51 Å (octahedral) and 1.64 Å (dodecahedral)
Stable isotopes: 39 (93.26%), 40 (0.011%), 41 (6.73%)
Atomic weight: 39.098
Long-lived isotope: 40 ($T^{\frac{1}{2}} = 1.25 \times 10^9$ a)
Condensation temperature: 1000 K
No significant complex in waters
Residence time in seawater: 12×10^6 years

Potassium is an alkali element, i.e. both volatile and lithophile. Its concentration in the Earth is therefore poorly constrained. As for other volatile elements, K is depleted in the Earth with respect to carbonaceous chondrites and enriched with respect to Mars and the Moon. The dual decay of the radioactive isotope ^{40}K into either ^{40}Ca by β^- emission or ^{40}Ar by electron capture makes this element, after U and Th, one of the three significant sources of heat in the Earth and accounts for about 20% of the radioactive heat production. As discussed in Chapter 11, it is not known whether K enters into the core.

The mineral phases that host K in the mantle are not well understood. Under upper mantle conditions, traces of high-K mineral phases, notably the Mg-rich mica phlogopite and occasionally alkali feldspar, may be present in those parts of the mantle that have been contaminated by subducted sediments or by fluids produced by the dehydration of the subducted oceanic crust. In the lower mantle, K may at least in part reside in the mineral hollandite, a high-pressure equivalent of alkali feldspar. The prime K repositories in igneous rocks are alkali feldspars, amphibole, and micas. In sedimentary rocks, most of the K inventory is sequestered in clay minerals, notably in smectites, illite, and in detrital feldspars.

The K^+ ion is very large. Although it may enter octahedral sites, it often fits 12-coordinated sites as in feldspars and micas. During melting in the mantle and basalt differentiation, K is strongly incompatible and follows other incompatible elements such as Th and U to the extent that the K/U ratio remains essentially constant in the mantle and the crust. Mid-ocean ridge basalts are far less concentrated (≈ 0.1 wt %) in K than ocean island basalts (1–2 wt %). Granitic melts typically contain 2–3 wt % K_2O. Felsic igneous melts are normally saturated in feldspar and biotite; K is therefore not incompatible in the genesis of granitic rocks.

At low temperature and therefore during weathering, feldspars react with water to produce clay minerals. The K-rich clay mineral illite is left as a residue and its ubiquitous presence keeps the concentration of K in low-temperature hydrous fluids (rivers, seawater) at very low levels. Potassium is therefore transported to the sea with the suspended load. In high-temperature hydrothermal fluids, K reacts with country rocks to form the K-feldspar crystals commonly observed in the metamorphic aureoles of granitic intrusions. Upon cooling, K is leached preferentially to Na, while the converse becomes true when the temperature of interaction between fluids and their ambient rocks falls below about 300 °C.

Solubility of KCl (sylvite, a major fertilizer) in water increases greatly with temperature. Through evaporation of brines in cool and cold (< 50 °C) landlocked or lagoonar environments, sylvite becomes saturated before NaCl (halite). In seawater, K is not biolimited and its concentration is essentially constant in the water column.

13.4 Sodium

Most common form: Na^+
Ionic radius: 1.02 Å (octahedral)
Stable isotope: 23 (100%)
Atomic weight: 22.990
Condensation temperature: 970 K
No significant complex in waters
Residence time in seawater: 83×10^6 years

Sodium is a volatile and lithophile alkaline element. As with K, the Na terrestrial abundance is therefore not well known. Breaking down the inventory of Na among mantle minerals is very difficult. Most major mineral phases do not accept this element but, under upper mantle conditions, Na solubility in clinopyroxene increases substantially with pressure. In the crust, Na is essentially hosted in the albite component of plagioclase feldspar. In contrast with potassium, there is no major Na-rich clay mineral of major geological importance and most sedimentary Na resides in detrital feldspar. Evaporitic rock salt NaCl represents another significant surface repository of sodium.

In silicates, the Na^+ ion normally occupies octahedral sites. It is a moderately incompatible element during mantle melting and the early stages of basalt differentiation. Plagioclase saturation in basalts differentiated at low pressure such as mid-ocean ridge and continental flood basalts (see Section 10.2) and in more felsic melts turns Na into a moderately compatible element. Basalts typically contain 2–3 wt% Na_2O and granitic melts 3–4 wt% Na_2O. Although albite may be stable under near-surface conditions, it rarely reaches saturation in hydrous fluids at ambient temperature. Authigenic albite is known in sediments, but is not ubiquitous. Sodium is therefore preferentially leached by low-temperature hydrothermal fluids in contrast to K, which dominates in higher-temperature fluids. Alteration of sea-floor basalts by warm percolating fluids rich in sodium and silica leads to the replacement of calcic plagioclase (anorthite) by albite:

$$3CaAl_2Si_2O_8 + 6Na^+ + 12SiO_2 \Leftrightarrow 6NaAlSi_3O_8 + 3Ca^{2+}$$

Basalts altered on the sea floor are therefore notably enriched in Si and Na with respect to their unaltered precursor. There is a stark contrast between K and Na behavior under surface conditions imposed by the lack of a major stable low-temperature Na-rich phase. Sodium ions set free by the weathering of feldspars are transported by rivers into the sea in the dissolved load. Sodium is the second most abundant element in seawater and the most abundant cation. Its concentration is limited mostly by interaction of seawater with the sea floor through diagenetic and hydrothermal fluids. Its concentration is constant throughout the water column. In warm landlocked or lagoonar environments, NaCl saturation may be reached and evaporitic halite becomes a prominent sedimentary deposit.

13.5 Magnesium

Most common form: Mg^{2+}
Ionic radius: 0.72 Å (octahedral)
Stable isotopes: 24 (78.99%), 25 (10.00%), 26 (11.01%)
Atomic weight: 24.305
Condensation temperature: 1340 K
Complexes in water: hydroxides, carbonates, sulfates
Reactions limiting solubility in water:

$$CaMg(CO_3)_2 \Leftrightarrow Ca^{2+} + Mg^{2+} + 2CO_3^{2-} \ (-\log K = 1.70) \ (dolomite)$$

Residence time in seawater: 13×10^6 years

Magnesium is a refractory and lithophile alkaline-earth element. It does not enter core composition in substantial concentration. After oxygen, magnesium is the second most abundant element in the mantle where it is hosted in most major minerals (olivine, pyroxenes, garnet, spinel, ringwoodite, etc.). Its concentration in the average crust is relatively low: as a consequence, amphiboles, micas, Mg-rich clays (smectite), and carbonates (dolomite) normally remain minor mineral phases. Magnesium-rich calcite is a major form of carbonate precipitated by marine organisms. After Na, Mg is the second most abundant cation of seawater in which it is partly complexed by carbonate and sulfate ions.

In silicates, Mg ions occupy octahedral sites. Many of its properties are better understood by observing that Fe and Mg have identical ionic charges and similar ionic radii. In sedimentary and igneous carbonates, Mg substitutes for Ca. During mantle melting and magma differentiation, Mg is strongly compatible. Primary basalts are rich in Mg and rapidly lose most of it by gravitational removal of olivine and clinopyroxene. The most common indices of magmatic differentiation involve Mg, either as the MgO content, the FeO/MgO ratio, or the fraction known as the mg# and which is the atom ratio $Mg/(Fe + Mg) \approx 100/(1 + 0.56 \, FeO/MgO)$. Relatively undifferentiated basalts may contain 8–16 wt % MgO with FeO/MgO ratios close to unity and mg# in excess of 65. Typical granitic melts only contain 2–4 wt % MgO. Probably the most striking property of Mg in hydrous fluids is its strong shift in solubility between ambient temperature and about 70 °C: seawater, cold groundwater, and run-off contain substantial amounts of Mg, while all hydrothermal and warm diagenetic fluids are Mg-free. A number of processes help explain this shift: precipitation of acid magnesium sulfate in the black smoker feeding zones (see Chapter 7), precipitation of magnesium–calcium carbonate in coastal environments (dolomitization), etc. The removal of seawater Mg in the ridge-crest hydrothermal systems approximately balances out riverine input of Mg to the ocean.

Magnesium is liberated from silicates by weathering (7.30) and transported to the sea in the dissolved load. Its abundance is not limited by a low-solubility mineral phase and is essentially constant down the water column. Partitioning of Mg into calcium carbonate seems to be dependent on the temperature conditions of precipitation: the Mg content of carbonates can therefore be used as a thermometer of ancient marine environments.

13.6 Calcium

Most common form: Ca^{2+}

Ionic radius: 1.00 Å (octahedral)

Stable isotopes: 40 (96.94%), 42 (0.65%), 43 (0.14%), 44 (2.09%), 46 (0.004%), 48 (0.19%)

Atomic weight: 40.078

Condensation temperature: 1518 K

Complexes in water: hydroxides, carbonates, sulfates

Reactions limiting solubility in water:

$$CaCO_3 \Leftrightarrow Ca^{2+} + CO_3^{2-} \ (-\log K = 8.22) \ \text{(aragonite)}$$
$$CaCO_3 \Leftrightarrow Ca^{2+} + CO_3^{2-} \ (-\log K = 8.22) \ \text{(calcite)}$$
$$CaSO_4.2H_2O \Leftrightarrow Ca^{2+} + SO_4^{2-} + 2H_2O \ (-\log K = 4.62) \ \text{(gypsum)}$$

Residence time in seawater: 1.1×10^6 years

Calcium is a refractory and lithophile alkaline-earth element. The abundance of the isotope 40 produced by radioactive decay of ^{40}Ar is occasionally used as a chronometer. The relative abundances of non-radiogenic isotopes have also been used as a tracer of certain biological processes. Just like Mg, Ca does not enter core composition in substantial quantities. In the mantle, Ca is stored in clinopyroxene and in its high-pressure equivalent Ca-perovskite. In igneous rocks, as in the crust in general, calcic plagioclase (anorthite) and amphibole are major hosts for calcium. There is no major Ca-rich clay mineral. The major low-temperature Ca-rich phase is calcium carbonate in its two forms of calcite and aragonite. Calcium phosphates are a ubiquitous form of Ca storage in igneous (fluorapatite) and sedimentary rocks (carbonate-apatite). Apatite is the essential ingredient of vertebrate hard parts (bones and teeth). Calcium sulfates (gypsum and anhydrite) are an essential component of evaporitic sequences.

In silicates, Ca ions occupy octahedral sites. During mantle melting, Ca is slightly incompatible. Its behavior during magmatic differentiation is controlled by the stability of plagioclase and clinopyroxene (see Section 11.1). High-pressure magmatic differentiation takes place in the presence of clinopyroxene but in the absence of plagioclase and there-fore decreases the Ca/Al ratio in the residual melt. Low-pressure fractionation of basalts in the stability field of clinopyroxene and plagioclase leaves the Ca/Al ratio essentially con-stant. Although the Ca content of olivine is low (a fraction of a percent), it increases with decreasing temperature. Olivine phenocrysts in basalts may contain up to 0.5 wt % CaO, while mantle peridotites normally are far more depleted. Basaltic melts typically contain 10–12 wt % CaO and a granitic melt 2–4 wt % CaO.

Calcium is a mobile element during water–rock interaction at any temperature. Pla-gioclase, clinopyroxene, and amphibole are easily weathered by water equilibrated with atmospheric CO_2 and by seawater alteration of submarine basalts. The Ca^{2+} ion goes into solution and is transported to the sea by run-off, while Ca-free mineral phases, silica and clay minerals, are left behind. Carbonate, sulfate, and phosphate complexes of Ca are strong. Calcium is the third most concentrated cation of seawater. The Ca concentration level in fresh water and seawater is controlled by the solubility of calcite and aragonite. Calcite is less soluble than aragonite. Surface oceanic water is saturated in calcite, while deep water is undersaturated. This results from the combined effect of temperature and pressure on the solubility product of calcium carbonate and the dissociation constants of carbonic acid together with the ΣCO_2 content of local seawater. Removal of calcium car-bonate from the ocean by biogenic carbonate precipitation is the most important control of seawater alkalinity; therefore it is the crux of the response of the ocean–atmosphere system to CO_2 fluctuations resulting from changes in volcanic activity, biological productivity, and erosion patterns.

In brines from landlocked and lagoonar environments, precipitation of calcium sulfate may occur as either hydrated gypsum at temperatures below about 35 °C or anhydrous anhydrite at higher temperatures. Gypsum normally converts to anhydrite during burial.

13.7 Iron

Most common forms: Fe^0, Fe^{2+}, and Fe^{3+}
Ionic radius: 0.61 Å for Fe^{2+} and 0.55 Å for Fe^{3+} (octahedral)
Stable isotopes: 54 (5.90%), 56 (91.72%), 57 (2.10%), 58 (0.28%)

Atomic weight: 55.847
Condensation temperature: 1336 K
Complexes in water: hydroxides, chlorides
Reactions limiting solubility in water:

$$Fe(OH)_3 \Leftrightarrow Fe(OH)_2^+ + OH^- \quad (-\log K = 16.5)$$
$$Fe(OH)_3 + H_2O \Leftrightarrow Fe(OH)_4^- + OH^- \quad (-\log K = 4.4)$$

Residence time in seawater: 55 years

Iron is the most abundant element in the Earth. It is refractory and, by definition, siderophile. It is the most abundant element of the core, both in the solid inner core and the liquid outer core in which convection generates the terrestrial magnetic field. In contrast with the core where Fe occurs in its metallic and most reduced form Fe^0, iron in the mantle is essentially in its Fe^{2+} form. Ferrous iron (Fe^{2+}) substitutes for Mg^{2+} in most silicate mineral phases. Iron is, after Si and Mg, the third most abundant cation in the mantle. In the upper mantle, ferrous iron is found in olivine, pyroxene, garnet, and amphibole. In the deep mantle, it enters with Mg into the perovskite structure of ringwoodite and also into the oxide structure of magnesio-wüstite $(Fe, Mg)\,O$. In igneous rocks, as in the crust in general, it is hosted in amphibole and biotite and also, together with Fe^{3+}, Al^{3+}, Cr^{3+}, and Ti^{4+}, in oxide minerals (magnetite, ilmenite). Ferric iron easily substitutes into the tetrahedral site of alkali feldspars, which is why so many granites turn reddish upon incipient weathering. When exposed to the atmosphere or seawater at low temperature, Fe is normally oxidized to Fe^{3+}. It is found in different forms of iron hydroxide (such as goethite, hematite, and limonite) that dominate soils, sediments, as well as ferromanganese nodules and encrustations from the deep sea. Iron-rich clay minerals and carbonates are uncommon. Organic compounds contain important Fe-rich proteins that have different functions, notably oxygen transport in the cell (porphyrins). Iron concentration in seawater is very low, again because of the very low solubility of hydroxides.

The sites occupied by Fe^{2+} and Fe^{3+} are normally octahedral, but Fe^{3+} can be found in tetrahedral sites, especially in feldspars. During mantle melting, Fe^{2+} has a neutral behavior (neither compatible nor incompatible) owing to its lack of octahedral-site preference energy (see Chapter 1). In contrast, Fe^{3+} is highly incompatible. Silicates are very poor electrical conductors so magmas cannot exchange significant amounts of electrons with country rocks. Upon removal of ferrous iron into cumulate minerals, the Fe^{3+}/Fe^{2+} ratio therefore increases in residual melts. The apparently more oxidizing conditions of differentiated rocks are therefore not the result of an externally imposed higher "fugacity" of oxygen, but simply reflect the increasing electron deficit in smaller and smaller quantities of melt. Most basalts would contain 10–12 wt % total iron as FeO and granites 3–4 wt %. Typically, about 15 wt % of the iron present in a primary basalt is in the form of Fe^{3+}. Even when small quantities of ilmenite and magnetite are present at the liquidus, iron concentration in melts does not change when the fractionating assemblage is dominated by olivine and pyroxene: this is the case for the high-pressure differentiation of ocean island basalts and for the wet differentiation of orogenic (calc-alkaline) magmas. When plagioclase is present, as in mid-ocean ridge and continental flood basalts (see Section 11.1), iron is significantly more incompatible. These contrasting trends are known in the petrological literature as the Fenner trend (constant Fe) and the Bowen trend (increasing Fe), respectively.

The behavior of iron during water–rock interaction must be understood with respect to the different properties of Fe^{2+} and Fe^{3+} in solution. Although Fe^{3+} is strongly complexed by Cl ions, its concentration is greatly limited by the solubility of ferric iron hydroxides. In contrast, Fe^{2+} is highly soluble. Reducing solutions can transport huge amounts of ferrous Fe and reprecipitate it when the conditions become more oxidizing. This is the case of the hydrothermal solutions of the submarine black smokers that precipitate enormous amounts of hydroxides and form the metalliferous sediments observed in the vicinity of the mid-ocean ridges. This is also the case of the Archean oceans that precipitated the banded iron formations (BIFs) that form our current prime iron ore: the low pressure of oxygen in the ancient atmosphere permitted the long-distance transport of ferrous iron until subtle changes in the marine oxidizing conditions triggered the massive precipitation of iron hydroxide. The modern atmosphere being oxygen-rich, iron is kept in its ferric form. Weathering of the continental crust leaves residues of hydroxide minerals that are transported to the sea with the riverine suspended load. Some of these hydroxides are in the form of colloids of very small dimension (smaller than the 0.45-micrometer pore dimension of common filtering devices). The very large surface area and the charged surface of these colloids allows their mutual repulsion and therefore maintains the colloids in suspension. Adsorption of many highly charged elements on these colloids is very efficient. This is the case for many transition elements such as Cr^{3+}, and also of all the rare-earth elements (3+), uranium, thorium, etc., that get their ride to the sea on these colloids. When rivers meet seawater in estuaries, the dielectric properties of the mixture change allowing the particles to flocculate and precipitate with clay minerals as estuarine sediments, entraining in this process a very large fraction of the highly charged elements before they reach the open ocean.

Iron is mostly introduced into the ocean as airborne particles of goethite and limonite transported by winds from arid areas (Sahara, Gobi). Its concentration profile in the water column is unusual (and is only matched by that of cobalt). It displays concentrations decreasing from the surface to the bottom, which is a result of the progressive dissolution of the falling airborne particles. Iron may play an important role in the control of biological productivity in the ocean, and, via CO_2 consumption by phytoplankton, in the control of the greenhouse effect and climates. Some scientists believe that Fe, which is a trace nutrient (micronutrient) indispensable to biological activity, is normally in short supply in surface waters. In this scenario, the introduction of Fe-rich airborne particles would fertilize the sea more efficiently during dry periods and quickly draw down part of the atmospheric CO_2.

The behavior of iron at the interface between seawater and the oceanic crust also represents a set of major geochemical processes. Upon circulation of seawater through the fractured oceanic crust, basaltic Fe^{2+} is oxidized into Fe^{3+} at the expense of dissolved O_2, then through the reduction of seawater SO_4^{2-} into sulfide S^{2-} (see Chapter 7). During diagenesis, a different process takes place in the oxygen-starved layers below the bioturbation layer: biological reduction of SO_4^{2-} into sulfide S^{2-} changes the redox conditions.

Ferric iron hydroxides (such as those composing ferromanganese nodules) are reduced to soluble Fe^{2+}, which then precipitates with S^{2-} to form the pyrite Fe_2S commonly found in reduced sediments.

13.8 Sulfur

Most common forms: S^0, S^{2-}, and SO_4^{2-}
Ionic radius: 0.31 Å (tetrahedral), 0.29 Å (octahedral), and 1.84 Å (S^{2-})
Stable isotopes: 32 (95.02%), 33 (0.75%), 34 (4.21%), 36 (0.02%)
Atomic weight: 32.07
Condensation temperature: 648 K
Dissociation of H_2S in water:

$$H_2S \Leftrightarrow HS^- + H^+ \ (-\log K = 7.02)$$
$$HS^- \Leftrightarrow S^{2-} + H^+ \ (-\log K = 13.9)$$

Reactions limiting solubility in water:

$$CaSO_4.2H_2O \Leftrightarrow Ca^{2+} + SO_4^{2-} + 2H_2O \ (-\log K = 4.62) \text{ (gypsum)}$$
$$FeS \Leftrightarrow Fe^{2+} + S^{2-} \ (-\log K = 18.1) \text{ (pyrrhotite)}$$

Residence time in seawater: not known but oceans are well mixed for sulfate

Sulfur is strongly chalcophile and volatile. It has been repeatedly suggested that very large quantities of this element are dissolved in the core and contribute to the relatively low seismic velocities of the core with respect to those of pure iron. Sulfur does not readily dissolve in silicates. Terrestrial sulfur is therefore stored in sulfides. At high temperatures, solid solutions of Ni and Fe dominate (monosulfide solid solution, MSS). At ambient temperature, sulfur enters a variety of sulfides. The major repository of sulfur in sediments is sulfides, notably pyrite. Because sulfates, the oxidized form of sulfur, are relatively soluble, these minerals play a minor role in the making of continental crust, with the exception of gypsum and anhydrite in evaporites, and barite in hydrothermal veins. In seawater, river, and rain water, in which substantial amounts of dissolved oxygen are present, the stable form of sulfur is the oxidized form, sulfate SO_4^{2-}, which is the third most abundant ion of seawater.

In magmas, sulfur is present as sulfide and also as sulfate in more oxidized granites. During mantle melting and magma differentiation, sulfur has a compatible behavior controlled by the exsolution and gravitational segregation of sulfide blebs (the melting point of sulfides is lower than average magma temperature) that are occasionally observed as mineral inclusions. Sulfides are the most abundant minerals of inclusions in diamond. The high-temperature form of sulfides precipitated out of mafic magmas are characteristically rich in nickel (pentlandite). The high-temperature forms of igneous sulfides are very unstable with respect to low-temperature alteration.

Weathering oxidizes sulfides into soluble sulfate and all the sulfur from the crust is transported to the sea, dissolved in the run-off. In landlocked and coastal environments, evaporation concentrates natural waters eventually to the point of gypsum and anhydrite saturation: these minerals are the most abundant form of evaporites. In seawater, $BaSO_4$ saturation is often reached, but most of it is redissolved shortly after deposition. Microbial activity during early diagenesis turns marine SO_4^{2-} from interstitial fluids into sedimentary pyrite. Marine SO_4^{2-} introduced into submarine hydrothermal systems (black smokers) is first precipitated as anhydrite whose solubility decreases significantly with temperature, reduced by ferrous iron from ambient basalt, and mixed with sulfide leached from the basalts. The reduced forms H_2S and HS^- are present in hydrothermal fluids, largely leached from the basalts, in proportions that vary largely with the pH. Upon cooling, hydrothermal sulfur commonly precipitates as sulfides of Cu and Fe (chalcopyrite, pyrrhotite) at temperatures in excess of 300 °C, and of Zn (sphalerite) at lower temperatures. Sulfur is abundant in volcanic fumaroles and is released as SO_2 in large quantities into the atmosphere by volcanic eruptions. An important aspect of sulfur atmospheric chemistry is the production of gaseous dimethyl-sulfide (DMS) generated in vast quantities by phytoplankton.

13.9 Phosphorus

Most common form: PO_4^{3-}
Ionic radius: 0.17 Å (tetrahedral)
Stable isotope: 31 (100%)
Atomic weight: 30.974
Condensation temperature: 1230 K
Dissociation of H_3PO_4 in water:

$$H_3PO_4 \Leftrightarrow H_2PO_4^- + H^+ (-\log K = 2.15)$$
$$H_2PO_4^- \Leftrightarrow HPO_4^{2-} + H^+ (-\log K = 7.20)$$
$$HPO_4^{2-} \Leftrightarrow PO_4^{3-} + H^+ (-\log K = 12.35)$$

Reaction limiting solubility in water:

$$Ca_5(PO_4)_3OH \Leftrightarrow 5Ca^{2+} + 3PO_4^{3-}$$
$$+ OH^- (-\log K = 55.6) \text{ (hydroxylapatite)}$$

Residence time in seawater: 70 000 years

Phosphorus is a lithophile and moderately siderophile element. Substantial amounts of this element are probably dissolved in the liquid core. It is almost exclusively hosted in calcium phosphate (apatite) (see Section 13.6). Apatite may be of igneous origin. Although apatite is certainly present in the upper mantle, P repository in the deep mantle is not well understood. Biogenic (fish teeth and bones) and diagenetic apatites are the essential repositories of sedimentary phosphorus. They occasionally form huge deposits, as in West Africa, that are actively mined to provide agricultural fertilizer. Some of these deposits, found in particular in the Late Precambrian of China, are chemical precipitates and seem to be associated with episodes of global glaciation. In low-temperature waters, phosphates form numerous complexes and, as indicated by the dissociation reactions above, speciation is pH-dependent. Phosphorus concentration in seawater and river water is limited by the very low solubility of apatite. Phosphate radicals often attach themselves to the surface of iron oxyhydroxide colloids when they precipitate in estuaries.

Phosphorus occurs principally in the center of oxygen tetrahedra. In seawater, P is one of the essential nutrients: in the Krebs cycle, adenosine triphosphate (ATP) is the major energy repository for cells, while ribose phosphates form the building blocks of nucleic acids DNA and RNA. Phosphate is brought to seawater by rivers and is removed, though only after many cycles across the thermocline, with the hard parts of biogenic debris. It is severely depleted in surface water but is regenerated during the dissolution of falling debris in deep water. It is depleted in young North Atlantic Deep Water with respect to the older Antarctic Bottom Water. Diagenetic dissolution, transport, and reprecipitation of apatite is common in sediments.

13.10 Carbon

Carbon is not an extremely abundant component of the Earth. It is both siderophile (in its reduced form) and atmophile (in its oxidized form). Substantial amounts of this element may be dissolved in the core. In the mantle, carbon occurs as graphite and, at depth in excess of about 120 km, as diamond when the conditions are reducing. In oxidizing conditions, carbon occurs as carbon dioxide, which, at depths of about 70–100 km, reacts with mantle silicates to form carbonates, e.g:

$$\underset{\text{(olivine)}}{Mg_2SiO_4} + CO_2 \Leftrightarrow \underset{\text{(magnesite)}}{MgCO_3} + \underset{\text{(pyroxene)}}{MgSiO_3} \tag{13.1}$$

Carbon dioxide solubility in magmas rapidly changes with pressure and therefore depth. Carbon dioxide outgassing from mantle-derived magmas starts at a depth of approximately 60 km and quickly strips the magma of many volatile species, such as rare gases, well before eruption. Carbon dioxide is found as fluid inclusions in olivine phenocrysts (the prime target for He isotope measurements) and makes up a very important component of the gas phases in mid-ocean ridge basalts and ocean island basalts.

Oxidized carbon in the crust occurs as sedimentary carbonates, mostly calcite which is two to three times more abundant than dolomite. The reduced forms are countless, from crystalline graphite to amorphous organic varieties (coal, oil, kerogen, methane). Carbon dioxide is an important component of metamorphic gases and, in particular, is the dominant species in fluids from the granulite facies. In ground-, river-, and sea-water, carbon occurs as the carbonate oxyanions CO_3^{2-} and HCO_3^-, but traces of soluble organic components (humic and fulvic acids) are ubiquitous. Carbon dioxide makes up about 350 ppm per volume of the atmospheric gases.

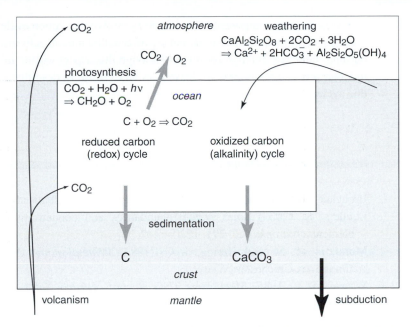

Figure 13.1 The long-term carbon cycles. The oxidized carbon cycle involves the production of alkalinity by weathering of silicates and its consumption by carbonate precipitation (carbonate weathering corresponds to recycling). The reaction shown on the right corresponds to the weathering of calcic plagioclase into dissolved ions and kaolinite. The modern residence time of alkalinity in the ocean is about 100 ka. Most of the reduced carbon is produced by photosynthetic plants which break atmospheric and dissolved CO_2 into reduced carbon and oxygen. Respiration returns some of the reduced carbon to carbon dioxide. The residence time of reduced carbon in the ocean is much shorter than the residence time of oxidized carbon. Both reduced and oxidized carbon are lost to the mantle by subduction of sediments, which leaves an excess of atmospheric oxygen. Atmospheric carbon dioxide is replenished by volcanism.

The geochemical cycle of carbon is of particular significance because it is the most essential component of life. The extremely diverse polymerization modes of carbon compounds and their easy binding to a number of other elements and molecules (nitrogen, phosphate, iron, magnesium, and scores of other metals) are unique in nature. A major source of carbon dioxide is volcanic outgassing from the mantle. Carbonate groups are essentially indestructible except by biological activity or in extremely reducing environments. The prime sites for production of reduced carbon from atmospheric CO_2 and oceanic carbonates are the ocean surface, continental shelves, and continental biosphere. This production, fuelled by photosynthetic processes, is called primary productivity. Igneous and biogenic reduced carbon is easily oxidized by atmospheric oxygen during weathering. The resulting carbon dioxide is distributed almost equally between atmospheric CO_2 and oceanic carbonates. Burial and subduction of sediments rich in organic carbon (reverse weathering) leaves unbalanced oxygen that accounts for the high proportion of this gas in the atmosphere. Figure 13.1 shows a simplified picture of the long-term oxidized and reduced carbon cycles.

The short-term components (10–100 000 years) of the carbon cycle are largely driven by fluctuations in biological activity and are affected by human activities such as coal and oil burning, land use, and deforestation. Although this aspect will be left to more specialized monographs and textbooks, we leave it as an exercise to describe a simplified version of this cycle.

References

Broecker, W. S. and Peng, T. H. (1982) *Tracers in the Sea*. Palisades: Eldigio

Lodders, K. (2003) Solar system abundances and condensation temperatures of the elements. *Astrophys. J.*, **291**, 1220–1247.

Morel, F. M. M. and Hering, J. G. (1993) *Principles and Applications of Aquatic Chemistry*. Chichester: Wiley.

Wasson, J. T. (1985) *Meteorites: Their Record of Early Solar System History*. Berlin, Springer.

A Appendix A Composition of the major geological units

Table A.1 The table below recaps the composition of the Earth's main reservoirs and of the CI carbonaceous chondrites, which serve as a reference for the composition of the Earth. The composition of the silicate Earth, representing the Bulk Silicate Earth less the core and fluid envelopes, is discussed in the text. The concentrations of many elements (nutrients) in the ocean are also highly variable and the values are indicative only. ppm = parts per million, ppb = parts per billion. BSE = Bulk Silicate Earth. A large amount of data can be found on the Geochemical Earth Reference Model (GERM) internet site http://www.earthref.org/GERM

Element	N	M at	Chondr. CI	Earth	Core	BSE	Cont. crust	Upper mantle	Unit	Ocean (μg l^{-1})	Rivers (μg l^{-1})
H	1	1.01	20000	67.5	600	100		2.5	ppm	1.11×10^8	
He	2	4.00								7.20×10^{-3}	
Li	3	6.94	1.5	1.08	0	1.6	11	1.568	ppm	1.78×10^2	3.0×10^0
Be	4	9.01	0.025	0.046	0	0.068		0.0442	ppm	2.25×10^{-4}	1.0×10^{-2}
B	5	10.81	0.900	0.203	0	0.3		0.075	ppm	4.39×10^3	
C	6	12.01	35000	309	2000	120		18	ppm	2.64×10^4	
N	7	14.01	3180	27	75	2		0.05	ppm	8.54×10^3	
O	8	16.00	46	30.5	0	44	45	44	%	2.40×10^3	
F	9	19.00	60	10.1	0	15		9.75	ppm	1.27×10^3	1.0×10^0
Ne	10	20.18								3.96×10^{-4}	
Na	11	22.99	0.510	0.180	0	0.267	2.37	0.214	%	1.08×10^7	6.0×10^3
Mg	12	24.31	9.65	15.39	0	23	2.65	22.8	%	1.26×10^6	4.1×10^3
Al	13	26.98	0.86	1.59	0	2.35	8.36	2.23	%	1.62×10^{-1}	5.0×10^1
Si	14	28.09	10.7	17.10	6	21	27.61	21	%	2.81×10^3	6.5×10^3
P	15	30.97	1080	1101	3500	90	873	54	ppm	6.00×10^1	2.0×10^1
S	16	32.07	54000	6344	19000	250		238	ppm	8.63×10^5	3.7×10^3
Cl	17	35.45	680	11.5	200	17		2.55	ppm	1.89×10^7	7.8×10^3
Ar	18	39.95								5.99×10^2	
K	19	39.10	550	162	0	240	15772	24	ppm	3.89×10^5	3×10^3
Ca	20	40.08	0.925	1.71	0	2.53	4.57	2.40	%	4.14×10^5	2×10^4
Sc	21	44.96	5.92	10.93	0	16.2	22	15.39	ppm	6.00×10^{-4}	4×10^{-3}
Ti	22	47.88	440	813	0	1205	4197	928	ppm	4.79×10^{-3}	3×10^0
V	23	50.94	56	94.4	150	82	131	82	ppm	1.78×10^0	9×10^{-1}

Table A.1 (cont.)

Element	N	M at	Chondr. CI	Earth	Core	BSE	Cont. crust	Upper mantle	Unit	Ocean (μg l^{-1})	Rivers (μg l^{-1})
Cr	24	52.00	2650	3720	9000	2625	119	2625	ppm	2.08×10^{-1}	1×10^{0}
Mn	25	54.94	1920	1680	3000	1045	852	1045	ppm	1.92×10^{-1}	7×10^{0}
Fe	26	55.85	18.1	30.3	85	6.26	5.13	6.26	%	5.59×10^{-2}	4×10^{1}
Co	27	58.93	500	838	2000	105	25	105	ppm	2.00×10^{-3}	1×10^{-1}
Ni	28	58.69	10500	17600	52000	1960	51	1960	ppm	4.80×10^{-1}	3×10^{-1}
Cu	29	63.55	120	60.9	125	30	24	29.1	ppm	2.00×10^{-1}	7×10^{0}
Zn	30	65.39	310	37.1	0	55	73	55	ppm	3.79×10^{-1}	2×10^{1}
Ga	31	69.72	9.20	2.70	0	4.000	16	3.8	ppm	6.97×10^{-4}	9×10^{-2}
Ge	32	72.61	31	7.24	20	1.100		1.1	ppm	5.08×10^{-2}	5×10^{-3}
As	33	74.92	1.850	1.669	5	0.050		0.01	ppm	1.72×10^{0}	2×10^{0}
Se	34	78.96	21	2.65	8	0.075		0.07125	ppm	1.18×10^{-1}	6×10^{-2}
Br	35	79.90	3.57	0.034	0.7	0.050		0.005	ppm	6.70×10^{4}	2×10^{1}
Kr	36	83.80								2.93×10^{-1}	
Rb	37	85.47	2.30	0.405	0	0.600	58	0.0408	ppm	1.24×10^{2}	1×10^{0}
Sr	38	87.62	7.25	13.4	0	19.865	325	12.935	ppm	8.00×10^{3}	7×10^{1}
Y	39	88.91	1.570	2.90	0	4.302	20	3.655	ppm	8.66×10^{-3}	4×10^{-2}
Zr	40	91.22	3.82	7.07	0	10.467	123	6.195	ppm	1.20×10^{-2}	
Nb	41	92.91	0.240	0.444	0	0.658	12	0.11186	ppm	4.65×10^{-3}	
Mo	42	95.94	900	1660	5000	50		30	ppb	1.10×10^{1}	6×10^{-1}
Ru	44	101.07	710	1310	4000	4.97		4.97	ppb		
Rh	45	102.91	130	242	740	0.910		0.91	ppb		
Pd	46	106.42	550	1015	3100	3.850		3.85	ppb	4.26×10^{-5}	
Ag	47	107.87	200	54.4	150	8		7.84	ppb	2.37×10^{-2}	3×10^{-1}
Cd	48	112.41	710	76.0	150	40		39.2	ppb	6.74×10^{-2}	1×10^{-2}
In	49	114.82	80	6.8	0	10		8.5	ppb	1.15×10^{-4}	
Sn	50	118.71	1650	251.1	500	130		97.5	ppb	4.75×10^{-4}	4×10^{-2}
Sb	51	121.76	140	46.2	130	5.5		1.1	ppb	1.22×10^{-1}	7×10^{-2}
Te	52	127.60	2330	285.7	850	12		11.76	ppb	1.02×10^{-4}	
I	53	126.90	450	6.8	0.13	10		1	ppm	5.71×10^{1}	7×10^{0}
Xe	54	131.29								6.56×10^{-2}	
Cs	55	132.91	0.190	0.014	0.065	0.021	2.6	0.504	ppm	3.00×10^{-1}	2×10^{-2}
Ba	56	137.33	2.410	4.455	0	6.600	390	0.449	ppb	1.17×10^{1}	2×10^{1}
La	57	138.91	0.237	0.437	0	0.648	18	0.080	ppm	3.88×10^{-3}	5×10^{-2}
Ce	58	140.12	0.613	1.131	0	1.675	42	0.538	ppm	7.01×10^{-4}	8×10^{-2}
Pr	59	140.91	0.093	0.171	0	0.254	5	0.114	ppm	7.72×10^{-4}	7×10^{-3}
Nd	60	144.24	0.457	0.844	0	1.250	20	0.738	ppm	2.60×10^{-3}	2×10^{-3}
Sm	62	150.36	0.148	0.274	0	0.406	3.9	0.305	ppm	5.10×10^{-4}	8×10^{-3}
Eu	63	151.97	0.056	0.104	0	0.154	1.2	0.119		1.34×10^{-4}	2×10^{-3}
Gd	64	157.25	0.199	0.367	0	0.544	3.6	0.430	ppm	7.94×10^{-4}	9×10^{-3}
Tb	65	158.93	0.036	0.067	0	0.099	0.056	0.080	ppm	1.81×10^{-4}	1×10^{-3}
Dy	66	162.50	0.246	0.455	0	0.674	3.5	0.559	ppm	9.47×10^{-4}	7×10^{-3}
Ho	67	164.93	0.055	0.101	0	0.149	0.76	0.127	ppm	2.79×10^{-4}	1×10^{-3}
Er	68	167.26	0.160	0.296	0	0.438	2.2	0.381	ppm	9.17×10^{-4}	4×10^{-3}
Tm	69	168.93	0.025	0.046	0	0.068	0.32	0.060	ppm	1.63×10^{-4}	6×10^{-4}
Yb	70	173.04	0.161	0.297	0	0.441	2	0.392	ppm	9.81×10^{-4}	4×10^{-3}
Lu	71	174.97	0.025	0.046	0	0.068	0.33	0.061	ppm	1.80×10^{-4}	6×10^{-4}
Hf	72	178.49	0.103	0.191	0	0.283	3.7	0.167	ppm	1.61×10^{-4}	
Ta	73	180.95	0.014	0.025	0	0.037	1.1	0.006	ppm	2.53×10^{-3}	
W	74	183.84	0.093	0.173	0.47	0.029		0.002	ppm	1.10×10^{-1}	3×10^{-2}

Table A.1 (*cont.*)

Element	N	M at	Chondr. CI	Earth	Core	BSE	Cont. crust	Upper mantle	Unit	Ocean (μg l^{-1})	Rivers (μg l^{-1})
Re	75	186.21	40	75.3	0.23	0.280		0.270	ppm	7.20×10^{-3}	
Os	76	190.23	490	900	2750	3.430		3.430	ppm	1.70×10^{-6}	
Ir	77	192.22	455	835	2550	3.185		3.185	ppb	1.15×10^{-6}	
Pt	78	195.08	1010	1866	5700	7.070		7.070	ppb	1.17×10^{-4}	
Au	79	196.97	140	164	500	0.980		0.960	ppb	9.85×10^{-6}	2×10^{-3}
Hg	80	200.59	300	23.1	50	10		9.800	ppb	1.40×10^{-3}	7×10^{-2}
Tl	81	204.38	140	12.2	30	3.500		0.350	ppb	1.23×10^{-2}	
Pb	82	207.20	2.470	0.232	0.4	0.150	12.6	0.018	ppb	5.00×10^{-4}	1×10^{0}
Bi	83	208.98	110	9.853	25	2.500		0.500	ppb	1.04×10^{-5}	
Th	90	232.04	0.029	0.054	0	0.079	5.6	0.006	ppm	5.00×10^{-5}	
U	92	238.03	0.0074	0.014	0	0.020	1.42	0.002	ppm	3.20×10^{0}	4×10^{-2}

Data sources: solid Earth: McDonough and Sun (1995); ocean: GERM website; rivers: mostly Martin and Meybeck (1979).

References

Martin, J.-M. and Meybeck, M. (1979) Elemental mass-balance of material carried by major world rivers. *Mar. Chem.*, **7**, 173–206.

McDonough, W. F. and Sun, S.-S. (1995) The composition of the Earth. *Chem. Geol.*, **120**, 223–253.

Let us write (2.6) describing the conservation of elements (or isotopes) A and B in the mixture of components $j = 1, 2, \ldots$:

$$C_0^A = \sum_j f_j C_j^A \qquad (B.1)$$

$$C_0^B = \sum_j f_j C_j^B \qquad (B.2)$$

and then divide one by the other:

$$\left(\frac{C^A}{C^B} \right)_0 = \frac{f_1 C_1^A + f_2 C_2^A + \cdots}{f_1 C_1^B + f_2 C_2^B + \cdots}$$

$$= \frac{f_1 C_1^B}{f_1 C_1^B + f_2 C_2^B + \cdots} \left(\frac{C^A}{C^B} \right)_1$$

$$+ \frac{f_2 C_2^B}{f_1 C_1^B + f_2 C_2^B + \cdots} \left(\frac{C^A}{C^B} \right)_2 + \cdots \qquad (B.3)$$

In the fractions of the right-hand side, we recognize proportions φ_1^B, φ_2^B, ..., of the denominator element B provided by components $1, 2, \ldots$, proportions that are written, for component 1, say:

$$\varphi_1^B = \frac{f_1 C_1^B}{f_1 C_1^B + f_2 C_2^B + \cdots} \qquad (B.4)$$

Summing the previous equation for all components, it can be verified that the sum of the various φ_j^B is equal to unity. We therefore arrive back at (2.9):

$$\left(\frac{C^A}{C^B} \right)_0 = \sum_j \varphi_j^B \left(\frac{C^A}{C^B} \right)_j \qquad (B.5)$$

For a binary mixture, when allowing for closure, this equation becomes:

$$\left(\frac{C^{A}}{C^{B}}\right)_{0} \quad \left(\frac{C^{A}}{C^{B}}\right)_{1} \quad \varphi_{2}^{B}\left[\left(\frac{C^{A}}{C^{B}}\right)_{2} - \left(\frac{C^{A}}{C^{B}}\right)_{1}\right] \tag{B.6}$$

If the element in the denominator is the same for two ratios, e.g. for A/B and X/B, or if the two denominators are proportional, φ_{2}^{B} can be eliminated between the two corresponding equations. We thus obtain a linear relation between the two ratios for the mixture. For two independent denominators, say for A/B and X/Y, the relation can be shown to be hyperbolic.

The first principle postulates the existence of a conservative quantity, energy. Energy exists in several forms: chemical bonds, thermal agitation of ions and molecules (the sum of these two giving the internal energy U), potential energy, kinetic energy, etc. Energy can be transferred in several ways, by heat transfer δQ (propagation of agitation) or mechanical work δW (exchange of a quantity of motion with a pressure agent, such as the atmosphere or the weight of a column of rock). The first principle can be written in simplified form as:

$$dU = \delta Q + \delta W = \delta Q - P dV \qquad (C.7)$$

where P is pressure and V is the volume of the system. Heat Q and work W are not themselves conservative properties.

The second principle of thermodynamics states that an isolated system drifts spontaneously toward the most probable state, i.e. a state in which a maximum number of equivalent microscopic configurations are available to it. The measure of the number of equivalent configurations accessible to a system is its entropy S. The entropy of an isolated system can therefore only increase. However, possible arrangements of a system can be added or removed by adding or subtracting energy, as in a game of checkers where the potential for varied situations depends on the number of pieces that can be arranged on the board. For a pure heat exchange at constant volume, we write:

$$dS = \frac{\delta Q}{T} + \sigma \, dt \qquad (C.8)$$

where the variable T, defined by this equation, refers to the absolute temperature (in kelvins) of the system, t to time, and σ to the entropy production by dissipative processes, such as shear heating and chemical diffusion. According to the second principle, σ must be non-negative. For an infinitesimal reversible transformation, σ is zero. For a system whose state is controlled by prescribing its entropy S and its volume V, the variation in energy is described by the variation in internal energy U such that:

$$dU \leq T dS - P dV \qquad (C.9)$$

the relation of which is reduced to an equality in the case of a reversible transformation. In (U, S, V) space, the entropy of a system whose energy remains constant evolves spontaneously toward a maximum (Fig. C.1). The system's geometry shows that an adiabatic system ($\delta Q = 0$) evolves spontaneously toward a minimum internal energy U.

For a system whose control variables are entropy and pressure, we use enthalpy $H = U + PV$. In a calorimeter ($dS = 0$) of constant volume, the change in thermal energy is

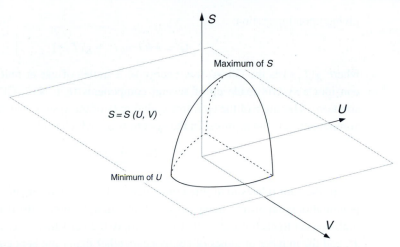

Figure C.1 The complementarity of the maximum entropy and minimum energy conditions are expressed by the single equation $S = S(U, V)$ representing the equation of a surface in (U, S, V) space. At constant energy U (i.e. for an isolated system), the volume V of the system evolves spontaneously so that its entropy is at a maximum. At constant entropy (no heat exchange with the outside), the same system evolves toward a minimum energy state.

ΔU, whereas in a calorimeter at constant pressure, this change is ΔH. For a system for which the temperature T and pressure P are now prescribed, there is nothing to preclude an energy exchange with the outside, whether thermal or mechanical. The intrinsic energy of such a system is measured by another conservative magnitude, free enthalpy or Gibbs' free energy G, so that:

$$G = U + PV - TS \tag{C.10}$$

Other forms of energy are used when control variables other than T and P are preferred. Differentiating G and allowing for the definition of U gives:

$$dG = -SdT + VdP \tag{C.11}$$

An apparently exotic property of G can be readily derived from the previous equation and proves very useful for the study of chemical equilibria:

$$\left[\frac{\partial (G/T)}{\partial (1/T)} \right]_P = H \tag{C.12}$$

For a perfect gas, the equation of state reads:

$$PV = nRT \tag{C.13}$$

where n is the number of moles of gas in the enclosure and R is the gas constant. The minimum energy (maximum entropy) corresponds to $dG = 0$. At constant temperature and constant mole number, we can write:

$$dG = VdP = nRT\frac{dP}{P} \tag{C.14}$$

giving upon integration:

$$G = nRT \ln P + g(T, n) \tag{C.15}$$

where $g(T, n)$ is the Gibbs' free energy of n moles of gas at unit pressure. Let us now consider a system made up of several components (e.g. Na^+, Cl^-, and H_2O for a salt solution). The share of the total free enthalpy G of the system that can be assigned to each component i having n_i moles in the system is μ_i, which is obtained by writing:

$$G = \sum_i \mu_i n_i \tag{C.16}$$

where the sum relates to all the components of the system; μ_i is known as the chemical potential of component i in the system. It is found, by taking the derivative of this relation with respect to each variable, that μ_i is simply the derivative of G relative to n_i, when T, P, and the number of moles of components other than i are kept constant. This yields the general expression of G for a system whose temperature, pressure, and composition are specified:

$$dG = -SdT + VdP + \sum_i \mu_i n_i \tag{C.17}$$

If we consider a mixture of ideal gases, i.e. gas molecules of species other than i which do not interact with each other, the equation of state can be written as:

$$PV = \sum_i (n_i) RT \tag{C.18}$$

where n is replaced by the sum of the numbers of moles of each species. The partial pressure P_i of gas i is defined as $(n_i / n)P$, and the Gibbs' free energy equation becomes:

$$G = \sum_i n_i \left[RT \ln P_i + \mu_i^0(T) \right] \tag{C.19}$$

where $\mu_i^0(T)$ is the chemical potential in the standard state $P_i = 1$ atm. By reference to (C.14), the chemical potential μ_i of gas i in the mixture of gases is defined as:

$$\mu_i = \mu_i^0(T) + RT \ln P_i \tag{C.20}$$

Let us now consider the following reaction in the gaseous state:

$$CH_4 + 2O_2 \Leftrightarrow CO_2 + 2H_2O \tag{C.21}$$

As the reaction progresses, the compounds are created and destroyed in proportions dictated by the stoichiometric coefficients v:

$$\frac{dn_{CH_4}}{v_{CH_4}} = \frac{dn_{O_2}}{v_{O_2}} = \frac{dn_{CO_2}}{v_{CO_2}} = \frac{dn_{H_2O}}{v_{H_2O}} = d\xi \tag{C.22}$$

where $v_{CH_4} = -1$, $v_{O_2} = -2$, $v_{CO_2} = +1$, and $v_{H_2O} = +2$. The parameter ξ measures the progress of the reaction and varies between 0 and 1. The reaction will be at equilibrium when the free energy of the reaction products (right-hand side) is equal to that of the reactants (left-hand side), i.e. when the transfer of matter from one side to the other occurs with no change in the energy of the system. This condition is written by differentiating the expression for Gibbs' free energy as:

$$\mu_{CH_4} dn_{CH_4} + \mu_{O_2} dn_{O_2} + \mu_{CO_2} dn_{CO_2} + \mu_{H_2O} dn_{H_2O} = 0 \qquad (C.23)$$

or alternatively:

$$\left(\mu_{CH_4} \nu_{CH_4} + \mu_{O_2} \nu_{O_2} + \mu_{CO_2} \nu_{CO_2} + \mu_{H_2O} \nu_{H_2O} \right) d\xi = \Delta G d\xi = 0 \qquad (C.24)$$

a condition which can only generally be observed when the content of the parentheses of the left-hand side cancels out ($\Delta G = \sum \mu_i \nu_i = 0$). By replacing the chemical potentials by their expression (C.14), we obtain:

$$RT \left[(-1) \ln P_{CH_4} + (-2) \ln P_{O_2} + (+1) \ln P_{CO_2} + (+2) \ln P_{H_2O} \right] =$$
$$- \left[(-1)\mu^0_{CH_4} + (-2)\mu^0_{O_2} + (+1)\mu^0_{CO_2} + (+2)\mu^0_{H_2O} \right] \qquad (C.25)$$

the relation of which can be compacted using the properties of the logarithms to the form of the "mass action law:"

$$\ln \frac{P_{CO_2} P^2_{H_2O}}{P_{CH_4} P^2_{O_2}} = \ln K(T, P) = -\frac{\Delta G_0(T)}{RT} \qquad (C.26)$$

In this equation, $K(T, P)$ is the equilibrium constant of the reaction and $\Delta G_0(T)$ the variation in Gibbs' free energy when all the components are in the standard state. Two essential equations accompany the mass action law and control the variation of the constant K with temperature and pressure. They are a consequence of the equations demonstrated above:

$$\left[\frac{\partial \ln K}{\partial (1/T)} \right]_P = -\frac{\Delta H_0(T, P)}{R} \qquad (C.27)$$

$$\left[\frac{\partial \ln K}{\partial P} \right]_T = -\frac{\Delta V_0(T, P)}{RT} \qquad (C.28)$$

This formalism established for ideal gases can be generalized to real gases by defining a parameter that satisfies the same equations as partial pressure; this is the gas fugacity.

It can also be transposed to liquid and solid solutions by replacing partial pressures by molar fractions $x_i = n_i/n$. For example, the substitution of rubidium (Rb) for potassium (K) between feldspar and mica can be described by the reaction:

$$Rb^+_{feld} + K^+_{mica} \Leftrightarrow Rb^+_{mica} + K^+_{feld} \qquad (C.29)$$

The chemical potential of rubidium in feldspar is written:

$$\mu^{feld}_{Rb} = \mu^{feld,0}_{Rb}(T, P) + RT \ln x^{feld}_{Rb} \qquad (C.30)$$

with similar expressions for other potentials. It will be noticed that the reference chemical potential $\mu^{feld,0}_{Rb} = 1$ is now defined for a unit molar concentration ($x^{feld}_{Rb} = 1$) and that it is dependent on temperature and pressure. The mass action law of equilibrium is written:

$$\ln \frac{x^{mica}_{Rb} x^{feld}_{K}}{x^{feld}_{Rb} x^{mica}_{K}} = \ln \frac{(Rb/K)_{mica}}{(Rb/K)_{feld}} = \ln K(T, P) = -\frac{\Delta G_0(T, P)}{RT} \qquad (C.31)$$

In many cases like this, the trace element (Rb, a few tens of ppm) displaces only a tiny fraction of the major element (K, 10% or more). Molar fractions of potassium may be considered constant and we can write:

$$\ln \frac{x_{Rb}^{mica}}{x_{Rb}^{feld}} = \ln K'(T, P) \tag{C.32}$$

where K' is the partition coefficient of rubidium between mica and feldspar. In practice, it is equivalent to an equilibrium coefficient of the mass action law, particularly with respect to the equation controlling its dependence on temperature and pressure.

Finally, let us consider a heterogeneous equilibrium, i.e. which involves solid, liquid, and gaseous phases at the same time. For example, let us examine the decarbonation of a siliceous limestone:

$$\begin{array}{cccccc} CaCO_3 & + & SiO_2 & \Leftrightarrow & CaSiO_3 & + CO_2 \\ (calcite) & & (quartz) & & (wollastonite) & \end{array} \tag{C.33}$$

to produce wollastonite $CaSiO_3$, a silicate similar to pyroxene, and carbon dioxide. The equality of the chemical potentials can be written:

$$\ln \frac{x_{CaSiO_3} P_{CO_2}}{x_{CaCO_3} x_{SiO_2}} = -\frac{\Delta G_0(T, P)}{RT} \tag{C.34}$$

Assuming the minerals to be relatively pure, the molar fractions, such as x_{CaCO_3}, are all very close to unity, and we obtain the equation:

$$\ln P_{CO_2} = -\frac{\Delta G_0(T, P)}{RT} \tag{C.35}$$

We say that the calcite–carbonate–silica assemblage buffers the CO_2 pressure. For constant confining pressure P (pressure imposed by the rock column), (C.12) and (C.35) can be transcribed as:

$$\left[\frac{\partial \ln P_{CO_2}}{\partial (1/T)} \right] = -\frac{\Delta H_0(T, P)}{R} \tag{C.36}$$

where ΔH_0 is the latent heat (enthalpy) of reaction of the equilibrium in question. If the latent heat varies little within the temperature range considered, the equation can be integrated as:

$$\ln P_{CO_2} = -\frac{\Delta H_0(T, P)}{RT} + constant \tag{C.37}$$

Similar equations can be written for the water vapor pressure (dehydration reactions), oxygen pressure (oxidation–reduction reactions), and for many other species. Similar equations are also valid for solubility of a pure solid phase in a solution:

$$\left[\frac{\partial \ln x_i}{\partial (1/T)} \right] = -\frac{\Delta H_i(T, P)}{R} \tag{C.38}$$

where $\Delta H_i(T, P)$ is the heat absorbed by dissolution of a mole of component i in the solution (melting-point depression). It should be remembered that the non-ideal character of mixtures of gases and of solutions has been ignored all along.

We have seen that when an equilibrium is established between several components that react together, the change in Gibbs' free energy ΔG of the reaction is zero. By definition, this is equal to:

$$\Delta G = \Delta H - T \Delta S \tag{C.39}$$

where $\Delta H \, d\xi$ is the change in enthalpy (heat of the reaction) and $\Delta S \, d\xi$ is the change in entropy when the reaction progresses by $d\xi$. Let us consider two equilibrium states of the system that are infinitely close and separated by temperature and pressure intervals dT and dP. We can write that between these states the ΔG of the reaction remains unchanged:

$$d\Delta G = -\Delta S dT + \Delta V dP \tag{C.40}$$

giving Clapeyron's equation:

$$\frac{dP}{dT} = \frac{\Delta S}{\Delta V} = \frac{\Delta H}{T \Delta V} \tag{C.41}$$

where the second equation results from the condition that, at equilibrium, $\Delta G = 0$. This equation is extremely important in the Earth sciences as it is used for connecting the slope of phase change or mineralogical reaction in the pressure–temperature space familiar to geologists with two measurable quantities: the change in density (and hence in molar volume) and the latent heat absorbed in the course of transformation.

The phase diagrams, which are representations of the stability field of the different phases that may appear within a system of given composition, are often approached qualitatively, but are also amenable to thermodynamic treatment. Let us consider a simplified granitic system, the quartz–albite (qz–ab) binary system, and ask which phases (quartz, albite, liquid) are stable at ambient pressure for a given proportion of albite/quartz and at a given temperature. If we consider that the latent heats of fusion of quartz and albite vary little with temperature, (C.38), applied to the solubility of quartz and albite, becomes after integration:

$$\ln x_{qz} = -\frac{\Delta H_{qz}(T, P)}{R} \left(\frac{1}{T} - \frac{1}{T_{qz}^f} \right) \tag{C.42}$$

$$\ln x_{ab} = -\frac{\Delta H_{ab}(T, P)}{R} \left(\frac{1}{T} - \frac{1}{T_{ab}^f} \right) \tag{C.43}$$

where the integration condition has been introduced so that, when the molar fraction x of each component is equal to unity, the temperature must be equal to the melting temperature T_{ab}^f and T_{qz}^f of the pure component. These two equations define the curve of the liquidus of the quartz–albite system (Fig. C.2). They cross at the eutectic point E. For each composition, the system has a stable branch at high temperature and a metastable branch at lower temperature.

If a liquid initially of composition x_A is cooled to temperature T_A, it allows some of the quartz to crystallize. Its composition evolves toward the eutectic point E where the albite begins to precipitate. If too little albite crystallizes, the system moves on to the metastable branch of quartz, on which it cannot remain without violating the minimum energy principle. If too much albite crystallizes, the system moves on to the metastable branch of albite,

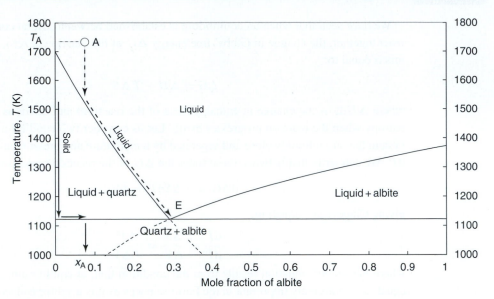

Simplified representation of the quartz–albite binary system. The melting temperatures of quartz and albite are 1700 K and 1373 K, and their latent heats of fusion are 9400 and 63 000 J mol^{-1}, respectively. The liquidus curves are calculated from (C.42) and (C.43). The metastable part of the curves is shown as thin dashed lines. Their point of intersection is the eutectic point E, where a solid and a liquid of the same composition co-exist. Evolution by cooling of a liquid of composition x_A and initially at temperature T_A, and that of a solid at equilibrium are shown by the thick solid line (solid) and the thick dashed line (liquid). The path for melting would be the exact reverse.

which is no more favorable. The system's composition is therefore locked at the eutectic point (constant temperature and composition). When a solid system of the same composition melts, the pathway is reversed. The first liquids are of eutectic composition until the albite disappears. At that point the liquid evolves toward quartz and total melting is achieved when its composition returns to that of the initial solid. We have chosen a binary system in which solids do not form a mutual solution. Other systems whose components crystallize in similar forms (Fe–Mg olivine, Na–Ca feldspar) have solid solutions that may be treated in similar ways at the cost of slight complications in form. This approach is not specific to solid–liquid equilibria. It can be extended to phase equilibria of all sorts (solid–gas, solid–solid). Reaction phase equilibria are processed in the same way, with the assemblages being referred to as peritectic.

Era	Period	Sub-period	Epoch	Ended–began (Ma)
	Quaternary		Holocene	0–0.018
			Pleistocene	0.018–1.6
Cenozoic	Tertiary	Neogene	Pliocene	1.6–5.1
			Miocene	5.1–24
		Paleogene	Oligocene	24–38
			Eocene	38–55
			Paleocene	55–65
Mesozoic	Cretaceous			65–144
	Jurassic			144–213
	Triassic			213–248
Paleozoic	Permian			248–286
	Carboniferous			286–360
	Devonian			360–408
	Silurian			408–438
	Ordovician			438–505
	Cambrian			505–570
Proterozoic				570–2700
Archean				2700–4000
Hadean				4000–4560

Appendix E **An overview of analytical methods**

The analytical methods of geochemistry are many and varied, but they can be grouped by family depending on what is to be analyzed. Setting aside the high-temperature and high-pressure experiments that involve methods often borrowed from mineralogy and petrology, these methods fall roughly into three groups:

- concentrations;
- isotopic ratios;
- speciation of elements in solutions, mineral phases, or organic matter.

We will omit the last item here, as the variety of methods would involve substantial developments with a large physics content about spectroscopic methods beyond the scope of this book.

Measurement of concentrations

Two general principles are commonly used. The first one uses comparison with a reference material by means of calibration curves and only requires off-the-shelf reagents, while the second one, isotope dilution, requires artificially altered nuclide mixtures. In the first case, the operator compares the response to physical stimulation (radiation, ionization) by means of a suitable detector upon the passage of a solution containing the dissolved sample and a set of reference solutions. The first step of most procedures is the dissolution of the powdered sample in hydrofluoric acid (HF), the only acid to dissolve silicates. Often this attack phase is replaced by melting of the sample powder in a lithium meta-borate "flux," the addition of which lowers the melting point of the sample–flux mixture for all the minerals, even the most refractory ones (zircon, oxides). The resulting glass can be dissolved in hydrochloric acid, which is far less dangerous than HF. The attack solution is then diluted so as to minimize problems of interference between the elements and the solution (matrix effects). Let us review four types of methods, each being associated with a different detector:

1. *Atomic emission spectroscopy* (AES) operates by light excitation of rays specific to an element. The method involves spraying the solution in a flame and is excellent for measuring the alkaline elements. As for all optical methods, the different wavelengths of the light beam are separated by an optical grating, a device which sorts the wavelengths by their interference patterns. Inductive plasma excitation, which we will

return to later, excites virtually all the elements. The large number of energy levels per element makes it easier to select rays with no interference with the matrix elements. The method is therefore relatively sensitive, but the interference problems between elements are complex.

2. *Atomic absorption spectroscopy* (AAS) relies on the opposite effect. Monochromatic light sources illuminate a vapor produced by heating the solutions in a graphite oven so as to excite electrons to higher energy levels. Excitation is particularly effective when the incident radiation energy corresponds to the transition energy (resonance). For a given element, the relative absorption energies of the sample and the standards are used to determine concentrations.

3. *X-ray fluorescence spectroscopy* (XRF) relies on the exposure of the sample powder to X-rays. The sample re-emits X-rays of lower energy by fluorescence and their wavelengths are analyzed. Each element appears as a peak whose intensity can be calibrated against standards. Measurements are precise, but below 100 ppm the method is not sensitive enough to produce good results.

4. *Inductively-coupled plasma mass spectrometry* (ICP-MS) relies on ionizing the solutions to be analyzed by inductive coupling of the solution nebulized in a stream of argon within a plasma torch. The ICP acronym, standing for inductively-coupled plasma, is a buzzword that the to-be geochemist must remember. Argon is chosen for its high ionization energy: once ionized as Ar^+ by induction in a coil, it captures the least firmly bonded electrons of the other ions vaporized in the plasma. The different masses are separated in the ion beam by a simple, lightweight mass spectrometer known as a quadrupole: the trajectory of ions injected between four bars with crossed potential is disturbed by radio frequencies so that only the ions of a specific mass cross the device and are counted by the detector. This type of device has no magnets and so allows very fast scanning across the mass range. Masses close to those of Ar, ArO, ArN (40–56) or to the atmospheric gases (N, O, C) cannot be measured, as the detector is saturated by the gas stream. For other masses, the respective heights of the signals are compared for the sample and the standard solutions. This method is both very sensitive (detection level of less than 10^{-11} g $= 10$ pg), and very precise (1–2%). Even so, it yields only poor isotopic ratios (at best 2%), as the tops of the mass peaks are not flat and the counting statistics marginal (see below). As an alternative to dissolution, a spot of the rock or mineral sample may be illuminated with a laser beam, preferably operated in the ultraviolet wavelength range to minimize elemental and isotopic biases due to partial evaporation, and the ensuing fine dust and vapor sprayed into the argon stream. The remainder of the protocol (ionization in the plasma torch and mass analysis) remains unchanged. This method, known as laser ablation, allows *in situ* analysis of concentrations with a beam size of less than 30 micrometers and a remarkably low detection level.

Isotope dilution is a complex, but accurate, reference method operating on a different principle; the equations will not be set out in detail here. The element in question must have more than one isotope, which, in particular, excludes phosphorus, aluminum, manganese, cobalt, etc. To cut a long story short, let us use another animal analogy and take as our job

The equation of isotope dilution

Let us set out to measure the number of moles of rubidium in a particular sample. Natural Rb has two isotopes, ^{85}Rb (72.16%) and ^{87}Rb (27.84%). A commercial supplier sells a ^{87}Rb spike containing 2.0% ^{85}Rb and 98.0% ^{87}Rb (numbers are atomic abundances). Note that, contrary to butterfly species, commercial spikes are never isotopically pure. Let us dissolve a sample, containing, by definition, natural Rb, and add a Rb spike to the solution. Let us call n the number of moles of any isotope in the mixture with atomic mass as superscript and use "nat" and "sp" to refer to the Rb contributed to the mixture ("mix") by the sample and the spike, respectively. Mass balance conditions on each isotope read:

$$n^{85}_{mix} = n^{85}_{nat} + n^{85}_{sp}$$
$$n^{87}_{mix} = n^{87}_{nat} + n^{87}_{sp} \qquad (E.44)$$

Dividing the second equation by the first and then dividing again by n^{85}_{sp} and using isotopic ratios, we get:

$$\left(\frac{^{87}Rb}{^{85}Rb}\right)_{mix} = \frac{\left(^{87}Rb/^{85}Rb\right)_{nat} n^{85}_{nat}/n^{85}_{sp} + \left(^{87}Rb/^{85}Rb\right)_{sp}}{n^{85}_{nat}/n^{85}_{sp} + 1} \qquad (E.45)$$

By collecting the terms containing the mole ratios, we get:

$$\frac{n^{85}_{nat}}{n^{85}_{sp}} = \frac{\left(^{87}Rb/^{85}Rb\right)_{sp} - \left(^{87}Rb/^{85}Rb\right)_{mix}}{\left(^{87}Rb/^{85}Rb\right)_{mix} - \left(^{87}Rb/^{85}Rb\right)_{nat}}$$

$$= \frac{98.0/2.0 - \left(^{87}Rb/^{85}Rb\right)_{mix}}{\left(^{87}Rb/^{85}Rb\right)_{mix} - 72.16/27.84} \qquad (E.46)$$

If we know n^{85}_{sp}, i.e., how many moles of ^{85}Rb from the spike were added to the mixture (typically by weighing an aliquot of a calibrated spike solution), the determination of the isotope composition of the mixture $(^{87}Rb/^{85}Rb)_{mix}$ with a mass spectrometer yields the number of moles n^{85}_{nat} of natural Rb present in the sample and therefore its Rb concentration. For better precision, $(^{87}Rb/^{85}Rb)_{mix}$ must be kept different enough from both the natural ratio (no under-spiking) and from the spike ratio (no over-spiking).

counting butterflies on a very small island, named Eye-Dee (Fig. E.3). Eye-Dee's forests are populated by red butterflies unfortunately not known by their pretty Latin name. We take advantage of a nearby island, which happens to be populated only by blue butterflies. Let us harvest a few buckets of blue butterflies, which we count, then quickly release them on Eye-Dee. A few wing flaps later, we collect a sample of the mixed population: the proportion of red and blue in our catch clearly gives us the number of red butterflies on Eye-Dee. The blue butterflies are the spike. If ^{85}Rb and ^{87}Rb are substituted for the red and blue butterflies, we have the measurement of rubidium concentrations by isotopic dilution. By measuring the proportion of each isotope, we gain access to the number of atoms of each of them initially present. Isotopic spikes are produced by specialist firms. A remarkable property of this method is that the size of the sample collected by the net

Figure E.3 Eye-Dee Island and the principle of isotope dilution. Blue butterflies are released on Eye-Dee in order to count its native population of red butterflies. If the number of foreign blue individuals (spike) is known, the blue/red ratio in a sample gives the number of red individuals on the island.

does not affect the result (provided a minimum number of individuals is taken): once the spike is added and has completely mixed with the sample, the mixing ratio can be obtained from any fraction of the mixture, however incomplete. Isotope dilution does not require a full recovery of the element to be analyzed. This absolute character of the method, offset unfortunately by its painstaking nature, makes it an outstanding reference method.

Measurement of isotope compositions

This measurement principle is very different from those governing the methods described above, as we seek to measure the ratio of two signals emitted simultaneously corresponding to two different isotopes of the same element. Because the ratio of two simultaneous signals is compared, the precision of isotopic measurements is far better than that of elemental analysis, for which the intensity of the signal itself constitutes the measurement. Figure E.4 is a diagram of a mass spectrometer. The different ion sources described in the following correspond to different physical processes and to different applications:

1. *Gas bombardment* sources are based on the principle of spraying the sample gas with electrons, which strip a peripheral electron from it. This type of source is used for measuring the isotopic compositions of hydrogen (H_2 gas), carbon and oxygen (CO_2 gas), sulfur (SO_2 gas), and also the inert gases (He, Ne, Ar, Xe). The use of lasers for *in*

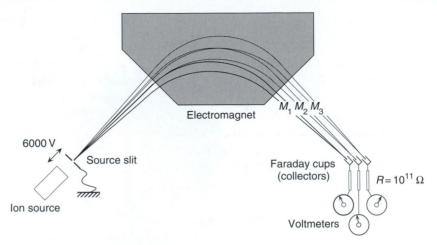

Figure E.4 Mass spectrometer. A source emits ions that are accelerated by a high voltage and then separated in a magnetic field according to their mass (M_1, M_2, etc.). Ion beams are collected in Faraday cages, converted to voltages by a high-value resistor, and these voltages, which are proportional to isotopic abundances, are analyzed by an array of voltmeters.

situ analysis of isotopic compositions, in particular of oxygen in minerals and of argon for geochronology, now allows the operator to select the domains to be analyzed so as to avoid mineral rims and altered material.

2. *Thermal ionization* (TIMS) sources are based on the probability of ions rather than neutral atoms being ejected from the surface of a sample placed on a filament – usually rhenium, tantalum, or tungsten – and heated to a high temperature (1200–1800 °C) in a vacuum. This method has dominated isotopic measurements of positive ions of Sr, Nd, Pb, and Th, and negative ions of Os for decades. However, many elements, especially those with high ionization potentials, cannot be analyzed in this way. In addition, the isotopic compositions of elements such as lead that do not have at least two stable isotopes cannot be corrected precisely for analytic mass bias.

3. The *inductively-coupled plasma* (ICP) sources have been described above. These most successful instruments couple this type of source with a magnetic sector and an array of multiple collectors (MC-ICP-MS). For many elements, this method, which emerged in the mid 1990s, has met with considerable success for two main reasons: the capacity of the plasma torch to ionize all elements, including those too refractory to be analyzed by TIMS; and the possibility of correcting analytical bias precisely for all elements. The outcome of this technique for radiogenic isotopes such as Hf, Nd, and Pb, as well as the "new" stable isotopes (B, Fe, Cu, Zn) makes this method the successor to thermal ionization.

4. *Secondary-ion sputtering* (SIMS), also known as the ion probe, uses a primary beam of ions, typically negative oxygen ions or positive cesium ions, to spray the sample surface and sputter the sample ions into the instrument. This very sensitive method is well adapted to *in situ* isotope measurement and is particularly successful for *in situ* U–Pb dating of zircons (especially the famous SHRIMP of Canberra, Australia). It is

poised to become the prime method for *in situ* analysis of isotopic compositions of carbon or oxygen.

Once the atoms from the sample are ionized, the ion beam is then accelerated at several thousand volts and spread in mass in a magnetic prism. The beam corresponding to each mass is normally collected in a Faraday cage, which often is a carbon-coated ceramic box attached to a high-value resistor, typically $R = 10^{11}$ Ω (ohms), and a voltmeter. By using Ohm's law, the reading of the flux of ionic charges I, i.e. the electric current, can be replaced by the reading of the potential difference $V = RI$, which is usually of the order of one volt. When the ion beam is too weak, the analog current reading is replaced by the counting of the individual arrivals of charged ions. Modern techniques allow this to be read simultaneously from several beams corresponding to different masses (multiple collection) and so make for more precise measurements.

Two remarks are called for so as to judge the relevance and quality of a method. First, for all the methods discussed, atoms are in competition for ionization. A mixture of atoms practically never gives an acceptable ionization yield. Except for ionic analysis, which does not allow it, we seek to isolate the element to be analyzed from the other components (matrix), while taking care to maintain a very high chemical separation yield. This is why complex separation procedures must be carried out (extraction lines for gases and stable isotopes, ion exchange on resins in solution for thermal ionization and plasma sources) before performing mass spectrometric analysis.

This brings us to the second point, which is the requirement of precision. While a precision of 1% can give good-quality information for the isotopic analysis of argon, precision better than 10^{-3} is required for meaningful lead and osmium isotopic data, and better than 10^{-4} for Hf, Nd, and Sr. How do we judge the required precision? A counting process is a Poisson process, i.e. the probability of an event (arrival of an ion) occurring depends only on the duration of measurement and not on the moment at which it is made. For this type of process, the standard deviation of measurement (which yields the precision) is equal to the square root of the number of events. If, say, we count a million ions, the standard deviation of the measurement will be 1000 counts and the relative precision of counting therefore will be one per 1000. Understandably then, this precision depends mainly on the number of ions analyzed. The flux of particles cannot be increased indefinitely as the ion, electron, and photon counting devices lose their linearity for a few hundred thousand strikes per second and are totally saturated at a few million. For isotopic compositions of lead, a significant measurement can be obtained on an impact of a few tens of micrometers on ionic analyzers such as the SHRIMP working at the extreme limit of experimental possibilities. Given the current state of sensitivity of instruments, it would be pointless, for example, to hope for a significant measurement on the same impact of the isotopic composition of Nd, as the quantity of the element required would be more than a hundred times greater than that required for lead. The elementary processes of physics must be comprehended and measurement can only be refined by improving the efficiency of ionization.

F Appendix F **Physical and geophysical constants**

Physical constants	
Avogadro number, $\mathcal{N}$ (mol^{-1})	$6.022\,137 \times 10^{23}$
Gas constant, R ($J\ mol^{-1}\ K^{-1}$)	8.314 51
Mean depth (km)	
Earth center	6371
Solid core	5150
Core–mantle boundary	2891
Upper mantle	660
Oceanic lithosphere	80
Oceanic crust	7.1
Continental crust	39
Ocean	3.77
Continental elevation	0.825
Surface areas (m^2)	
Whole Earth	5.10×10^{14}
Land	1.48×10^{14}
Masses (kg)	
Whole Earth	5.9736×10^{24}
Inner core	9.84×10^{22}
Outer core	1.841×10^{24}
Mantle	4.002×10^{24}
Lower mantle	2.940×10^{24}
Upper mantle	1.062×10^{24}
Continental crust	2.97×10^{22}
Oceans	1.40×10^{21}
Polar ice and glaciers	4.34×10^{19}
Atmosphere	5.1×10^{18}
Mean densities ($kg\ m^{-3}$)	
Whole Earth	5526
Core	10 957
Lower mantle	4903
Upper mantle	3604
Fluxes	
Oceanic crust at ridge crests ($km^2\ a^{-1}$)	3.5
Rivers and run-off ($kg\ a^{-1}$)	3.6×10^{16}
Mechanical erosion ($kg\ a^{-1}$)	2.0×10^{13}

Appendix G **Some equations relative to residence time**

First proposition: The relaxation time τ^i is the mean residence time of element i in the reservoir. For an upstream concentration of zero, this essential property can be demonstrated from (6.2) as follows. The mean time $\langle t \rangle$ that element i spends in the reservoir is the mean algebraic value of variable t summed for all increments of mass $dM^i(t)$ of the element passing through the reservoir between $t = 0$ and $t = \infty$:

$$\langle t \rangle = \frac{1}{MC^i(\infty) - MC^i(0)} \int\limits_{MC^i(0)}^{MC^i(\infty)} t \, \mathrm{d}\left[MC^i(t) \right] \qquad (G.47)$$

where the masses of element i contained in the reservoir at those times are given by simply multiplying the concentrations by the mass M of the reservoir. Under the chosen conditions, concentration $C^i(\infty)$ is zero, mass M is constant and cancels out. By taking the derivative of $C^i(t)$ obtained from (6.3), and by introducing the variable $u = t/\tau^i$, we obtain:

$$\langle t \rangle = \tau^i \int\limits_{0}^{\infty} u e^{-u} \mathrm{d}u \qquad (G.48)$$

Integrating by parts, i.e. using:

$$\int u e^{-u} \mathrm{d}u = -u e^{-u} + \int e^{-u} \mathrm{d}u \qquad (G.49)$$

shows that the sum in (G.48) equals unity and demonstrates the proposition. A similar demonstration applies to non-zero upstream concentrations. Observe that if the mass of the reservoir varies with time, the demonstration is no longer valid.

   Second proposition: Extend (6.3) to the case of a constant upstream concentration. When the concentration C_{in}^i is constant, (6.2) can be re-written as:

$$\tau^i \frac{\mathrm{d}\left[C^i - C_{\mathrm{in}}^i / \left(1 + \beta^i \right) \right]}{\mathrm{d}t} = -\left(C^i - \frac{C_{\mathrm{in}}^i}{1 + \beta^i} \right) \qquad (G.50)$$

a homogeneous first-order equation that integrates simply as:

$$\left(C^i - \frac{C_{in}^i}{1 + \beta^i} \right)_t = \left(C^i - \frac{C_{in}^i}{1 + \beta^i} \right)_0 e^{-t/\tau^i} \tag{G.51}$$

or

$$C^i(t) = C^i(0)e^{-t/\tau^i} + \frac{C_{in}^i}{1 + \beta^i}\left(1 - e^{-t/\tau^i} \right) \tag{G.52}$$

where the right-hand side contains the sum of the relaxation term and the forcing term.

Appendix H **The adiabatic atmosphere**

Let us consider an adiabatic atmosphere in which movements are fast enough for heat transfer between different parts to be negligible.

$$dU \quad dQ - PdV \quad -PdV \tag{H.53}$$

We also assume that it is made of perfect gases for which the internal energy is only a function of the temperature and use C_V as the heat capacity at constant volume:

$$dU \quad C_V dT \tag{H.54}$$

Combining the two equations and using the law of perfect gases to eliminate P, we get

$$C_V \frac{dT}{T} \quad R \frac{dV}{V} \quad 0 \tag{H.55}$$

The gas density ρ relates to the volume through $\rho V \quad M$ (molar mass), and therefore:

$$\frac{d\rho}{\rho} \quad \frac{R}{C_V} \frac{dT}{T} \tag{H.56}$$

which we can be integrated as:

$$\frac{\rho}{\rho_0} \quad \left(\frac{T}{T_0}\right)^{\frac{C_V}{R}} \tag{H.57}$$

where the subscript 0 refers to surface conditions (elevation $z \quad 0$). The next task is to find a relationship between temperature and elevation in the atmosphere. To this end, we will use the law of perfect gases one more time to eliminate V from (H.55):

$$R \frac{dP}{P} \quad (R \quad C_V) \frac{dT}{T} \tag{H.58}$$

and combine it with the hydrostatic condition:

$$dP \quad -\rho g dz \quad -\frac{M}{V} g dz \tag{H.59}$$

to obtain

$$-M \frac{RT}{PV} dz \quad (R \quad C_V) dT \tag{H.60}$$

or, using $R = C_P - C_V$:

$$\frac{dT}{dz} \quad -\frac{Mg}{C_P} \tag{H.61}$$

an expression known as the adiabatic gradient. This equation can be integrated to give the variation of temperature in the atmosphere:

$$T \quad T_0 \left(1 - \frac{Mg}{C_P T_0} z \right) \tag{H.62}$$

and for the density:

$$\rho \quad \rho_0 \left(1 - \frac{Mg}{C_P T_0} z \right)^{\frac{C_V}{R}} \tag{H.63}$$

For a diatomic gas such as N_2 or O_2, $C_V = \frac{5}{2} R$, so the density of the atmosphere decreases with the elevation much faster than its temperature.

Further reading

Level of difficulty: (A) armchair reading; (B) for students; (C) serious stuff. This list of references is not exhaustive. It is only intended to guide the reader into more specialized fields.

Albarède, F. (1995) *Introduction to Geochemical Modeling.* Cambridge: Cambridge University Press. Geochemical modeling methods with plenty of examples. C.

Atkins, P. and de Paula, J. (2006) *Atkins' Physical Chemistry* 8th edn. The European classic, very clear and nicely printed. C.

Allègre, C. J. (2008) *Isotope Geology.* Cambridge: Cambridge University Press. A textbook about isotopes and global geochemistry. The focus is on mass balance. B.

Anderson, D. L. (2007) *New Theory of the Earth.* Oxford: Blackwell. The original ideas of one of the most influential geophysicists who does not hesitate to cross the line between geophysics and geochemistry. B.

Aris, R. (1999) *Elementary Chemical Reactor Analysis.* Mineola: Dover. A reprint of the 1989 classic textbook on chemical engineering. Outstanding value. C.

Berner, R. A. (1980) *Early Diagenesis.* Princeton: Princeton University Press. A common sense but powerful explanation of early diagenesis. B.

Bethke, C.M. (2008) *Geochemical and Biogeochemical Reaction Modeling,* 2nd edn. Cambridge: Cambridge University Press. A rigorous presentation of geochemical concepts with a focus on environmental issues and water–rock interaction. A useful introduction to a well-distributed software package. B.

Broecker, W. S. (1994) *Greenhouse Puzzles.* Palisades: Eldigio. The greenhouse effect will never seem the same to you again. A.

Broecker, W. S. (1995) *The Glacial World According to Wally.* Palisades: Eldigio. Quaternary climates by the undisputed champion. A.

Broecker, W. S. and Peng, T. H. (1982) *Tracers in the Sea.* Palisades: Eldigio. Not read it yet? Still fascinating 25 years later. B.

Brownlow, A. H. (1986) *Geochemistry.* Upper Saddle Review, Prentice Hall. One of the most popular textbooks on geochemistry. B.

Cottingham, W. N. and Greenwood, D. A. (1986) *An Introduction to Nuclear Physics.* Cambridge: Cambridge University Press. If you desperately need to get into this difficult field, this book tries hard to make your life less miserable. C.

Criss, R. E. (1999) *Principles of Stable Isotope Distribution.* Oxford: Oxford University Press. An enjoyable reference with examples. C.

Denbigh, K. (1981) *The Principles of Chemical Equilibrium*. Cambridge: Cambridge University Press. Many have tried to do better. I keep patching up the cover of my copy. B.

Dickin, A. (2005) *Radiogenic Isotope Geology,* 2nd edn. Cambridge: Cambridge University Press. A very well written reference work. B.

Faure, G. and Mensing, T.M. (2005) *Isotopes. Principles and Applications*. Chichester: Wiley. A scholarly treatise emphasizing the methods with abundant references. B.

Greenwood, N. N. and Earnshaw, A. (1995) *Chemistry of the Elements*. Oxford: Butterworth–Heinemann. A best-seller on the inorganic chemistry of the elements.

Killops, S. and Killops, V. (2005) *Introduction to Organic Geochemistry,* 2nd edn. Blackwell: Oxford. An enormous amount of information, well-organized and easy to read. B.

Konhauser, K. (2007) *Introduction to Geomicrobiology*. Blackwell: Oxford. A great introduction to the interface between microbiology and geochemistry. B.

Krauskopf, K. B. and Bird, D. K. (1995) *Geochemistry*. New York: McGraw-Hill. One of the most popular textbooks on geochemistry. The third edition was extensively re-written. B.

Lasaga, A. (1998) *Kinetic Theory in the Earth Sciences*. Princeton: Princeton University Press. An outstanding account of the physics that underlies geochemistry. C.

Lerman, A. (1988) *Geochemical Processes*. Malabar, Krieger. An amazing book on low-temperature geochemistry with a unique approach to physical processes. Never out of fashion. B.

Lunine, J.I. (1999) *Earth. Evolution of a Habitable World*. Cambridge: Cambridge University Press. The Earth from the Enterprise. Delightful. A.

Maczek, A. (1998) *Statistical Mechanics*. Oxford: Oxford University Press. A bargain book that leads the reader swiftly through the arcanes of elemental and isotopic fractionation. B.

McBirney, A. R. (2006) *Igneous Petrology,* 3rd edn. Boston: Jones and Bartlett. An elegant and non-conventional introduction to magmatic processes. B.

McBride, N. and Gilmour, I. (2005) *An Introduction to the Solar System*. Cambridge: The Open University Press and Cambridge University Press. A very well written tutorial, very pleasant to read. A.

McDougall, I. and Harrison, T. M. (1999) *Geochronology and Thermochronology by the ^{40}Ar–^{39}Ar method*. Oxford: Oxford University Press. An exhaustive but difficult reference of an essential geochronology technique. C.

McSween, H. Y., Jr (1999) *Meteorites and Their Parent Planets*. Cambridge: Cambridge University Press. An excellent modern textbook on these weird and fascinating objects. B.

McSween, H. Y., Richardson, S.M., and Uhle, M.E. (2003) *Geochemistry. Pathways and Processes.* 2nd edn. Columbia University Press, New York. An excellent alternative to the book you are reading, great! B.

Millero, F. J. (2005) *Chemical Oceanography,* 3rd edn. Boca Raton, CRC Press. An authoritative reference. B.

Morel, F. M. M. and Hering, J. G. (1993) *Principles and Applications of Aquatic Chemistry*. Chichester: Wiley. A particularly insightful introduction to usually difficult concepts of water chemistry. C.

The Oceanography Course Team of the Open University (1991) Oxford: Pergamon. These seven titles of very clear and wonderfully illustrated bargain books supersede any introductory text in Oceanography. I recommend particularly two of these: *Ocean Circulation* and *Seawater: Its Composition, Properties, and Behavior*. A.

Ottonello, G. (2000) *Principles of Geochemistry*. New York: Columbia. An excellent textbook focused on thermodynamics, now in an affordable paperback edition. B.

Ozima, M. and Podosek, F. (2002) *Noble Gas Geochemistry,* 2nd edn. The reference by two pioneers of the field. B.

Schubert, G., Turcotte, D.L., and Olson, P. (2001) *Mantle convection in the Earth and Planets*. Cambridge: Cambridge University Press. If you can't find something about the interior of the Earth and planets, it will surely be in this book. Difficult but luminous, indispensable. C.

Sharp, Z. (2007) *Principles of Stable Isotope Geochemistry.* Prentice Hall: Upper Saddle River. An insightful and inspiring book with a modern spirit. B.

Silbey, R.J., Alberty, R.A., and Bawendi, M.G. (2005) *Physical Chemistry,* 4th edn. Wiley: Chichester. A powerful book. Beats all my teachers. Your ignorance won't survive. B.

Stumm, W. and Morgan, J. J. (1995) *Aquatic Chemistry*. Chichester: Wiley. The undisputed bible with plentiful examples. C.

Taylor, S. R. (2001) *Solar System Evolution: A New Perspective*. Cambridge: Cambridge University Press. An extremely well-written textbook by one of the leading scholars of geochemistry. A.

Taylor, S. R. and McLennan, S. M. (1985) *The Continental Crust: its Composition and Evolution*. Blackwell: Oxford. The ultimate read on the geochemistry of continental crust. B.

Valley, J. W. and Cole, D. R. (2001) *Stable Isotope Geochemistry.* Washington: Mineralogical Society of America. A set of varied and very high-quality review papers. B.

Wayne, R.P. (2000) *Chemistry of Atmospheres,* 3rd edn. Oxford: Oxford University Press. An enormous amount of information on modern issues without the hassle of atmospheric dynamics. B.

Wulfsberg, G. (2000) *Inorganic Chemistry.* Sausallito: University Science Books. Illuminating on some of the most difficult concepts of the field. B.

Index